Using this book

D1198081

The *Philip's Astronomy Dictionary* contai
nearly 200 colour illustrations covering a
has been designed to be easy to use by r

The alphabetical order of headwords
spaces between words. So, for example, '**G star**
and '**Gum Nebula**'. The exception is that '**H α**', '**H I region**', and '**H II region**' appear at the beginning of the 'H' section. Terms beginning with a number are alphabetized as if the number were spelled out; an example is '**Sixty-one Cygni**'.

Cross-references are used extensively, and appear in **bold type**. A cross-reference usually indicates that the entry referred to provides secondary information on the topic under discussion. Sometimes it means that the term in **bold type** is defined or discussed at greater length in its own entry. An entry which is just a cross-reference means that the term is defined in another entry, where it will be found in italic type. A heading or cross-reference followed by a number, for example '**aberration** (2)', is used whenever the same term has more than one entry.

Most entries are fairly concise, but a number of subjects are treated at greater length, for example 'spectral classification' and the entries for the major planets. The term 'planetary body' is used in this dictionary to refer to large satellites and asteroids in addition to the major planets. For the purposes of the values given for the diameters of the giant planets, their 'surfaces' have been taken to be the levels in their atmospheres at which the pressure is 1 bar, i.e. the average pressure at the surface of the Earth.

In the past decade numerous small satellites of the outer planets have been discovered, and many have been named. To avoid overburdening the dictionary with the names of minor objects, only named satellites larger than a certain diameter have been added for this edition (see the note to the table on pp.206–207).

Particularly in tables, large numbers are sometimes given in exponential form. Thus 10^3 is a thousand, 2×10^6 is two million, and so on. 'Billion' means a thousand million, or 10^9.

Dates in this book have been expressed in the order day, month, year, although it is customary in astronomy to show dates in the order year, month, day.

In biographical entries, if the first forename of a person is not the one by which they were most commonly known, it is enclosed in parentheses.

John Woodruff *April 2005*

A

aberration (1) The small apparent displacement of a star's position produced by the combination of the motion of the Earth and the finite velocity of light.

Annual aberration is the displacement resulting from the motion of the Earth in its orbit around the Sun. This is seen by an observer as an apparent motion of a star in an ellipse over the course of the year. The maximum displacement of the star, given by the **semimajor axis** of this ellipse, is just over 20 arc seconds. This angle (α) is the *constant of annual aberration* and is determined from the relationship

$$\tan \alpha = v/c,$$

where v is the Earth's orbital velocity and c is the speed of light. For a star at the pole of the ecliptic the ellipse becomes a circle; for a star on the ecliptic the ellipse becomes a straight line.

Diurnal aberration is a very much smaller displacement of a star, to the east, resulting from the rotation of the Earth on its axis. It reaches a maximum of 0.32 arc seconds, for an observer at the equator, and is zero at the poles. The Earth's rotation carries an observer in a circle with a speed that is greatest at the equator and zero at the poles.

The aberration of starlight was discovered by James **Bradley**, who published his findings in 1729. In demonstrating that the Earth is not fixed in space – thus furnishing the first observational evidence to support Copernicus's theory – and that light travels at a finite velocity, the discovery is important in the development of cosmological ideas.

aberration (2) An optical defect, such as blurring, distortion or false coloration that can occur in the image produced by a lens or curved mirror. The major aberrations are **chromatic aberration**, **spherical aberration**, **coma** (1) and **astigmatism**.

ablation Wearing away of the surface of an object. An example is the erosion of a meteoroid during its high-speed passage through the Earth's atmosphere: atmospheric molecules striking the meteoroid vaporize its surface layers.

absolute magnitude (symbol M) The apparent magnitude a star would have if it were at a standard distance of 10 parsecs from the Earth. The absolute magnitude may be derived from the **apparent magnitude** (m) and the **parallax** (π) by the formula

$$M = m + 5 + 5 \log \pi,$$

where π is in arc seconds. Most stars have an absolute magnitude between -5 and $+15$. The Sun has an absolute magnitude of $+4.8$. SEE ALSO **magnitude**.

absolute temperature Temperature as measured on a scale whose zero point is *absolute zero*, the point at which almost all motion at the molecular level ceases. The unit of absolute temperature is the degree kelvin; an interval of one kelvin (K) is equivalent to an interval of one degree Celsius (°C). The relationship between an absolute temperature (T) and the same temperature as measured on the Celsius scale (t) is

$$T = t + 273.15,$$

absolute zero being -273.15°C. Absolute temperatures are widely used in astronomy. (SEE ALSO the table of conversion factors on page 263.)

absorption line A dark line or band in a **continuous spectrum**. Absorption lines are produced when particular wavelengths of radiation from a hot source are absorbed by a cooler, intervening medium. They are typical of the spectra of stars – radiation from the hot interior is absorbed by a star's cooler outer layers, producing absorption lines. The absorption lines in the Sun's spectrum are called **Fraunhofer lines**. The wavelengths at which absorption lines occur are characteristic of the chemical elements present in the radiating source. This enables a star's chemical composition to be determined from its spectrum.

accelerating Universe An increase in the speed of expansion of the Universe. It had been expected that the expansion of the Universe imparted to it by the **big bang** would be slowed by the effects of gravity, but recent measurements of the distances of Type Ia **supernovae** and of X-ray emissions from clusters of galaxies have shown that about 5 to 6 billion years ago the rate of expansion started to increase. This acceleration has been attributed to the effects of **dark energy**. SEE ALSO **Hubble constant**.

accretion The coalescence of small particles in space as a result of collisions, and the gradual building up of larger bodies from smaller ones by gravitational attraction; also, the accumulation of matter by a star or other celestial object. Matter can be transferred from one component of a close binary system to another by an accretion process, via an **accretion disk**. Accretion is an important factor in the evolution of stars, planets and comets.

accretion disk A disk of gas in orbit around an object, formed by inflowing matter. In a binary star system in which the two component stars are very close, gas can be transferred from the larger to the smaller star, creating an accretion disk. If the accreting star has a strong gravitational field, as it will have if it is a neutron star or black hole, friction will heat the orbiting gas to millions of degrees, hot enough to emit X-rays. Accretion disks also occur at the centre of galaxies.

Achernar The star Alpha Eridani, magnitude 0.46, distance 144 light years, luminosity 1150 times that of the Sun. It is a main-sequence star of spectral type B3 and the ninth-brightest star in the sky.

Achilles Asteroid no. 588, one of the **Trojan asteroids** sharing Jupiter's orbit, and the first to be discovered, by Max Wolf in 1906. It has a diameter of 147 km (91 miles).

achondrite A type of **stony meteorite**, consisting of rock that was once molten, and which therefore was once part of a planetary body that underwent **differentiation**. Like all stony meteorites, achondrites contain no uncombined metallic elements, and they also lack chondrules – millimetre-sized globules of silicate rock. The principal types are *howardites*, *eucrites* and *diogenites* (collectively known as *HED achondrites*), differing in chemical make-up and structure but all with properties, such as radiation damage, that suggest they were once part of a planetary surface. A main

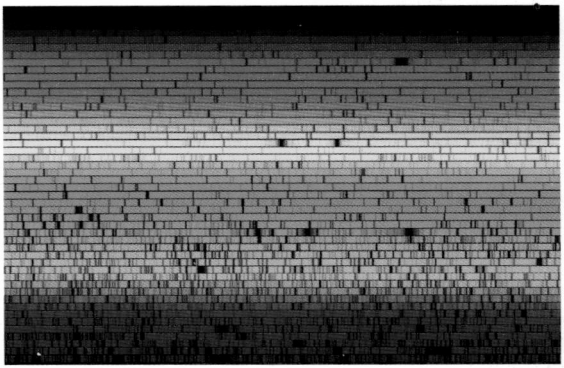

► **absorption line** The visible light spectrum of the cool giant star Arcturus (α Boötis) is shown here. The dark vertical lines in the spectrum are caused by atoms in the star's atmosphere absorbing radiation. Because each element absorbs radiation at characteristic wavelengths, the spectrum of a star can be used to determine which elements are present.

◀ **active optics** The active optics actuators on the reverse of the primary mirror of the WIYN telescope at Kitt Peak National Observatory allow the mirror to be flexed continually to compensate for the effects of gravity as the telescope moves. This system means that far thinner mirrors can be used without risking distortion of the images and data obtained.

candidate for the HED achondrite parent is the asteroid **Vesta**. Achondrites include a handful of meteorites known to have originated on the Moon, and the **SNC meteorites**, believed to have originated on Mars. SEE ALSO **chondrite**.

achromatic lens (achromat) A lens designed to reduce **chromatic aberration**, cutting down the amount of false colour in the image. This is usually achieved by using two lens elements made from different types of glass, for example one element of crown glass and the other of flint glass. The combination is chosen so that two selected wavelengths of light come to the same focus, and at the same time the residual aberration is reduced to a minimum. SEE ALSO **apochromat**.

active galactic nucleus (AGN) The region, at the heart of an *active galaxy*, in which huge amounts of energy are generated by processes other than those that power stars. The energy is believed to be released as matter is accreted by a supermassive **black hole**, of up to a few billion solar masses, at the galaxy's centre. **Jets** are often emitted by AGNs. Active galaxies show a range of behaviour, depending on the amount of energy generated in their nuclei: SEE **Seyfert galaxy**, **BL Lac object**, **blazar**, **quasar**.

active optics A technique for maintaining the primary mirror of a large telescope in accurate shape and alignment, countering the distortions caused by temperature changes or tilting of the telescope. The mirror rests on numerous supports which are automatically adjusted by computer to keep the mirror's figure smooth under all conditions. The technique can also compensate for any slight imperfections that remain after the manufacturing of the mirror. Mirrors with active supports can be made thinner, and thus lighter and cheaper, than ordinary telescope mirrors. The term is sometimes used to refer to the more complicated technique more properly known as **adaptive optics**.

Adams, John Couch (1819–92) English astronomer. While still an undergraduate at Cambridge he set out to analyse the motion of Uranus. The planet's observed path was not in agreement with its calculated orbit, and Adams believed that the discrepancies could be accounted for by the gravitational influence of an undiscovered planet. He calculated an orbit for the new planet, but no search was mounted from England. It was Johann **Galle** who located Neptune, as it was subsequently called, near a position predicted by the French astronomer Urbain **Le Verrier**. After a lengthy priority dispute, the honours of predicting the existence of Neptune came to be shared between Adams and Le Verrier. However, it now seems that neither Adams's nor Le Verrier's calculations were that accurate, and that the two predicted orbits and the actual orbit roughly coincided at the point where Neptune happened to be when Galle found it.

adaptive optics A technique to counteract the blurring of images caused by **seeing** conditions in the Earth's atmosphere. Sensors measure the distortions in a telescope's image caused by atmospheric effects. The image falls onto a small, wafer-thin subsidiary mirror, which can be flexed in a controlled fashion hundreds of times a second to cancel out the imperfections in the image. The resulting image is almost as sharp as if the telescope were in space.

Adhara The star Epsilon Canis Majoris, magnitude 1.50, 431 light years away, and 3700 times as luminous as the Sun. It is a blue giant of spectral type B2.

Adrastea One of the small inner **satellites** of Jupiter discovered in 1979 during the Voyager missions. It is irregular in shape, averaging 16 km (10 miles) in diameter, and orbits near the outer edge of Jupiter's main ring.

aeon (US **eon**) A period of a billion (10^9) years.

Agena Another name for **Beta Centauri**.

age of the universe The length of time for which the Universe has been in existence. Assuming, as seems likely, that we inhabit a **flat universe**, then the results from the **Wilkinson Microwave Anisotropy Probe** give the time from the **big bang** as 13,700±300 million years. The oldest stars are about 12,100 million years old, so the 'age problem', whereby some earlier measurements appeared to imply that the Universe is younger than some of its stars, has been resolved. SEE ALSO **Hubble constant**.

AGN ABBREVIATION FOR **active galactic nucleus**.

airglow A faint glow in the Earth's atmosphere. Molecules in the tenuous upper atmosphere, where the solar radiation is strong, are ionized (split up into charged particles) by ultraviolet radiation or by collision with other charged particles. These particles are short-lived and soon recombine to form the original molecules, emitting a faint light as they do so. The recombination of oxygen molecules is a principal source of emission, which continues long after sunset. From a dark site on a clear, moonless night the airglow is visible as a grey luminous background to the stars, but for many observers **light pollution** makes the airglow impossible to see. The **ashen light** observed on Venus's night side may be a similar phenomenon.

Airy, George Biddell (1801–92) English astronomer and geophysicist. He became Astronomer Royal and Director of the Royal Greenwich Observatory, and re-established the Observatory's standards and authority, equipping it with new instruments. Airy determined the mass of the Earth by taking gravity measurements at the top and bottom of a mineshaft.

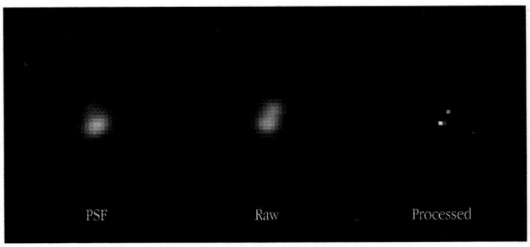

► **adaptive optics** The Very Large Telescope (VLT) uses adaptive optics. Here the light from a close binary pair has been reflected from the primary mirror onto a subsidiary mirror, which is continually adjusted to compensate for variations in the Earth's atmosphere; it is then computer processed.

Airy disk The central spot in the image of a star formed in a telescope. The effects of **diffraction** make it impossible for the image to be formed as a point. Instead, it is spread out into an Airy disk, surrounded by diffraction rings and spikes. (After George Biddell Airy.)

al-Battānī, Muhammad ibn Jābīr (Latinized name **Albategnius**) (858–929) Arab astronomer. His 40 years of observations yielded a catalogue of star positions more accurate than those given in Ptolemy's ***Almagest***, and he introduced trigonometry into Arab astronomy. He made accurate determinations of the precession of the equinoxes and the obliquity of the ecliptic, and showed that the Earth's distance from the Sun varies over the course of the year.

albedo A measure of the reflectivity, or brightness, of a material or body. The albedo scale goes from 1, for a perfectly reflecting (white) surface, to 0, for a perfectly absorbing (black) surface. In astronomy, albedo is used to indicate the fraction of sunlight reflected by Solar System bodies. Rocky bodies such as Mercury and the Moon have low albedos, while cloud-covered bodies like Saturn, or those consisting of a high proportion of water ice – like many of the satellites of the outer Solar System, and comets – have high albedos. Albedo can be defined in various ways. The one most used now is *geometric albedo*, defined as the ratio of the reflectance of the body to that of a flat white surface of the same surface area at the same position.

Albiorix A small outer **satellite** of Saturn, discovered in 2000.

Aldebaran The star Alpha Tauri, magnitude 0.87 (but slightly variable, of irregular type), 65 light years distant, luminosity about 150 times that of the Sun. Aldebaran is a giant of type K5. It appears to lie in the Hyades cluster, but is actually a foreground object at about half the cluster's distance.

Algol The star Beta Persei, the prototype of a class of eclipsing binaries known as Algol-type variables. The first recorded observation of Algol's variability was made by Geminiano Montanari of Bologna in 1669. John **Goodricke** in 1782 was the first to suggest that the variability resulted from two stars in binary motion, the fainter and larger one periodically passing in front of the brighter one. There are two stars, of spectral types B8 and G, orbiting each other in 2.867 days. The B8 star is a dwarf and is the visible component; the fainter star is a subgiant. During eclipses, Algol's brightness drops from 2.1 to 3.4. There is also a third star in the system, but it does not take part in the eclipses. Algol lies 93 light years away.

Algonquin Radio Observatory A radio astronomy observatory at Lake Traverse, Ontario, run by the National Research Council of Canada. Its principal instrument is a 46-m (150-ft) fully steerable radio dish which has been in operation since 1966.

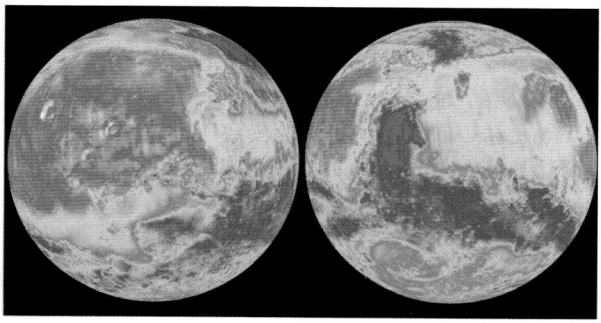

◄ **albedo** This albedo map of Mars was produced by NASA's Mars Global Surveyor. Red areas are bright and show where there is dust. Blue areas show where the underlying, darker rocks have been exposed.

Almagest An encyclopedia of astronomy compiled by **Ptolemy** in about AD 140. It is thought to be based mainly on the work of **Hipparchus**, whose star catalogue it incorporates. It was the ancient world's standard accurate description of the heavens and remained the recognized authority until the middle of the 16th century.

almanac A yearbook containing predicted positions of celestial objects, and other astronomical and calendrical data. For astronomical and navigational purposes the leading publication is *The Astronomical Almanac*, produced each year jointly by Her Majesty's Nautical Almanac Office and the US Naval Observatory.

Alpha Centauri The brightest star in the constellation Centaurus, and the third-brightest star in the sky, magnitude −0.28; also known as Rigil Kentaurus. It is a visual binary, with G2 and K1 main-sequence components of magnitudes −0.01 and 1.35, orbital period 80 years, distance 4.4 light years. SEE ALSO **Proxima Centauri**.

al-Sūfi, Abu'l-Husain (903–86) Arab astronomer. In his *The Book of the Fixed Stars* he presented a detailed revision of the star positions in Ptolemy's ***Almagest***, based on his own observations. It contains the first recorded reference to the Andromeda Galaxy. He measured the length of the year, and attempted to determine the length of a degree of the meridian.

Altair The star Alpha Aquilae, magnitude 0.76, distance 17 light years, luminosity 11 times that of the Sun. It is a main-sequence star of type A7.

altazimuth A mounting for an instrument such as a telescope that allows the instrument to be moved in both altitude and azimuth – that is, about a horizontal and a vertical axis. The disadvantage of this mounting for an amateur telescope is the problem of moving it simultaneously in azimuth and altitude in order to counteract the Earth's diurnal motion, and so keep the object being observed in the field of view. This is made easier in the **Dobsonian telescope**, and circumvented in the EQUATORIAL TELESCOPE. Since the introduction of computer-controlled drive mechanisms, the altazimuth has been the preferred mounting for all large professional instruments.

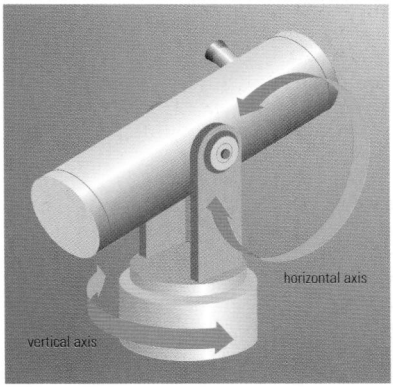

▲ **altazimuth** This simple form of telescope mount allows free movement in both horizontal and vertical axes. An altazimuth mounting is not suitable for use with motor drives, however, unless they are computer controlled.

altitude The angular distance of a celestial body above the observer's horizon. It is measured in degrees from 0 (on the horizon) to 90 (at the zenith) along the **great circle** passing through the body and the zenith. If the object is below the horizon, the altitude is negative. SEE ALSO **coordinates**.

Amalthea The largest of Jupiter's small inner **satellites**, discovered in 1892 by E. E. Barnard. It is an irregular body, averaging 167 km (104 miles) in diameter. Amalthea is dark red, which may indicate that its surface is covered with sulphur from the volcanoes of Io, the innermost of Jupiter's large satellites.

Amor asteroid One of a class of small **asteroids** whose **perihelion** distances lie inside the orbit of Mars (specifically, those

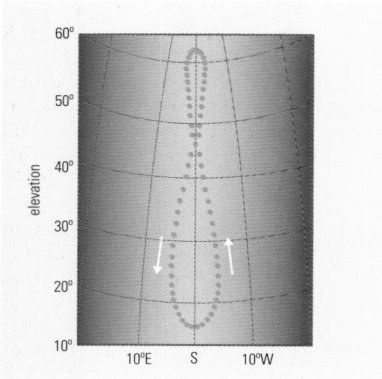

▲ **analemma** A plot of the Sun's apparent position from 52°N, looking south at midday, at 5-day intervals throughout the course of a year. The Sun is at the top of the figure 8 at the summer solstice and at the bottom at the winter solstice.

with perihelion between 1.017 and 1.3 AU). They are named after the 1-km (0.6-mile) diameter asteroid 1221 Amor. The first to be discovered, the most studied and the largest is 433 **Eros**. Over 1200 were known by the end of 2004.

Some of these small bodies may have been perturbed out of the main asteroid belt by Mars, while others may be the nuclei of extinct comets. Close encounters with Mars or Earth can further perturb Amor asteroids into Earth-crossing **Apollo asteroid** orbits; the two groups are often classified together as *Apollo–Amor objects*. Like other small bodies in the inner Solar System, Amor asteroids have limited lifetimes, typically about 10 million years, and many of them will eventually impact with the Earth or Mars. SEE ALSO **Aten asteroid**.

analemma A long, thin figure-of-eight formed by plotting the position of the Sun in the sky at the same time of day throughout the year. The variation in the Sun's position along the long axis of the '8' – highest at the summer **solstice**, lowest at the winter solstice – is caused by the Earth's axial inclination. The variation across the short axis is a result of the eccentricity of the Earth's orbit.

Ananke An outer **satellite** of Jupiter discovered in 1951 by Seth Nicholson. It is in a **retrograde** (1) orbit and may be a captured asteroid.

anastigmatic lens (anastigmat) A type of lens designed to compensate for **astigmatism**.

Andromeda A large constellation of the northern hemisphere, adjoining the Square of Pegasus. The main stars lie in a line leading away from Pegasus, and the star Alpha Andromedae (Sirrah or Alpheratz, mag. 2.1) actually forms one corner of the Square. In mythology, Andromeda was the princess rescued by Perseus. Gamma (Almach or Alamak) is a fine double of magnitudes 2.3 and 4.8, spectral types K3 and B9. The open cluster NGC 752 is an easy binocular object, and NGC 7662 is a planetary nebula visible in small telescopes. The most famous object in the constellation is the **Andromeda Galaxy**.

Andromeda Galaxy A spiral galaxy 2.5 million light years away in the constellation Andromeda, the most distant object clearly visible to the naked eye, also known as M31 or NGC 224. The Andromeda Galaxy has a mass of over 300,000 million Suns. Cepheid and long-period variable stars, novae and a supernova have been observed in it. To the naked eye it appears as a smudge of magnitude 3.4; through binoculars and small telescopes it shows an elongated structure some 1° × 3° in extent, and its plane is tilted at 13° to our line of sight. Its true diameter is about 150,000 light years, somewhat larger than our own Galaxy. The Andromeda Galaxy is accompanied by two small satellite elliptical galaxies, M32 (NGC 221) and M110 (NGC 205).

Andromedids A meteor shower, last seen in 1907, which gave spectacular displays on 27 November in 1872 and 1885, following the break-up in the 1840s of its parent comet, **Biela's Comet**. The shower, also known as the *Bielids*, had its radiant in the constellation Andromeda, near the star Gamma. The Andromedid meteor stream no longer intersects the Earth's orbit but should do so again in the 22nd century.

Anglo-Australian Observatory An observatory at Siding Spring Mountain, near Coonabarabran, New South Wales, jointly funded by the UK and Australian governments. It contains the 3.9-m (150-inch) Anglo-Australian Telescope and the 1.2-m (48-inch) UK Schmidt Telescope, and has unique facilities for wide-field observations.

angstrom (angstrom unit) (symbol Å) A unit of length, equal to 10^{-10} m, formerly widely used in spectroscopy for wavelengths of light. It is no longer recommended for use in astronomy, having largely been superseded by the nanometre, symbol nm (1 nm = 10^{-9} m, 1 Å = 0.1 nm). It is named after the Swedish physicist and astronomer Anders Ångström (1814–74), who introduced it in his atlas of the solar spectrum, published in 1868.

angular diameter The apparent diameter of a celestial body expressed in angular measure (usually in **arc minutes** and **seconds**). It is the angle subtended at the observer by the actual diameter of the body under observation. If the distance of the celestial body from the observer is known, its true diameter can be calculated.

annual parallax (heliocentric parallax) The angle subtended at a celestial object by the radius of the Earth's orbit, which is 1 astronomical unit. It is measured by determining the **semimajor axis** of the parallactic ellipse traced by the star on the celestial sphere. The reciprocal of the annual parallax in arc seconds is the distance of the object in **parsecs**. SEE ALSO **parallax**.

annular eclipse A type of **solar eclipse** in which the Sun's disk is not completely hidden by the Moon, a thin ring of light (*annulus* is Latin for 'ring') remaining visible around the dark body of the Moon. A solar eclipse is annular if it occurs when the Moon is close to **apogee**, at its farthest point from the Earth in its elliptical orbit. The tip of its shadow-cone does not reach down to the Earth's surface, and its apparent diameter is a little less than the Sun's.

anomalistic month The time taken for one revolution of the Moon around the Earth relative to its **perigee**. It is equal to 27.55455 days of **mean solar time**.

anomalistic year The time taken for one revolution of the Earth around the Sun relative to the orbit's **perihelion**. It is equal to 365.25964 days of **mean solar time**. This is about 4m 43.5s longer than a **sidereal year**, because the perihelion point moves eastwards.

anomaly An angular measurement used in determining the position of a body in an elliptical orbit. It is reckoned from periapsis (SEE **apsides**), in the direction of the object's motion. There are three types of anomaly. *True anomaly*, in the case of a planet, is the angle between the planet's radius vector and its perihelion. In the diagram it is the angle PSX. *Mean anomaly* is

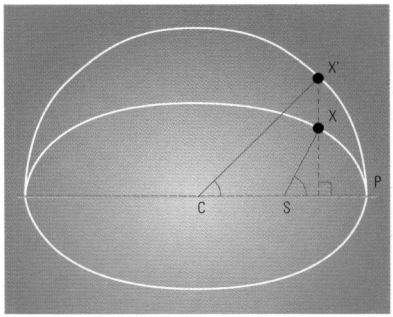

▲ **anomaly** The true anomaly is described by the angle PSX in the diagram here. The eccentric anomaly, PCX', is given by projecting the body's position onto a circumscribing circle, and can be used as a matter of convenience in orbital calculations. C, centre of ellipse; S, Sun; P, perihelion; X, position of planet.

the same angle, but for an imaginary planet moving along its orbit at constant angular speed, rather than at non-uniform angular speed as in reality. *Eccentric anomaly* relates to the motion of a point along a circle whose diameter is the long axis of the planet's orbit. In the diagram, the eccentric anomaly is the angle PCX′. The point X′ is where a perpendicular from the long axis through the planet crosses the circle that circumscribes the ellipse.

ansae (singular **ansa**) The outer parts of Saturn's rings, which through a telescope can look like handles projecting from the planet's disk. The word is Latin for 'handles'.

antapex SEE **apex**.

Antares The star Alpha Scorpii, a supergiant of type M1 that varies irregularly from about magnitude 0.9 to 1.2. Its distance is 604 light years, and it is more than 10,000 times as luminous as the Sun. Antares has a main-sequence companion star of magnitude 5.4, type B2, orbiting it with a period of about 900 years.

anthropic principle The argument that life in the Universe is somehow related to the physical properties of the Universe. Its simplest form is the *weak anthropic principle*, which states that the Universe we inhabit has properties that have made it possible for life to develop. This is not controversial, but the *strong anthropic principle* goes further. It argues that the laws of physics that have shaped the Universe are such that the emergence of intelligent life is an inevitable outcome.

Antlia A small southern constellation representing an air pump, added to the sky by **Lacaille**. It has no star brighter than magnitude 4.3.

Antoniadi, Eugène Michael (1870–1944) French astronomer, born in Turkey of Greek parents. He was a highly skilled observer, particularly of the inner planets. His maps of the surface features of Mars and Mercury were not bettered until those planets were visited by spacecraft. The **canals** of Mars were shown by Antoniadi to be illusory. He devised a system known as the **Antoniadi scale**, used to rate seeing.

Antoniadi scale The scale devised by Eugène Antoniadi, widely used by amateur astronomers to rate the **seeing** conditions (image quality as affected by air movements) under which lunar and planetary observations are made. It distinguishes five gradations of seeing:

I Perfect seeing, without a quiver

II Slight undulations, with periods of calm lasting several seconds

III Moderate seeing, with larger air tremors

IV Poor seeing, with constant troublesome undulations

V Very bad seeing, scarcely allowing a rough sketch to be made.

ap-, apo- Prefixes referring to the farthest point in an object's orbit around another body, as in **aphelion** and **apogee**.

Apache Point Observatory An observatory in the Sacramento Mountains of New Mexico. Its main instrument is a 3.5-m (138-inch) reflector, and the **Sloan Digital Sky Survey**'s 2.5-m (100-inch) reflector is also sited there. The observatory is operated by a group of US universities called the Astrophysical Research Consortium.

apastron The point in the orbit of one star in a binary system at which it is farthest from the other star.

aperture The clear diameter of an optical telescope's objective lens or primary mirror. More generally, for other types of telescope, the aperture is the size of the principal radiation-collecting element – for example, the antenna of a radio telescope. Aperture is the most significant parameter of any telescope, for the amount of radiation collected from distant objects is all-important.

aperture synthesis A technique in radio astronomy in which the signals from a number of radio dishes are used to synthesize the resolution of a single, much larger telescope. The output from numerous small antennae, which are often arranged in a linear or two-dimensional array many kilometres long, are combined electronically. As the Earth rotates, the array sweeps out an area equivalent to a single large dish with a diameter equal to the maximum dimension of the array. An example is the **Very Large Array**. SEE ALSO **radio interferometry**.

apex The point on the celestial sphere towards which the Sun and the entire Solar System appear to be moving, at a velocity of 19–20 km/s (12 mile/s) relative to the nearby stars; it is also known as the *solar apex*. The apex lies in the constellation Hercules at about RA 18h, dec. +30°. The point on the celestial sphere opposite the apex is the *antapex*, in the constellation Columba.

aphelion The point in the orbit of a planet or a comet at which it is at its farthest from the Sun.

apochromat A lens especially corrected to a high degree to eliminate **chromatic aberration**. By using three elements, all of different types of optical glass, a greater amount of correction is possible than with the two-element **achromat**.

apogee The point in the orbit of a body around the Earth at which it is at its farthest from the Earth.

Apollo asteroid One of a class of small **asteroids** whose **perihelion** distances lie inside the Earth's orbit (specifically, those with perihelion less than 1.017 AU and **semimajor axis** greater than 1 AU). They are named after the 2-km (1.25-mile) diameter object 1862 Apollo. Almost 1500 were known by the end of 2004.

Like the **Amor asteroids**, with which they are often grouped as *Apollo–Amor objects*, some may be extinct cometary nuclei, but others may have been perturbed inward by Jupiter from one of the **Kirkwood gaps** in the main asteroid belt. Many Apollo asteroids could impact the Earth within a few million years. SEE ALSO **Aten asteroid**.

Apollo programme A US space project that landed astronauts on the Moon. The Apollo spacecraft had three main parts: a pressurized *command module* for the three crew members; a *service module* containing oxygen and electrical supplies and the main manoeuvring engine; and the *lunar module*, in which two astronauts made the actual Moon landing. The craft was launched by a Saturn V rocket. After four preparatory crewed flights, Apollo 11 made the first Moon landing in July 1969. Neil Armstrong and Edwin Aldrin became

► **Apollo programme** The Lunar Roving Vehicles allowed the astronauts on the final three Apollo lunar missions to range farther from the landers and so collect more varied samples from the Moon's surface. Here, James B. Irwin works next to the Vehicle during the Apollo 15 mission. The photograph was taken by David R. Scott.

the first men to step on the Moon, remaining on the surface for nearly a day and bringing back 22 kg (49 lb) of lunar rock and dust. Subsequent Apollos extended the exploration of the Moon's surface, bringing back further samples and leaving behind instruments which transmitted data about moonquakes. The last mission of the series, Apollo 17, was in December 1972.

apparent magnitude (symbol *m*) The brightness of an object as seen from Earth. The apparent magnitude differs from the **absolute magnitude** because of the object's distance. SEE ALSO **magnitude**.

apparent solar time Local timescale, based on the apparent daily movement of the Sun. It is the time shown on a sundial. Noon occurs when the Sun reaches its maximum altitude, i.e. when it crosses the observer's meridian. Time reckoned in this way is subject to considerable variations because of the Sun's non-uniform motion, a result of both the elliptical shape of the Earth's orbit and the fact that the Sun moves along the ecliptic, not the celestial equator. The time kept by clocks is **mean solar time**, based on the motion of a hypothetical mean Sun.

▲ **Arecibo Observatory** As well as its more famous role in the SETI project, the radio telescope is a powerful radar transmitter and receiver, and is used to map small bodies such as asteroids and to examine the Earth's upper atmosphere.

apparition The appearance in the sky of a comet, or a period during which a planet or satellite is best observed.

appulse The apparent close approach of two celestial bodies, such as a planet and a star, whose directions of motion as seen on the celestial sphere cause them to converge (though in reality they are remote from each other). SEE ALSO **occultation**.

apsides (singular **apsis** or **apse**) The two points in an orbit that are respectively closest to (*periapsis*) and farthest from (*apapsis*) the primary body. For example, the apsides of the Earth's orbit are its perihelion and aphelion; in the Moon's orbit the apsides are its perigee and apogee. The *line of apsides* is the line connecting these two points.

Apus A far-southern constellation representing a bird of paradise, first shown in Johann **Bayer**'s atlas of 1603. Its brightest star is of magnitude 3.8.

Aquarids Either of two **meteor showers**, the **Delta Aquarids** and the **Eta Aquarids**.

Aquarius A constellation of the zodiac, representing a figure pouring water from a jar. Alpha (Sadalmelik, magnitude 3.0) and Beta (Sadalsuud, 2.9) are its brightest stars. Zeta is a fine binary with a period of 856 years, and components of magnitudes 4.4 and 4.6. R Aquarii is a **symbiotic variable** with a range from 5.8 to 12.4 and a mean period of 387 days. There are two important planetary nebulae: the **Saturn Nebula** and the **Helix Nebula**. Aquarius also contains two globular clusters, M2 and M72.

Aquila A distinctive constellation, led by first-magnitude **Altair**; most of it lies in the northern hemisphere of the sky, although the celestial equator crosses it. Mythologically it represents the eagle sent by Zeus to collect the shepherd-boy Ganymede, who was to be cup-bearer to the gods. Gamma (Tarazed), magnitude 2.7, and Beta (Alshain), 3.7, stand either side of

Altair. Eta, a Cepheid variable, has a period of 7.18 days and a range from 3.5 to 4.4; it lies between Theta (3.2) and Delta (3.4), which make good comparison stars. The Mira variable R Aquilae has a range from 5.5 to 12.0 and a period of 284 days. The Milky Way runs through the constellation and is very rich in this region.

Ara A southern constellation between Scorpius and Triangulum Australe, representing the altar of the gods. The brightest stars are Alpha and Beta, both magnitude 2.8. The brightest variable in the constellation is the eclipsing binary R Arae (6.0 to 6.9; period 4.43 days, Algol type). There is also a prominent open cluster, NGC 6193, and a scattered globular, NGC 6397.

arc minute, second (symbols ′, ″) Units of angular measure, equal to 1/60 of a degree and 1/3600 of a degree, respectively. These units are used to express the apparent diameter or separation of astronomical objects and the **resolving power** of a telescope. Often abbreviated to arcmin, arcsec.

Arcturus The star Alpha Boötis, the fourth-brightest in the sky at magnitude −0.05, distance 37 light years, luminosity 100 times that of the Sun. It is a giant of type K2.

Arecibo Observatory A radio and radar telescope near Arecibo in Puerto Rico, diameter 305 m (1000 ft), the largest single-bowl type of instrument in operation. It is built into a natural hollow in the limestone terrain and cannot be steered. Instead, an ingenious feed system enables sources as far as 20° from the zenith to be tracked. It was opened in 1963 and is operated by Cornell University. An upgrade completed in 1998 corrected the telescope's spherical aberration.

areo- Prefix referring to the planet Mars, as in *areography* – the description and mapping of the surface features of Mars. (From *Areos*, Greek for Mars.)

▲ **Ariel** Voyager 2 captured this image of Uranus's satellite Ariel in 1986. Ancient faults and valleys are overlain by abundant impact craters, indicating that the moon has not been geologically active for billions of years.

Argelander, Friedrich Wilhelm August (1799–1875) German astronomer. In 1852 he began the vast undertaking of preparing an atlas and catalogue of all stars down to magnitude 9.5 in the northern hemisphere. This immense work, covering over 300,000 stars, was published as the ***Bonner Durchmusterung***. Argelander introduced the system of nomenclature for variable stars in which they are assigned one or two capital roman letters, beginning with R and RR, and also the subdivision of magnitudes into tenths.

Ariel One of the five main **satellites** of Uranus, diameter 1158 km (720 miles), discovered in 1851 by William Lassell. It is composed largely of a mixture of rock and water ice, with a density of 1.6 g/cm^3. Ariel has been geologically active in the past. Its characteristic features are wide, steep-sided troughs that wind across a landscape of cratered terrain, ridged terrain and plains.

Tectonics accounts for the ridges, and volcanic flooding for the plains; the troughs formed when the water interior froze and the satellite expanded, splitting the crust.

Aries The first constellation of the zodiac – although, since **precession** has now shifted the **vernal equinox** into Pisces, Aries has technically lost this distinction. In mythology it represents the lamb with the golden fleece. Its brightest stars are Alpha (Hamal), magnitude 2.0, and Beta (Sheratan), magnitude 2.6.

Aristarchus of Samos (3rd century BC) Greek mathematician and astronomer. He tried to calculate the distances of the Sun and Moon from the Earth, as well as their sizes. His method was sound but the results were inaccurate. He explained the immobility of the fixed stars by their great distance from the Earth's orbit. He put the Sun at the centre of the Universe, making him the first to propose a heliocentric theory, but the idea was not taken up because it did not seem to make the calculation of planetary positions any easier.

Aristotle (384–322 BC) Greek philosopher and encyclopedist. He developed a view of the Universe based on the four 'elements' (earth, air, fire and water) and the system of concentric spheres proposed by Eudoxus of Cnidus upon which various celestial bodies were carried. Aristotle added more spheres to account for the motion of all celestial bodies. The celestial bodies were, he said, made of a substance called aether, and are perfect and incorruptible, unlike the Earth. He attempted to estimate the size of the Earth, which he maintained was spherical but unmoving – the centre of the Universe. This 'Aristotelian' view of the world remained almost unchallenged until Copernicus.

Arizona Meteor Crater SEE **Meteor Crater**.

array An arrangement of antennae used in radio astronomy. SEE **radio telescope**.

artificial satellite A spacecraft, in particular one that is uncrewed, in orbit around the Earth or some other body, as opposed to a natural satellite. The first artificial satellite, Sputnik 1, was successfully put into orbit by the USSR on 4 October 1957.

ascending node The point at which an orbit crosses from south to north of, for example, the celestial equator or the ecliptic. SEE **node**.

ashen light A faint luminosity on the night side of Venus when the planet appears as a thin crescent, close to inferior conjunction. The ashen light is occasionally reported by observers, but has never been imaged. Its appearance is similar to **earthshine** on the Moon, although that is produced in a different way. If it is a real phenomenon, and not illusory, it is probably the result of electrical disturbances in the planet's ionosphere, and may therefore be similar in nature to the **airglow** in the Earth's atmosphere.

aspect The position of a planet or the Moon relative to the Sun as seen from the Earth. SEE **conjunction**, **elongation**, **opposition**, **quadrature**.

association, stellar A group of stars that have formed together, but are more loosely linked than the members of a star cluster. Many are young and define portions of the spiral arms of our Galaxy. SEE **OB association**, **T association**.

A star A star of spectral type A, characterized by extremely strong hydrogen absorption lines in its spectrum. On the main sequence, A stars range in surface temperature from 7500 K at type A9 up to 9900 K at type A0. Their corresponding masses and radii are, respectively, 1.8 and 1.4 times the Sun's at A9, increasing to 3.2 and 2.5 at A0. Examples are Sirius, Vega, Altair and Deneb. Roughly 10% of A stars have strong magnetic fields. In these, certain chemical elements become concentrated in their atmospheres and produce unusually strong spectral lines. These are classified as Ap ('p' for peculiar) stars. Another kind of spectral anomaly produces the metallic-line (Am) stars, of which Sirius is an example.

asterism A group of easily recognizable stars, part of a constellation. Examples are the seven stars that form the shape of the Plough or Big Dipper in Ursa Major, and the 'Teapot' in Sagittarius. The term is also used for distinctive groupings of stars visible in binoculars or small telescopes, some of which have been given descriptive names by observers, such as the Coathanger (in Vulpecula) and Kemble's Cascade (in Camelopardalis).

asteroid (minor planet) A small Solar System body in an independent orbit around the Sun. The majority move between the orbits of Mars and Jupiter. The largest asteroid, and the first to be discovered, on 1 January 1801 by Giuseppe **Piazzi**, is **Ceres**, which has a diameter of 934 km (580 miles). The smallest that have been measured are the so-called **near-Earth asteroids**, some of which are just a few metres across. Some of the larger asteroids are spherical, but most are irregularly shaped. Since the 1970s, polarization, infrared and radar techniques have allowed improved measurements to be made of asteroids' albedos, sizes and shapes. There are thought to be over a million asteroids with a diameter greater than 1 km (0.6 mile); below this, they extend in size down to dust particles. Objects less than about 10 m (30 ft) in diameter are best classed as **meteoroids**. Some of these very small objects have found their way to Earth as **meteorites**.

The second asteroid to be found was **Pallas**, in 1802, followed by **Juno** in 1804 and **Vesta** in 1807. It was not until 1845 that the fifth asteroid, Astraea, was discovered, but from then on the pace of discovery quickened. In 1891 Max **Wolf** made the first asteroid discovery by photography, and went on to find hundreds more. By late 2004 nearly 100,000 asteroids had been catalogued and had their orbits calculated, and this figure is now increasing by thousands each year. Tens of thousands more have been observed, but not often enough for an accurate orbit to be calculated.

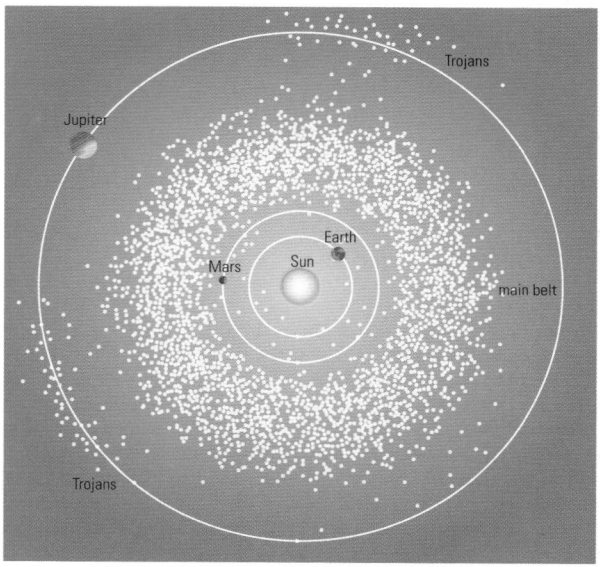

► **asteroid** Plots of the larger known asteroids: most of them lie in the main belt between the orbits of Mars and Jupiter, while the Trojans move in the same orbit as Jupiter, 60° in front of or behind it.

ASTEROIDS

Name	Number	Discovery	Diameter (km)	Av. distance from Sun (million km)	Orbital period (years)	Rotation period (hours)
The first discovered						
Ceres	1	1801, Piazzi	934	413.6	4.60	9.08
Pallas	2	1802, Olbers	525	414.0	4.61	7.81
Juno	3	1804, Harding	268	399.3	4.36	7.21
Vesta	4	1807, Olbers	510	353.2	3.63	5.34
Astraea	5	1845, Hencke	120	385.2	4.13	16.81
Other large asteroids						
Hygeia	10	1849, De Gasparis	408	469.2	5.55	27.6
Davida	511	1903, Dugan	326	475.4	5.67	5.13
Interamnia	704	1910, Cerulli	316	458.1	5.36	8.72
Europa	52	1858, Goldschmidt	302	463.3	5.46	5.63
Sylvia	87	1866, Pogson	260	521.5	6.52	5.19
Some Trojan asteroids						
Hektor	624	1907, Kopff	225	775.3	11.80	6.92
Patroclus	617	1906, Kopff	149	782.7	11.97	—
Achilles	588	1906, Max Wolf	147	774.4	11.78	—
Some Amor asteroids						
Ganymed	1036	1924, Baade	41	398.7	4.35	—
Eros	433	1898, Witt	21	218.6	1.76	5.27
Amor	1221	1932, Delporte	1	287.2	2.66	—
Some Apollo asteroids						
Phaethon	3200	1983, Green	5.0	190.2	1.43	4
Daedalus	1864	1971, Gehrels	3.3	218.5	1.77	8.57
Apollo	1862	1932, Reinmuth	2.0	220.1	1.78	3.07
Some Aten asteroids						
Ra-Shalom	2100	1978, Helin	3.5	124.5	0.76	19.79
Aten	2062	1976, Helin	1	144.6	0.95	—
Hathor	2340	1976, Kowal	0.5	126.2	0.78	—

Diameters and rotation periods of asteroids are often uncertain, and different measurement techniques give different results. For irregular asteroids, the diameter given is an average.

Most asteroids orbit within the main *asteroid belt*, a torus-shaped region of space between about 2.2 and 3.3 AU from the Sun. Within the main belt are the **Kirkwood gaps** at distances from the Sun at which **resonances** rule out stable orbits. The typical asteroid orbit is elliptical but more eccentric than those of the major planets, and with a greater inclination to the ecliptic. Some asteroids have orbits that bring them into the inner Solar System (SEE **Amor asteroid**, **Apollo asteroid** and **Aten asteroid**), while others, such as **Hidalgo** and **Chiron**, have orbits that lie partially or entirely in the outer Solar System.

The great majority of asteroids almost certainly originate from small **protoplanets** left over from the formation of the Solar System, prevented by Jupiter's gravitational influence from forming into a major planet. A few may be the extinct nuclei of **comets**. The present asteroid population does not represent the remnants of a single large planet that disintegrated, as was once thought: put together, they would make up a body only about 1500 km (950 miles) in diameter.

New asteroid discoveries are designated by the year of discovery, and a pair of letters and subscripted numbers to denote the half-month and order of discovery (e.g. 2004NB denotes the second asteroid discovered in the first half of July in that year; the letter I is not used). A permanent number is given when the orbit has been accurately calculated, and the discoverer may then add a name to the number (as with, for example, 1707 Chantal), subject to the approval of the International Astronomical Union.

SEE ALSO **Centaur**, **Hirayama family**, **Kuiper Belt**, **Trojan asteroid**.

astigmatism An **aberration** (2) in lenses and mirrors that affects light from objects away from the centre of the observer's field of view. Instead of being brought to a point focus, these oblique rays of light form an image which is either an ellipse or a straight line, depending on the distance from the lens at which it is viewed. A lens designed to minimize astigmatism is called an *anastigmat*.

astrobiology (exobiology) The study of the possible existence of life beyond the Earth. This includes looking for life forms on other planetary bodies in the Solar System, such as Mars and Europa; the study of organic molecules in nebulae, meteorites and comets; the study of planetary surfaces (and subsurfaces) and atmospheres; questions relating to the origin of life and the range of conditions under which it can survive; and the search for signs of life in space, such as incoming messages at radio or other wavelengths and investigating **extrasolar planets**. SEE ALSO **SETI**.

astrolabe An early astronomical instrument for showing the appearance of the celestial sphere at a given moment, and for determining the altitude of celestial bodies. The basic form consisted of two concentric disks – one with a star map and one with a scale of angles around its rim – joined and pivoted at their centres (rather like a modern **planisphere**), with a sighting device attached. Astrolabes were used from the time of the ancient Greeks down to the 17th century for navigation, measuring time, and terrestrial measurement of height and angles.

astrology A pseudoscience professing to assess people's personality traits and to predict events in their lives and future trends in general from aspects of the heavens, in particular the positions of the planets. Astrology is based on ideas that are scientifically unsound and dismissed by the majority of rational people, but horoscopes are still published in the popular press, though mainly for their entertainment value. In many ancient cultures it was for astrological reasons that observational records first came to be kept. Such records from ancient China have proved of great historical importance in research into past eclipses, novae and comets. In later times, casting horoscopes provided astronomers such as Johann **Kepler** with the livelihood to sustain them in their astronomical work.

astrometric binary A star that is recognized as a **binary star** from its wave-like **proper motion**, caused by its orbital motion with an unseen companion.

astrometry The branch of astronomy concerned with the measurement of precise positions of celestial objects, usually determined from images obtained specially for the purpose. **Parallaxes** and **proper motions** of stars are derived from positions obtained at widely differing times. The most accurate positional measurements available were made by the astrometry satellite **Hipparcos**.

Astronomer Royal An honorary title bestowed upon a prominent British astronomer (until 1972, an additional title of the Director of the Royal Greenwich Observatory). The first, John Flamsteed, was appointed by Charles II when he founded the Greenwich Observatory. SEE the table on page 20.

astronomical twilight The dawn and dusk periods when the Sun is less than 18° below the horizon. SEE ALSO **twilight**.

astronomical unit (symbol AU) The mean distance between the Earth and the Sun, used as a fundamental unit of distance,

ASTRONOMERS ROYAL

Name	Held office
John **Flamsteed** (1646–1719)	1675–1719
Edmond **Halley** (1656–1742)	1720–42
James **Bradley** (1693–1762)	1742–62
Nathaniel Bliss (1700–64)	1762–64
Nevil **Maskelyne** (1732–1811)	1765–1811
John Pond (1767–1836)	1811–35
George Biddell **Airy** (1801–92)	1835–81
William Christie (1845–1923)	1881–1910
Frank Watson Dyson (1868–1939)	1910–33
Harold Spencer **Jones** (1890–1960)	1933–55
Richard Woolley (1906–86)	1956–71
Martin **Ryle** (1918–84)	1972–82
Francis Graham-Smith (1923–)	1982–90
Arnold Wolfendale (1927–)	1991–94
Martin Rees (1942–)	1995–

particularly for distances in the Solar System. It is equal to 149,598,000 km (92,956,000 miles). There is a more complicated 'official' definition, which gives a value that differs from the mean Earth–Sun distance by a few kilometres.

astrophotography The application of photography to astronomy. Astrophotography makes it possible to obtain a permanent record of the positions and appearance of objects, while long exposures show faint objects such as remote galaxies invisible to the naked eye. It differs from normal photography in that exposures can be very long, often an hour or more, and the camera or telescope must accurately track the object as the Earth rotates during the exposure.

In 1850 the first star images, of Vega and Capella, were obtained by W. C. Bond at Harvard College Observatory. Henry Draper obtained a spectrogram of Vega in 1872 and photographed the Orion Nebula in 1880. The 1880s marked photography's breakthrough in astronomy, as techniques improved and plates became more sensitive. The first photographic atlases of the sky were made, and libraries of stellar spectra were built up. In the 20th century astrophotography rapidly supplanted visual observation for many purposes, but has itself now been superseded by electronic detectors particularly **CCD**-based systems. Photography using conventional film is now practised only by amateur astronomers, and many of those are now using CCDs to image astronomical objects.

astrophysics The study of the physical properties of celestial bodies. It is based mainly on the study of radiation from these bodies, and has developed since the 19th century through the application of photography, photometry and spectroscopy. New observing techniques at wavelengths from radio and infrared to X-rays and gamma-rays enormously increased the scope of astrophysics in the second half of the 20th century. Theoretical astrophysics applies mathematical methods and physical laws to explain or predict the behaviour of the Universe and objects within it.

Atacama Large Millimeter Array (ALMA) A millimetre-wavelength telescope being built by a US–European consortium at Llano de Chajnantor, in the Chilean Andes. The instrument, at an elevation of 5000 m (16,400 ft), will consist of at least 64 antennas 12 m (39 ft) in diameter. The high altitude will give the telescope access to wavebands between 350 μm and 10 mm. ALMA is scheduled to begin operation in 2007 and to be fully completed in 2012.

Aten asteroid One of a class of small **asteroids** whose orbits lie largely within the Earth's (specifically, those with **semi-major axis** less than 1 AU). They are named after the 1-km (0.6-mile) diameter object 2062 Aten. More than 250 were known by late 2004. SEE ALSO **Amor asteroid**, **Apollo asteroid**.

Atlas One of the small inner **satellites** of Saturn, discovered in 1980 during the Voyager missions. Atlas orbits just outside Saturn's A Ring, and seems to act as the **shepherd moon** to the whole bright ring system.

atmosphere The envelope of gases around a star, planet or satellite. The ability of a planet or satellite to retain an atmosphere depends on its outer temperature, determined by its distance from the Sun, and its escape velocity, determined by its mass. Smaller bodies in the inner Solar System, such as Mercury and the Moon, have almost no atmosphere or none at all, while the giant planets have deep, massive atmospheres. Far from the Sun, even small worlds like Pluto and Triton have clung on to a tenuous atmosphere. Venus, Earth and Mars were too small to hang on to the lighter gases, such as hydrogen, from their original atmospheres, which had the composition of the solar nebula from which the Solar System formed. Subsequent geological processes such as **volcanism** have modified their atmospheres considerably, and they now contain carbon dioxide, oxygen and nitrogen (although in very different proportions). The giant planets have their original, largely hydrogen, atmospheres. Saturn's largest moon, **Titan**, is the only satellite with a substantial atmosphere, mainly of nitrogen.

The outer layers of a star, which is where the lines in the star's spectrum originate, are referred to as its atmosphere.

▼ Atacama Large Millimeter Array The Chajnantor region in the Chilean Andes has been chosen for the new millimetre-wavelength telescope.

atmospheric extinction The reduction in the brightness of light from celestial bodies when it passes through the Earth's atmosphere. The effect is more pronounced for objects at low altitude, for their light has then had to pass through a greater volume of atmosphere. Observations of, for example, a variable star have to be corrected for extinction, which can dim the star's light by half a magnitude when viewed at an altitude of 20°, and by a whole magnitude at 10°. The main cause of atmospheric extinction is the **scattering** of light by molecules in the atmosphere. As scattering affects wavelengths at the blue end of the visible spectrum more than it does the red, the light becomes reddened, as is most obvious at sunrise or sunset.

atmospheric refraction The small apparent change in altitude of a celestial object, seen by an observer at the Earth's surface, caused by its light passing through the Earth's atmosphere. Just as light is refracted when it passes from air into glass, so it is refracted when it passes from the vacuum of space into the atmosphere. Atmospheric refraction makes a body appear to be at a higher altitude than it actually is. The effect is greater the closer the body is to the horizon, and a body can even become visible before it has risen. All

astronomical observations have to be corrected for atmospheric refraction in order to obtain the true position from the observed apparent position.

AU ABBREVIATION FOR **astronomical unit**.

Auriga A large northern constellation, containing the first-magnitude star **Capella**. Mythologically it represents Erichthonius, a king of Athens said to have invented the four-horse chariot. Beta (Menkalinan), magnitude 1.9, is an eclipsing variable of very small range. The stars Eta and Zeta are popularly called the Haedi or Kids. **Epsilon Aurigae** is a remarkable eclipsing binary. Zeta is also an eclipsing binary, consisting of a K5 giant orbited by a B7 main-sequence star with a period of 972 days; the visual range is magnitude 3.7 to 4.0. There are three prominent open clusters: M36, M37 and M38.

aurora (plural **aurorae**) An illumination of the night sky, also known as the *aurora borealis* in the northern hemisphere and as the *aurora australis* in the southern. Aurorae can take many forms, including *arcs* and *bands*, which may be *homogeneous* or show vertical *ray* structure; folded rayed bands resemble rippling curtains. Green and red colours predominate, resulting from excitation of atmospheric oxygen and nitrogen by electrons accelerated in Earth's **magnetosphere**. Emissions are seen at altitudes typically between 100 and 1000 km (60 and 600 miles), sometimes reaching as low as 80 km (50 miles) during strong displays. Auroral activity is present more or less continuously in oval regions surrounding either geomagnetic pole at high latitudes. Aurorae are commonest at lower latitudes at times of high solar activity, when solar **flares** and **coronal mass ejections** deliver powerful bursts of charged particles via the **solar wind**. Aurorae have been observed on all the giant planets, which have strong magnetic fields.

Australia Telescope National Facility A large **aperture synthesis** telescope in New South Wales, the most powerful radio astronomy instrument in the southern hemisphere. With three separate parts, it is an example of the use of **very long baseline interferometry**. One part consists of a 6 km line of six dish aerials, each 22 m (72 ft) in diameter, at Culgoora. These can be linked to another 22-m dish, 100 km to the south at Mopra, and also the large radio dish 200 km farther south at **Parkes Observatory**. The Australia Telescope came into operation in 1988.

autumnal equinox SEE **equinox**.

averted vision Looking slightly to one side of a faint object in order to see it better. The part of the retina in the human eye most sensitive to low light levels is not right at the centre, but in a small region surrounding the centre. Using averted vision when observing causes the light from the object being observed to fall in this region, making it easier to see.

axis (plural **axes**) The imaginary line, joining the north and south poles of a celestial body, about which it rotates. The *magnetic axis* of a body possessing a magnetic field is the line joining the north and south poles of the field. A body's rotational and magnetic axes do not necessarily coincide; with **Uranus** and **Neptune**, for example, they are markedly different.

azimuth The angle measured westward from north in a horizontal plane to the vertical circle (the meridian) that runs through a celestial object. SEE ALSO **coordinates**.

B

Baade, (Wilhelm Heinrich) Walter (1893–1960) German-American astronomer. From Mount Wilson Observatory in the 1943 wartime blackout he was able to observe individual stars in the Andromeda Galaxy and distinguish the younger, bluer Population I stars from the older, redder Population II stars (SEE **stellar populations**). He went on to improve the use of **Cepheid variable** stars as distance indicators, and showed that the Universe was older and larger than had been thought.

background radiation SEE **cosmic microwave background**.

Baily's beads A very short-lived phenomenon seen during a solar eclipse, just before or after totality, in which the extreme edge of the Sun's disk appears to break into a string of bright lights. It is caused by rays of light from the Sun shining through the valleys on the Moon's limb, while other rays are blocked off by mountains on the limb. It was first described by English astronomer Francis Baily (1774–1844), who observed it in 1836. SEE ALSO **diamond-ring effect**.

Balmer series The distinctive series of lines in the **hydrogen spectrum**. The Balmer series is a dominant characteristic of the spectra of A stars. The first Balmer line is called the **hydrogen alpha line** ($H\alpha$), and is at 656.3 nm; the second line is $H\beta$, at 486.1 nm; the third is $H\gamma$, at 434.2 nm; and so on. The formula governing the wavelengths of lines in the spectrum of hydrogen was worked out in 1885 by Swiss mathematics teacher Johann Balmer (1825–98).

Barlow lens An extra lens used in conjunction with a telescope's eyepiece in order to increase the magnification, usually by a factor of 2. It can also improve definition. (After English physicist Peter Barlow, 1776–1872.)

Barnard, Edward Emerson (1857–1923) American astronomer. He was a skilled visual observer, and also a pioneer of photographic astronomy. Barnard discovered many comets, the fifth satellite of Jupiter (Amalthea) and **Barnard's Star**. He speculated (correctly) that what appear to be regions containing few stars are dark nebulae, and published a catalogue of them.

Barnard's Loop A large but faint emission nebula, spanning 14°, in the form of an arc centred on the Orion **OB association**. It appears to be the result of gas and dust blown away from the association by radiation pressure. It was discovered by E. E. Barnard.

Barnard's Star A red dwarf star 6 light years away in the constellation Ophiuchus. It has the largest known **proper motion**, at 10.4 arc seconds per year, and is the closest star to the Sun after the Alpha Centauri system. It was discovered in 1916 by E. E. Barnard. Barnard's Star is magnitude 9.5, with a luminosity 0.0004 times the Sun's, and type M4.

barred spiral galaxy A spiral **galaxy** in which the arms extend from the ends of a bar, an elongated distribution of stars

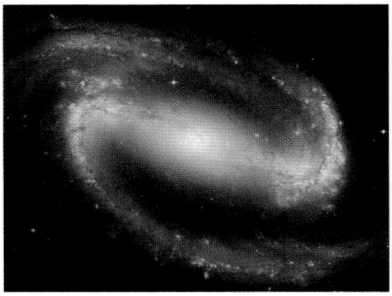

▲ **barred spiral galaxy** NGC 1300 in Eridanus is a typical barred spiral galaxy, with arms connected to the ends of a bar, which has the nucleus at its centre.

extending either side of the centre of the galaxy's disk. Most galaxies – including our own – are thought to possess a bar, which plays an important role in their evolution and in triggering star formation.

Barringer Crater SEE **Meteor Crater**.

barycentre The **centre of mass** of two or more celestial bodies and the point around which they revolve. The term is used in particular for the Earth–Moon system. As the Earth is so much more massive than the Moon, the barycentre lies within the Earth.

Bayer, Johann (1572–1625) German astronomer. In 1603 he published a star atlas, *Uranometria*, in which he introduced 12 new constellations. In it he allocated letters of the Greek alphabet (SEE the table on page 263) to the stars in each constellation: alpha (α), beta (β), gamma (γ), and so on, usually in order of brightness. This system of what came to be known as *Bayer letters* is still in use.

BD ABBREVIATION FOR ***Bonner Durchmusterung***.

Beijing Astronomical Observatory An observatory with its main optical/infrared observing site at Xinglong, about 150 km (95 miles) north-east of Beijing. It houses China's largest telescope the Large Sky Area Multi-Object Fiber Spectroscopic Telescope (LAMOST), due for completion in 2006. LAMOST is an unusual variant of the Schmidt telescope design. The main mirror, 6.7 × 6.0 m (265 × 236 inches), is fed light by a reflective corrector plate which moves to track objects and transmits light via fibre-optic tubes to stationary spectrographs.

Belinda One of the small inner **satellites** of Uranus discovered in 1986 during the Voyager 2 mission.

Bell Burnell, (Susan) Jocelyn (1943–) English astronomer. While a graduate student she ran a survey of radio galaxies which led to the discovery of the first **pulsar**. She is now generally credited, with her then supervisor, Antony **Hewish**, as the co-discoverer of pulsars, but she received no share of the Nobel prize subsequently won by Hewish.

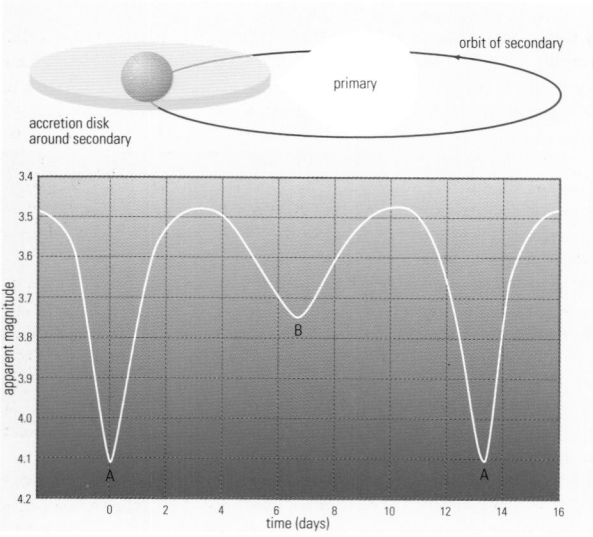

◀ Beta Lyrae star
The continuous light curve of Beta Lyrae stars is believed to be caused by an accretion disk of material which surrounds the secondary star and eclipses the primary. This theory also explains why the secondary star appears to be dimmer than the primary, despite being far more massive.

Bennett, Comet One of the brightest comets of the 20th century, discovered by South African amateur astronomer John Caister Bennett (1914–90) in 1969, and a first-magnitude object in April 1970. Its tail extended for 20°.

Bepi Columbo A European Space Agency spacecraft scheduled to enter orbit around Mercury in 2011. After launch in 2009, the craft will fly past Venus en route. Bepi Colombo will deploy a small lander, which will be the first craft to touch down on Mercury, and also a small sub-satellite to study the planet's magnetosphere.

Bessel, Friedrich Wilhelm (1784–1846) German mathematician and astronomer. His catalogue of 63,000 accurate star positions, based on refinements of previous observations as well as his own, marks the beginning of modern astrometry. In 1838 Bessel became the first to announce that he had determined the parallax of a star (61 Cygni). From the 'wobbles' in the motions of Sirius and of Procyon he deduced the existence of their companion stars, which were not detected until many years after his death.

Beta Centauri The second-brightest star in the constellation Centaurus, magnitude 0.61 (slightly variable), also known as Hadar or Agena. It is a giant of spectral type B1, and lies 525 light years away.

Beta Lyrae star A type of **eclipsing binary** in which gas escapes from a bloated primary star and falls onto an **accretion disk** that surrounds the secondary star. Beta Lyrae stars are members of a broad class of double stars known as *semi-detached binaries* so called because only one of the two stars fills its **Roche lobe**. In Beta Lyrae itself the primary star is overflowing its Roche lobe, and gases stream onto the disk at the rate of 0.00001 solar masses per year; some gas escapes from the system altogether. In contrast to **Algol**-type stars, Beta Lyrae variables do not show intervals of constant (or near-constant) light over part of their orbital period. The lower half of the diagram shows the light curve of Beta Lyrae; the primary minimum (A) is when the secondary star passes in front of the brighter primary, and the shallower minimum (B) is when the opposite occurs.

▲ **Betelgeuse** The Hubble Space Telescope's first direct image of a star other than our Sun revealed that Betelgeuse (α Orionis) has a bright spot more than 2000 K hotter than the rest of the star's surface. The exact nature of the spot is not yet understood.

Beta Pictoris SEE **Pictor**.

Betelgeuse The star Alpha Orionis, the second-brightest star in Orion. It is an orange-red supergiant of type M2 and a semiregular variable that ranges from magnitude 0 to 1.3 with a period of about $6\frac{1}{2}$ years. Betelgeuse lies 427 light years away. It has a diameter about 500 times the Sun's (about 700 million km/400 million miles) and is about 10,000 times as luminous.

Bethe, Hans Albrecht (1906–) German-American physicist. In 1938 he and Carl von **Weizsäcker** independently proposed a detailed theory to account for the production of energy in the Sun and other stars (SEE **carbon–nitrogen cycle**) in which hydrogen is converted into helium by nuclear fusion; this is essentially still accepted today. In recognition, he was awarded the 1967 Nobel Prize for Physics. SEE ALSO George **Gamow**.

Bianca One of the small inner **satellites** of Uranus discovered in 1986 during the Voyager 2 mission.

Biela's Comet A now disintegrated comet of orbital period 6.62 years, discovered by Wilhelm von Biela (1782–1856) in 1826. At its return in 1846 it was observed to have split into two parts. At the next return in 1852 the two parts were about 2.3 million km (3.7 million miles) apart. The comet has not been seen since, but in 1872 there was a magnificent display of the **Andromedid** meteor shower, at precisely the time when the Earth crossed the orbital path of the vanished comet. This was a dramatic demonstration that the particles that produce meteors are the debris of disintegrated comets.

big bang The theory that a giant explosion about 14 billion years ago created the Universe and started its expansion. It was developed in the 1940s by George **Gamow** from the ideas of Georges **Lemaître**.

According to the big bang theory, everything in the Universe once constituted an exceedingly hot and compressed gas with a temperature exceeding 10,000 million degrees K. When the Universe was only a few minutes old its temperature would have been 1000 million degrees K. As it cooled, nuclear reactions took place that led to the material emerging from the fireball consisting of about 75% hydrogen and 25% helium by mass, the composition of the Universe as we observe it today. Small quantities of deuterium, helium-3 and lithium were also produced, and their abundances, together with the lack of heavier elements, enable a limit to be placed on the amount of ordinary matter in the universe (SEE **critical density**). There were fluctuations from place to place in the density or expansion rate. Slightly denser regions of gas whose expansion rate lagged behind the mean value collapsed to form galaxies when the Universe was perhaps 10% of its present age.

The **cosmic microwave background** radiation, first detected in 1965, is considered to be the residual radiation of the big bang explosion. The *ripples* in this background radiation discovered in 1992 by the **Cosmic Background Explorer** and later confirmed by the **Wilkinson Microwave Anisotropy Probe** are evidence of the initial density variations from which galaxies formed.

The expansion of the Universe, the abundance of the light elements and the **cosmic microwave background** radiation are the three strongest observational supports for the big bang theory. However, there remain some problems that the basic or standard version of the big bang does not explain.

We can trace the Universe back at least to an age of 10^{-5} seconds after the big bang, when it contained matter, antimatter, and radiation at a temperature of around 10^{12} K. In the standard big bang theory we cannot extrapolate back to the moment of creation itself because the theory predicts the existence of a singularity, where the laws of physics break down. Suggestions for an origin have arisen from progress in the unification of the fundamental forces of physics, known as **grand unified theories**. This, though, leads back to one of the three main problems with the big bang theory: the so-called causality problem, of why the Universe should have expanded from a singularity (SEE ALSO **oscillating universe**).

Other observations that the standard big bang theory fails to explain have led some cosmologists to postulate that there was a period of **inflation** very soon after the big bang occurred. The observed amount of ordinary matter in the Universe is only 5% of that needed for the **critical density**, leading to the **missing mass** problem and the possible existence of **dark energy** and **dark matter**. Finally, there is the antimatter problem: the question of why there are not equal amounts of matter and antimatter in the Universe, the solution to which may lie with particle physics.

Big Bear Solar Observatory A solar observatory built on an artificial island in Big Bear Lake, in the San Bernardino

► **Big Bear Solar Observatory** The suite of telescopes in the dome of the Big Bear Solar Observatory observe the Sun in visible, near-infrared and ultraviolet light, thus giving astronomers detailed information about different parts of the solar surface and atmosphere. This information helps them to monitor and understand the way in which features on the Sun's surface form and evolve.

Mountains, California. The surrounding expanse of water ensures that there is none of the bad seeing caused by heat haze over land. The main instrument is a 0.65-m (26-inch) reflector, opened in 1973.

Big Dipper Popular US name for the group of stars in Ursa Major known in the UK as the *Plough*.

binary pulsar A **pulsar** orbiting another star, forming a binary. The first to be discovered was PSR 1913+16 in 1974; its companion is also thought to be a neutron star, but one that does not emit pulses. The orbital period is nearly 8 hours. Observations show that the pulsar's orbit is gradually contracting, caused by energy being emitted in the form of **gravitational waves**, as predicted by Einstein's General Theory of Relativity. American radio astronomers Joseph Taylor and Russell Hulse were awarded the 1993 Nobel Prize for Physics for the discovery of PSR 1913+16. In 2000, a binary system was discovered in which both stars are pulsars.

binary star Two stars in orbit around a common **centre of mass**. They are usually classified as visual, spectroscopic or eclipsing binaries, according to the means by which they are observed. Their orbital periods vary enormously, from hours to years for spectroscopic and eclipsing binaries, and from a decade to many centuries for visual binaries. A star with more than two components is called a *multiple star*, **Castor** being an example.

In a *visual binary*, of which about 50,000 are known, the components are sufficiently far apart, with an angular separation greater than about 0.1 arc seconds, for the two to be seen separately through a telescope. The first visual binary discovered was **Mizar**, in 1650 by Giovanni Battista Riccioli, who saw that it had a close companion (as distinct from its wider companion, Alcor).

In an **eclipsing binary**, one star periodically passes in front of the other so that the total light output appears to fluctuate. Most eclipsing binaries are also spectroscopic binaries. **Algol** was the first eclipsing binary to be identified, by John **Goodricke** in 1782.

A **spectroscopic binary** is a system whose components are too close for their separation to be measured visually. Instead, the binary nature is revealed by a Doppler shift in the absorption lines of the stars' spectra as they move around their orbits. The brighter component of Mizar was the first star shown to be a spectroscopic binary, by Edward **Pickering** in 1889.

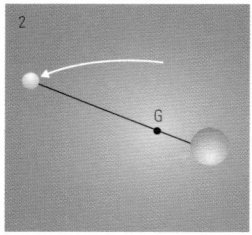

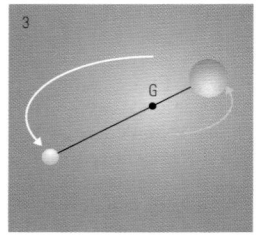

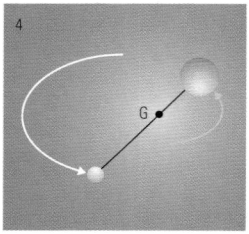

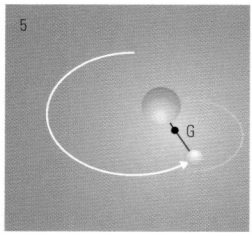

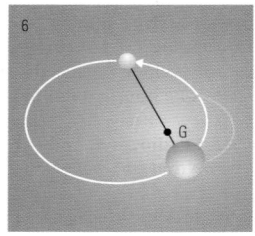

▲ **binary star** The two components of a binary system move in elliptical orbits around their common centre of gravity (G), which is not halfway between them but nearer to the more massive component. In accordance with Kepler's second law of motion, they do not travel at uniform rates. In (2) they are greatly separated and moving slowly. In (6) they are closer and moving more rapidly.

The study of irregularities in the **proper motions** of stars has in some cases revealed the existence of invisible companions; such a star is known as an *astrometric binary*.

binoculars Two small telescopes joined by a hinge, giving the observer a low-magnification view with both eyes simultaneously. The hinge allows the binoculars to be adjusted for the distance between the eyes, and there is usually a means of focusing the eyepieces independently. Compactness is achieved by using prisms to fold the light-paths within the instrument. Binoculars have designations such as 8 × 40, where 8 is the magnification and 40 is the diameter of each object lens in millimetres.

bipolar flow The flow of material from a star in two streams in opposite directions. In young stars this can produce **Herbig–Haro objects**; in old stars, bipolar planetary nebulae. A star that generates bipolar flow is usually surrounded by a dusty envelope that is oriented perpendicularly to its rotation axis. The interplay of the star's rotation and mass outflow, and its magnetic field, is thought to create bipolar flows.

black-body radiation The radiation that would be emitted by a *black body*, a hypothetical object that absorbs all thermal radiation (heat) falling on it, and is a perfect emitter of thermal radiation. The spectrum of the radiation emitted by a black body is a **continuous spectrum**, and the intensity of the radiation is greatest at a wavelength that depends only on the body's temperature. Although the black body is only hypothetical, its importance in astronomy is that the spectra of stars can be interpreted by assuming that they approximate to black bodies. SEE ALSO **effective temperature**.

black drop An optical effect, visible at **second contact** or **third contact** during a transit of Mercury or Venus across the Sun's disk, in which a dark ligament – the 'drop' – appears to connect the limb of the planet with the limb of the Sun. The phenomenon limits the accuracy of the timing of transits.

Black Eye Galaxy A spiral galaxy (M64, NGC 4826) in the constellation Coma Berenices which contains a dark cloud of dust near the nucleus, creating the 'black eye' effect. The Black Eye Galaxy lies 24 million light years away.

black hole A localized region of space from which neither matter nor radiation can escape – in other words, where the escape velocity exceeds the velocity of light. The boundary of this region is called the *event horizon*. Its radius, the **Schwarzschild radius**, depends on the amount of matter that has fallen into the region and it increases linearly as the mass increases.

A black hole of stellar mass is thought to form when a massive star undergoes total gravitational collapse. For stars up to about 1.4 solar masses, gravitational collapse can be physically halted to produce a **white dwarf**. A slightly more massive object will collapse to form a **neutron star**. If, however, the mass exceeds about 3 solar masses, even after a supernova explosion has blasted away the outer layers of the star, the collapse continues beyond even the neutron star stage. As it contracts below its Schwarzschild radius, the object becomes a black hole and effectively disappears. A star of 3 solar masses has a Schwarzschild radius of about 9 km. Inside the event horizon of the black hole, space and time are highly distorted and the stellar matter is increasingly compressed until it forms an infinitely dense **singularity**.

Black holes can have an immense range of sizes. *Supermassive black holes* with up to a billion solar masses are believed to be the source of energy in **quasars** and other **active galactic nuclei**. At the other end of the scale, some primordial black holes (SEE below) could be truly microscopic.

Because no light or other radiation can escape from black holes, they are extremely difficult to detect. Any matter encountered by a black hole would probably first go into orbit rather than being drawn directly into it. A rapidly spinning disk of matter, known as an **accretion disk**, forms around the object and heats up through friction to such high temperatures that it emits X-rays. Black holes may therefore appear as X-ray sources in binary stars. An example is **Cygnus X-1**.

Not all black holes result from stellar collapse. During the big bang, some regions of space might have become so compressed that they formed so-called *primordial black holes*. Such black holes would not be completely black, because radiation could still 'tunnel out' of the event horizon at steady rate, leading to the evaporation of the hole. (Such small-scale effects are not important for larger black holes.) Primordial black holes could thus be very hot.

blazar A term, compounded from **BL Lacertae object** and **quasar**, for a galaxy with an exceptionally **active galactic nucleus**. Blazars show variable optical brightness, strong and variable optical polarization, and strong radio emission. The optical variations may be on timescales as short as days. The explanation for their exceptional activity is that in these galaxies we are viewing relativistic **jets** end-on.

BL Lacertae object (BL Lac object) A highly luminous **active galactic nucleus**, the prototype of which, BL Lacertae, was originally classified as a variable star. BL Lacertae objects are variable at all wavelengths from radio to X-rays, sometimes on a timescale of just a few hours, and their optical spectra are unusual in being completely featureless, containing neither absorption nor emission lines. This is consistent with them being jets from **quasars** or **radio galaxies** emitted along our line of sight.

blue giant A massive star that has exhausted the hydrogen at its core and initiated helium burning, causing it to expand and leave the **main sequence**. Blue giants have high surface temperatures, of about 30,000 K, and luminosities about 10,000 times that of the Sun. At a later stage of evolution, they expand still further and cool, becoming red **supergiants**.

blue moon An occasional blue coloration of the Moon caused by particles injected into the upper atmosphere by, for example, a large volcanic eruption or an extensive forest fire. The particles scatter light, making it appear bluer. The term has also come to mean a second full Moon in the same calendar month, which happens about seven times in every 19 years, and is used in everyday speech to denote a rare event.

blue straggler A bright, blue star in a globular cluster that remains on the **main sequence** long after stars of similar or lower mass have evolved into red giants. It is thought that they 'straggle behind' because they are close binaries in which mass has been transferred from one to the other, or, in some cases, the two stars have merged.

Bode, Johann Elert (1747–1826) German astronomer. He published the theoretical relationship between the distances of the planets that had previously been pointed out by Johann Titius (1729–96). On the basis of this relationship (SEE **Bode's law**), Bode suggested the existence of an undiscovered planet between Mars and Jupiter. In 1801 he published the beautifully illustrated star atlas *Uranographia*, and a catalogue of 17,000 stars and nebulae.

Bode's law More properly called the *Titius–Bode law* (SEE **Bode**), a numerical sequence announced by Bode in 1772 that matches the distances from the Sun of the then known planets. It is formed by adding 4 to each number in the sequence 0, 3, 6, 12, 24, 48, 96 and 192, giving 4, 7, 10, 16, 28, 52, 100 and 196. Taking the Earth's distance from the Sun as 10, Mercury, Venus and Mars fell into place quite well (at 3.9, 7.2 and 15.2), as did Jupiter and Saturn (52 and 95). The discovery in 1781 of Uranus at a distance corresponding to 192 prompted a search for a major planet to fill the gap at 28, but led instead to the discovery of the asteroids. The next number in the sequence is 388, but although Neptune (301) does not fit the bill, Pluto (395) does.

Although, in the form in which it was originally proposed, Bode's law has no real basis, it does give an indication of how **res-**

◀ **Bok globule** The dark cloud Barnard 68 (B68), shown here in an image obtained by the Very Large Telescope, is a Bok globule. It is 410 light years away.

onances and tidal interactions (SEE **tides**) between bodies in the Solar System can result in them having orbits which are **commensurable**.

Bok, Bartholomeus Jan ('Bart') (1906–83) Dutch-American astronomer. He made extensive studies of the Milky Way, particularly of its star-forming regions (SEE **Bok globule**). Bok showed that stellar **associations** consist of young stars.

Bok globule A round, dark cloud of gas and dust which is a likely precursor of a **protostar**; named after Bart **Bok**. Small globules have diameters of 0.1 light year (6000 AU) and can be seen in front of bright nebulae. Larger globules, up to 3 light years across, are seen as dark patches against the stellar background of the Milky Way. Estimated globule masses range from about 0.1 solar masses for the smallest to about 2000 solar masses for those in the Rho Ophiuchi dark cloud.

bolide An exceptionally bright meteor – a major **fireball** – that is accompanied by a sonic boom. Objects whose passage through the atmosphere produce bolides may survive to reach the surface as meteorites.

bolometer An instrument for measuring all the electromagnetic radiation from a source, used in astronomy to measure the total radiation from a star, by registering the change in electrical resistance that results when radiation falls on a thin conductor.

bolometric magnitude (symbol m_{bol}) A measure of the total amount of energy radiated by a body at all wavelengths and expressed on the visual magnitude scale.

Bondi, Hermann (1919–) British cosmologist and mathematician, born in Austria. With Thomas **Gold** and Fred **Hoyle** he developed the **steady-state theory**, in which the continuous creation of matter drives the expansion of the Universe. A by-product was Bondi and Hoyle's elucidation of the production of heavy elements inside stars. Bondi later demonstrated the reality of gravitational waves.

Bonner Durchmusterung (BD) A catalogue containing data on over 300,000 stars, published in 1859–62 by F. W. A. **Argelander** and extended in 1886 to over 450,000 stars by Eduard Schönfeld. The stars are numbered in declination zones from +90° to −23°, and are cited in the form 'BD +52° 1638'.

Boötes A prominent northern constellation containing **Arcturus**, the brightest star north of the celestial equator. In mythology Boötes represents a herdsman. Epsilon (Izar or Pulcherrima) is a beautiful double star, with K0 and A0 components, magnitudes 2.6 and 4.8, separation 2.9 arc seconds. Mu (Alkalurops) is a wide double, magnitudes 4.3 and 6.5.

Borrelly, Comet A periodic comet discovered by Alphonse Borrelly (1842–1926) in 1904. It has an orbital period of 6.9 years, and became the fourth comet to be visited by a spacecraft when the **Deep Space 1** probe flew past at a distance of 2200 km (1380 miles) in 2001. During the flyby, the nucleus was imaged as a dark (albedo 0.03) elongated body with a long axis of 6 km (4 miles) from which jets of gas were emerging.

bow shock A curved region ahead ('upwind') of a planet's **magnetosphere** where the solar wind is decelerated and deflected, rather like the effect set up in front of a ship's bow as it moves through the water. SEE ALSO **heliosphere**.

Bradley, James (1693–1762) English astronomer. In 1742 he succeeded Edmond Halley as Astronomer Royal. He discovered the **aberration** (1) of light – the first observational evidence for Copernicus's heliocentric theory – and the **nutation** of the Earth's axis. His accurate catalogue of 3000 stars was published posthumously, and formed the basis of Friedrich **Bessel's** catalogue.

Brahe, Tycho (1546–1601) Danish astronomer. In 1572 he observed the supernova in Cassiopeia, and his report of it, *De nova stella*, soon made him famous. Under the patronage of King Frederick II of Denmark he built and equipped two observatories – Uraniborg and Stjerneborg – on the island of Hven in the Baltic. He became the most skilled observer of the pre-telescopic era, expert in making accurate naked-eye measurements of the positions of stars and planets. He calculated the orbit of the comet seen in 1577, and this, together with his study of the supernova, showed that **Aristotle**'s picture of the unchanging heavens was wrong. He could not, however, accept the world system put forward by **Copernicus**. In his own planetary theory (the *Tychonian system*), the planets move round the Sun, and the Sun itself, like the Moon, moves round the stationary Earth. In 1597 he left Denmark and settled in Prague, where Johann **Kepler** became his assistant. The accurate series of planetary observations made by Tycho Brahe were used by Kepler in deriving his laws of planetary motion.

breccia A type of rock made up of coarse, angular fragments set in a finer-grained material. Impacts produced the fragments, which were later incorporated in a younger rock formation. Most of the lunar rock samples returned by the Apollo missions were breccias, and the surface rocks of many other cratered, geologically inactive bodies in the Solar System are expected to be made up largely of breccias.

bremsstrahlung (German, literally 'braking radiation') Electromagnetic radiation produced when energetic electrons are decelerated, for example when approaching the nucleus of an atom. Also known as *free–free radiation*, it is generated in ionized gas clouds (SEE **H II region**).

Bright Star Catalogue A catalogue of naked-eye stars (above magnitude 6.5) first published by Yale University in 1930. The current (5th) edition is in electronic form.

brown dwarf An object with a mass less than 0.08 solar masses, in which the core temperature does not rise high enough to initiate thermonuclear reactions. Such a *substellar object* is, however, luminous, because as it slowly shrinks in size it radiates away its gravitational energy. The sur-

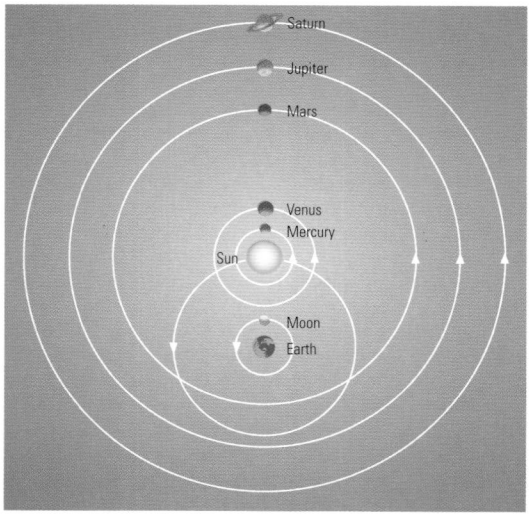

◀ **Brahe, Tycho** In the Tychonian model of the Solar System, there is a static Earth at the centre. The Moon and Sun orbit the Earth, and the other planets move around the Sun.

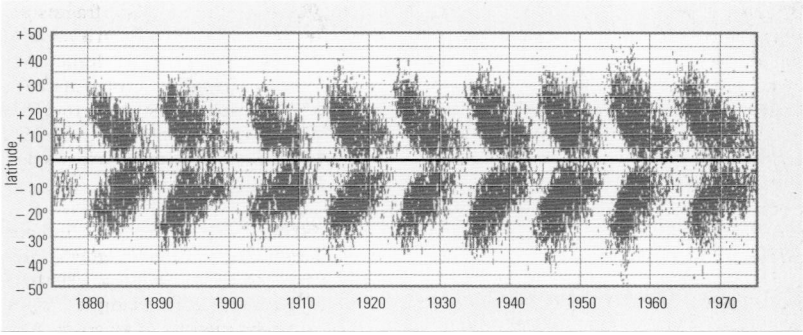

▲ **butterfly diagram** During the solar cycle, the latitude at which sunspots form changes. This is related to changes in the Sun's magnetic field during its cycle. Cycles usually overlap, with the first spots of a new cycle forming concurrently with the last spots of the old cycle.

face temperature of a brown dwarf is below the 2500 K lower limit of a **red dwarf**. The very coolest exhibit specific spectral features, and two new spectral classes – **L stars** and **T stars** – have been introduced to describe these stars. Over a hundred brown dwarfs had been found by 2004, both singly and in orbit around larger stars. One of the coolest, at 900 K, is the first brown dwarf to be discovered, designated Gliese 229B.

B star A star of spectral type B, characterized by strong neutral helium absorption lines and the presence of hydrogen lines (although not as strong as in the **A stars**). On the **main sequence**, the temperatures of B stars range from 10,500 K at B9 to 31,000 K at B0. Their masses and sizes range correspondingly from 2.3 to 14 solar masses, and from 2.5 to 10 solar radii. At their hottest, B stars have 25,000 times the luminosity of the Sun. Examples of B stars are Achernar, Regulus, Rigel and Spica. A B9 star stays on the main sequence for about 800 million years, but a B0 star remains there for only 11 million years. The hottest B stars (and the even hotter **O stars**) form loose groupings known as **OB associations**. O and B stars evolve to become supergiants with luminosities up to 100,000 times that of the Sun.

butterfly diagram A diagram illustrating the changing distribution of sunspots in solar latitude over the course of the 11-year cycle. At the start of a cycle there are spots at latitudes of up to around 30° north and south, but very few near the equator. As the cycle progresses, spots appear nearer the equator. When all the spots in a cycle are plotted on the diagram, a characteristic pattern in the shape of a butterfly appears. The diagram was devised by Edward Maunder in 1904.

C

Caelum A constellation of the southern sky, adjoining Columba, introduced by **Lacaille**; it represents an engraving tool. It has no star above magnitude 4.4. One variable, R Caeli, of Mira type, ranges from magnitude 6.7 to 13.7 with a period of 391 days.

Calar Alto Observatory A German–Spanish observatory near Almeria in southern Spain. Its main telescope is a 3.5-m (138-inch) reflector, opened in 1984.

calendar A list of the days of the year, grouped into months, and based on the apparent motions of the Sun or Moon, or both. The ancient Egyptians used a calendar based on a solar year. The Babylonians (and the modern Hebrews and Muslims) used a lunar year of 12 months, which is 11 days shorter than a solar year, so an extra month is added every third year.

The present *Gregorian calendar* was introduced by Pope Gregory in 1582 to correct a discrepancy of 10 days that had accumulated in the earlier *Julian calendar*, which had a year of 365 days and added a leap day every four years. This gave an average calendar year of 365.25 days, longer than a **tropical year** (the interval between two successive passages of the Sun through the vernal equinox) by over 11 minutes. The Gregorian calendar reduces the average length of the calendar year by not inserting a leap day in century years that are not divisible by 400. Thus the years 1600 and 2000 are leap years, but 1700, 1800 and 1900 are not. This gives a calendar year of 365.2425 days, which is accurate to one day in 3300 years. For religious and political reasons many countries did not adopt the Gregorian calendar for many years or even centuries. It was adopted in Great Britain and its territories in 1752, when 2 September was followed immediately by 14 September. Dates before and after this 'gap' are sometimes referred to as *Old Style* and *New Style*, respectively. SEE ALSO **Julian Day**.

Caliban An outer **satellite** of Uranus in a **retrograde** (1) orbit, discovered by Brett J. Gladman and Philip D. Nicholson in 1997.

California Nebula An emission nebula, designated NGC 1499, in the constellation Perseus. Discovered by E. E. **Barnard** in 1884–5, it takes its name from the similarity between its outline in long-exposure images to that of the American state. The nebula is part of a complex of gas and dust, 1000–1600 light years distant, from which stars are forming in an adjacent spiral arm of our Galaxy.

Callisto The second-largest of Jupiter's **Galilean satellites**, with a diameter of 4821 km (2996 miles), and the outermost of them. It has the lowest average density, at 1.86 g/cm^3, indicating that it contains a high proportion of water ice. The surface of the satellite is dark and very heavily cratered – in fact it is the most densely cratered object known. Callisto's ancient surface bears scars from the heavy bombardment phase of the Solar System's formation, ending 3.8 billion years ago. As well as the dense craters, there are several large, multiringed **impact features**, the largest being **Valhalla**, with an overall diameter of 4000 km (2500 miles) and Asgard (1600 km; 1000 miles). The **Galileo** probe found evidence for a 10-km (6-mile) deep subsurface ocean 100 km (60 miles) below Callisto's surface, and detected a weak magnetic field and a thin carbon dioxide atmosphere. Callisto orbits just outside Jupiter's main **radiation belt**. SEE ALSO **satellite**.

Caloris Basin (Caloris Planitia) The largest **impact feature** on Mercury. About half of it was photographed by the Mariner 10 probe in 1974–5. It is 1300 km (800 miles) in diameter and, like a number of similar features elsewhere in the Solar System, has a multiringed structure.

Calypso A small **satellite** of Saturn, discovered by Bradford Smith and others in 1980 between the two Voyager encounters. Calypso is irregular in shape, with an average diameter of 19 km (12 miles). It is **co-orbital** with Tethys and Telesto; Calypso and Telesto orbit near the L_5 and L_4 **Lagrangian points**, respectively, of Tethys's orbit around Saturn.

Camelopardalis A barren northern constellation representing a giraffe. Its brightest star, Beta, is of magnitude 4.0 and has a wide magnitude 8.6 companion. It contains the open cluster NGC 1502.

canals A network of dark linear markings on the surface of Mars, reported to exist by observers from the 1870s to the early 20th century. They were first described by Angelo **Secchi** and Giovanni **Schiaparelli** as *canali*, the Italian word for 'channels'. In some English reports *canali* was mistranslated as 'canals' – with the inference that they were artificial constructions. This idea was seized upon by, in particular, Percival **Lowell**, who was convinced that the canals were artificial waterways built by intelligent beings to irrigate a dying planet. Despite the demonstration in 1909 by Eugène **Antoniadi** that the canals were illusory, this romantic notion of Mars persisted in many minds until spacecraft visited Mars and sent back pictures that showed no sign of a canal network.

Cancer A dim zodiacal constellation, between Gemini and Leo. In legend it represents the crab sent by Juno, queen of the gods, and trodden on by Hercules during his fight with the multi-headed Hydra. Beta, magnitude 3.5, is its brightest star. Zeta is a multiple system; the main components are of magnitudes 5.0 and 6.2, and the brighter star is itself a close double. R Cancri is a Mira variable, range 6.1 to 11.8, period 362 days. RS Cancri is a semiregular variable (5.1 to 7.0, period about 120 days), as is X Cancri (5.6 to 7.5, period about 195 days). The large open cluster **Praesepe**, M44, is flanked by Gamma, 4.7, and Delta, 3.9, known as the Asses. Also in Cancer is M67, a very old open cluster visible in binoculars.

Canes Venatici A constellation introduced by **Hevelius** on his map of 1690, representing the dogs Asterion and Chara, held by the herdsman Boötes. The only star above fourth magnitude is Cor Caroli ('Charles's Heart'), magnitude 2.9. This name was given to it by Edmond Halley, in honour of King Charles I of England. The star is the prototype of the class of *magnetic variable* stars whose spectra contain lines that vary in intensity. It is a double, with a companion of magnitude 5.6, separation 20 arc seconds, excellent for small telescopes. Y is a semiregular variable, ranging in magnitude from 5.0 to 6.5 in a period of about 160 days. The famous **Whirlpool Galaxy**, M51, lies in Canes Venatici. Other spiral galaxies are M63, M94 and M106. There is also the 6th-magnitude globular cluster M3, which is distinct in binoculars.

Canis Major A constellation representing one of Orion's hunting dogs. It is distinguished by the presence of **Sirius**, the brightest star in the sky. Other bright stars are Epsilon (Adhara), magnitude 1.4; Delta (Wezen), 1.8; and Beta (Mirzam), 2.0. Beta is variable over a very small range and is a prototype of the class of variables known either as *Beta Canis Majoris stars* or as *Beta Cephei stars*. Other variables are R (5.7 to 6.3, 1.14 days, Algol type), W (6.4 to 7.9, irregular) and UW (4.8 to 5.3, 4.39 days, Beta Lyrae type). There is one prominent open cluster, M41, of 5th magnitude, south of Sirius.

Canis Minor A constellation representing the smaller of Orion's two dogs. The first-magnitude star **Procyon** makes it easy to locate. Beta (Gomeisa), magnitude 2.9, is a slightly variable **shell star**. There are no other objects of note, although three Mira variables (V, R and S) can rise to above magnitude 8 at maximum.

Cannon, Annie Jump (1863–1941) American astronomer. Working at the Harvard College Observatory, initially as an assistant to E. C. **Pickering**, she reorganized the classification of stars into spectral types, and classified the 225,000 stars in the ***Henry Draper Catalogue*** of stellar spectra.

Canopus The star Alpha Carinae, the second-brightest star in the sky, magnitude −0.62. It is a bright giant of type F0, more than 10,000 times as luminous as the Sun, and lies 313 light years away.

Capella The star Alpha Aurigae, magnitude 0.08 (making it the sixth-brightest star in the sky), distance 42 light years. It is a spectroscopic binary, consisting of two giant stars of types G6 and G2 with an orbital period of 104 days.

Capricornus One of the less prominent of the zodiacal constellations; it has been identified with the Greek god Pan. Its brightest star is Delta (Deneb Algedi), magnitude 2.9. Alpha (Algedi) is a wide optical double, magnitudes 3.6 and 4.2. The semiregular variable RT has a range from 6th to 9th magnitude and a period of 393 days. The most important cluster is M30, a globular of magnitude 7.5.

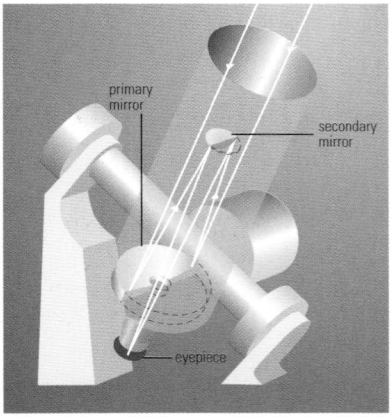

▲ **Cassegrain telescope** Light collected by the primary mirror is reflected from a hyperboloidal secondary mirror to the eyepiece through a hole in the centre of the primary.

captured rotation SEE **synchronous rotation**.

carbonaceous chondrite A rare type of chondritic meteorite (SEE **chondrite**) with a higher-than-average content of carbon, in the form of organic compounds. Not counting the **volatiles**, the proportions of chemical elements in a carbonaceous chondrite are very similar to the composition of the Sun. These meteorites are therefore believed to have come from bodies that formed very early in the Solar System's history, which makes them important in the study of its origin.

carbon–nitrogen cycle A sequence of nuclear reactions that accounts for the production of energy inside main-sequence stars that are more massive and hence hotter than the Sun. It was first described in 1938 by Hans **Bethe**. The reactions involve the fusion of four hydrogen nuclei into one helium nucleus with the release of a huge amount of energy. Nitrogen and oxygen are formed as intermediate products. The presence of carbon is essential, but it behaves as a catalyst and remains unchanged at the end of the cycle. The process takes place at temperatures above 15 million degrees K. It is also known as the *carbon cycle* and the *carbon–nitrogen– oxygen cycle*.

carbon star (C star) A cool red giant star whose surface composition shows carbon-based molecules such as carbon monoxide (CO), cyanogen (CN), molecular carbon (C_2) and other compounds of carbon, and also the presence of much lithium. A star of about the Sun's mass in the red-giant phase has carbon, produced by nuclear fusion at its core, brought to the surface by convection currents. The carbon reacts to form the molecules visible in its spectrum. Carbon stars are allocated spectral type C in the Morgan–Keenan system of **spectral classification**.

Carina A prominent southern constellation, once part of the larger grouping of Argo Navis, the ship Argo, and representing the ship's keel. It contains **Canopus**, the second-brightest star in the sky. Other bright stars are Beta (Miaplacidus), magnitude 1.7, and Epsilon (Avior), magnitude 1.9. Epsilon and Iota (2.2) form part of the **False Cross**. Upsilon is a double, 3.0 and 6.3. Carina's variable stars include the unique **Eta Carinae**, which lies within the nebula NGC 3372. Other variables with maximum above magnitude 6.5 are the Cepheids U (range 5.7 to 7.0, period 38.77 days) and l ('ell', 3.3 to 4.2, 35.54 days), as well as the Mira-type stars R (3.9 to 10.5, 309 days) and S (4.5 to 9.9, 150 days). There are several open clusters visible to the naked eye, principally NGC 2516, NGC 3114 and IC 2581. The open cluster IC 2602 around Theta is spectacular when viewed with binoculars. There is also a 6th-magnitude globular cluster, NGC 2808.

Carme An outer **satellite** of Jupiter discovered in 1938 by Seth Nicholson. It is in a **retrograde** (1) orbit, and may be a captured asteroid.

Cassegrain telescope A **reflecting telescope** in which the light received by the primary mirror is reflected back from the secondary mirror through a hole in the primary. The light is brought to a focus – the *Cassegrain focus* – on the axis of the telescope, where the image can be observed by the eye or by instruments. The combination of a concave, **paraboloidal** primary and a convex (usually **hyperboloidal**) secondary tends to cancel **aberrations** (2), such as coma, of the separate mirrors. The arrangement is used for both amateur and professional telescopes. It was devised in 1672 by the French priest Laurent Cassegrain (1629–93). SEE ALSO **Schmidt–Cassegrain telescope**.

Cassini French family of Italian origin that produced four generations of astronomers and cartographers, all of whom ran the Paris Observatory. The two most prominent were:

Cassini, Giovanni Domenico, known as **Jean Dominique** (1625–1712) after he moved to Paris in 1669. He was the first to accurately measure the dimensions of the Solar System. He discovered the division in Saturn's rings that now bears his name, and also four of that planet's satellites. He improved the **ephemerides** of Jupiter's satellites, which helped Ole Römer in his determination of the velocity of light.

Cassini, Jacques (1677–1756), son of Giovanni. He accurately measured an arc of a meridian running through France. The results helped to show that the Earth is flattened at the poles, although he and his father had believed there to be a polar elongation. He determined the proper motion of the star Arcturus.

▲ **Cassini-Huygens** Saturn and two of its moons, Titan and Rhea, can be seen in this image captured by the Cassini spacecraft on 5 February 2005, when it was 3.4 million km (2.1 million miles) from the planet. Titan is at lower left, and Rhea above centre.

Cassini-Huygens A joint space mission between NASA and the European Space Agency to explore Saturn and its satellites, in particular Titan. It was launched on 15 October 1997 and reached Saturn in July 2004. The main Cassini craft entered orbit around Saturn, on a four-year mission, while the Huygens sub-probe landed on **Titan** in January 2005 and returned data and images from the surface.

Cassini Division The gap between Rings A and B in the ring system of **Saturn**, discovered by Giovanni Cassini in 1675. The Voyager spacecraft showed that it is not a true gap, but contains several narrow rings.

Cassiopeia A distinctive northern constellation, representing the mother of Andromeda. The five leading stars make up a W or M pattern, on the opposite side of Polaris from Ursa Major. Alpha (Schedar) is its brightest star, magnitude 2.2. Gamma, magnitude range 1.6 to 3.0, is the prototype of the variables known as **shell stars**. Rho is a highly luminous semi-regular variable, usually of around magnitude 5 but which on rare occasions falls to below 6. Other variables are R (4.7 to 13.5, 430 days, Mira type), RZ (6.2 to 7.7, 1.2 days, Algol type) and SU (5.7 to 6.2, 1.95 days, Cepheid). Eta is a fine coloured binary, magnitudes 3.5 and 7.5; the orbital period is just under 500 years. Cassiopeia contains several open clusters resolvable in binoculars: M52, NGC 457, NGC 559, NGC 663 and M103.

Cassiopeia A The strongest radio source in the sky apart from the Sun, and the remnant of a supernova that exploded around 1660 but was not recorded by observers on Earth. It lies about 10,000 light years away and appears optically as a faint nebula.

Castor The star Alpha Geminorum, magnitude 1.58, distance 52 light years.

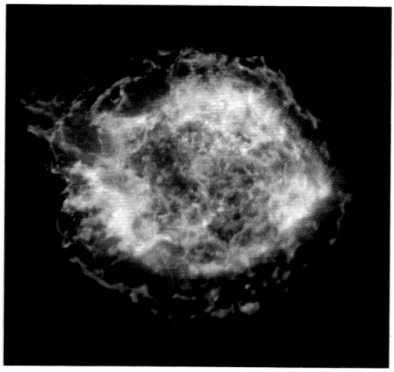

▲ **Cassiopeia A** This image of Cassiopeia A was obtained by the Chandra X-ray Observatory. The green ring marks the position of a shock wave generated by the supernova explosion. The blue areas are iron gas.

Castor is one of the finest visual binaries in the sky for small telescopes, consisting of main-sequence components of magnitudes 2.0 and 2.9, types A2 and A5, period 470 years. Each of the stars is itself a spectroscopic binary. A distant companion, Castor C, orbits the system. It is an eclipsing binary also known as YY Geminorum, made up of two faint red dwarfs, range 9.2 to 9.6.

cataclysmic variable A term used for a diverse group of stars that erupt in brightness, including **supernovae**, **novae**, **dwarf novae**, **symbiotic variables**, nova-like variables and some X-ray objects. With the exception of one type of supernova, these objects are very close binaries whose outbursts are caused by interaction between the two components. A typical system of this type has a low-mass secondary which fills its **Roche lobe** so that material is transferred over to the primary, which is usually a white dwarf. The transferred material has too much angular momentum to fall directly onto the primary, and so forms an **accretion disk**. A hot spot may arise on the outer edge of the accretion disk where the infalling material hits it. Outbursts occur when the flow rate varies, and take place at irregular, unpredictable intervals from weeks to many years – thousands of years, it is thought, for novae. In dwarf novae the thick disk brightens in outburst, but in novae (in particular) outbursts occur when the material that has accreted onto the white dwarf is ignited. SEE ALSO **eruptive variables**.

catadioptric A type of telescope in which a lens–mirror combination forms the image. The **Schmidt–Cassegrain telescope** is the commonest type of catadioptric design for amateur use; another is the **Maksutov telescope**. A catadioptric telescope is basically a reflector with a full-aperture lens at its top end. The lens corrects for **aberration** (2), and produces a divergent light-beam which is reflected onto the secondary mirror by a short-focus **spheroidal** or **paraboloidal** primary mirror. The design is compact and often used for portable instruments.

Cat's Eye A bright planetary nebula in the constellation Draco. Designated NGC 6543, it has a pronounced blue-green colour. It was from observations of this object that William **Huggins** was able to demonstrate, in 1864, that planetary nebulae shine by emission rather than reflection. The Cat's Eye lies at a distance of 3000 light years.

CCD Abbreviation for *charge-coupled device*, a solid-state electronic imaging device. A CCD is an integrated circuit whose upper layer contains a square array of closely spaced electrodes. Photons of light from a celestial object falling on an electrode are converted into an electronic charge, and the charges from all the electrodes are 'coupled' together and read off as a pattern – an instantaneous image of the object. A series of these images are fed to a computer and combined to create a picture, just as a photographic emulsion builds up a picture over the course of its exposure.

Images that formerly required hours of exposure on traditional photographic emulsion can be recorded in minutes by using a CCD. Fainter objects can be detected, the effects of **light pollution** can be compensated for, image processing by computer can be used to bring out detail, and images taken though coloured filters can be combined to give true-colour pictures. In professional astronomy CCD imaging has replaced traditional **astrophotography** techniques, and they are becoming increasingly cheaper and readily accessible to amateur astronomers.

celestial equator The projection of the Earth's equator onto the celestial sphere, dividing the sky into the northern and southern hemispheres.

celestial latitude, longitude SEE **coordinates**.

celestial mechanics The branch of astronomy dealing with the motions of heavenly bodies. It uses the laws of physics to explain and predict the orbits of planets, satellites and other celestial bodies.

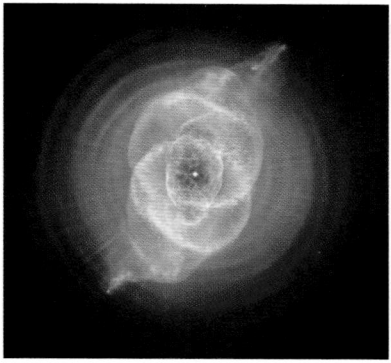

▲ **Cat's Eye** This detailed image was taken by the Hubble Space Telescope's Advanced Camera for Surveys. Concentric gas shells are clearly visible.

celestial meridian The **great circle** passing through the observer's zenith, the nadir, and the celestial poles, cutting the horizon at the north and south points.

celestial poles SEE **celestial sphere, pole**.

celestial sphere The imaginary sphere, of infinite radius and with the Earth at its centre, upon which all the celestial bodies appear to be projected as though they were at a uniform distance. The most obvious aspect of the celestial sphere is its apparent daily east-to-west rotation, caused by the axial spin of the Earth. The celestial sphere appears to rotate about the *celestial poles*, as though on an extension of the Earth's own axis, which points towards the celestial poles. Although the positions of the two celestial poles seem to be fixed, they are in fact slowly drifting because of the effect of **precession**.

The celestial sphere is divided into two halves by the *celestial equator*, which is equidistant from the poles. One can thus visualize a set of **great circles**, analogous to circles of latitude on Earth, called *declination circles*, each parallel to the equator. Declination circles allow the position of an object to be specified in terms of its angular distance north or south of the celestial equator.

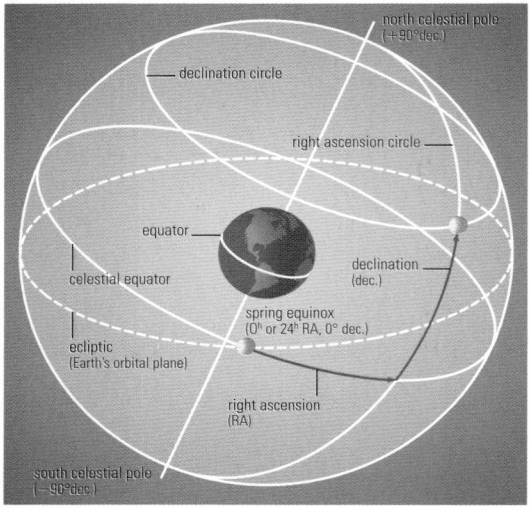

◀ **celestial sphere** A useful concept for describing positions of astronomical bodies is the celestial sphere. As shown, key points of reference – the celestial pole and celestial equator – are projections onto the sphere of their terrestrial equivalents. An object's position can be defined in terms of right ascension and declination (equivalent, respectively, to longitude and latitude).

Right ascension, the celestial equivalent of longitude, is also an extension of the terrestrial system. The zero of right ascension is related to the apparent motion of the Sun around the sky during the year, caused by the Earth's orbital motion around the Sun. The track traced by the Sun on the celestial sphere, called the **ecliptic**, indicates the plane of the Earth's orbit. The ecliptic is not the same as the celestial equator because the Earth's axis is tilted by about $23\frac{1}{2}°$.

For half of each year the Sun is in the northern hemisphere of the sky, and for the other half it is in the southern hemisphere. On two dates each year the Sun crosses the celestial equator; these points are known as the **equinoxes**. It is the *spring equinox*, when the Sun crosses into the northern hemisphere, that is chosen as the zero of right ascension. Right ascension is measured eastward from this point but, unlike terrestrial longitude, it is measured in units of time rather than degrees, minutes and seconds of arc (SEE **sidereal time**).

Centaur One of a class of small objects orbiting the Sun between Jupiter and Neptune. The first to be discovered was **Chiron**, and about 150 were known by late 2004. Typically 100 km (60 miles) in diameter, the Centaurs are probably icy **planetesimals** that have been perturbed inwards from the **Kuiper Belt** and now occupy short-lived, unstable orbits.

Centaurus A brilliant southern constellation representing a centaur, with 13 stars above magnitude 3.5. The brightest are **Alpha Centauri** and **Beta Centauri** (also known as Hadar or Agena), which indicate the direction of the Southern Cross. Gamma is a close binary with an 85-year period; the components are equal, at magnitude 2.9, and their combined brightness is magnitude 2.2. R is a Mira-type variable, range 5.3 to 11.8, and the period is unusually long, 546 days on average. T is a semi-regular variable (5.5 to 9.0, 90 days). The globular cluster **Omega Centauri** is the finest in the entire sky. There are also two bright open clusters, NGC 5460 and NGC 3766, and a fine planetary nebula, NGC 3918. The elliptical galaxy NGC 5128 is the radio source Centaurus A.

Centaurus A A strong radio source about 10 million light years away in the constellation Centaurus, identified with the galaxy NGC 5128. Optically it appears

▲ **Centaurus A** Crossed by a prominent dark lane, this elliptical galaxy (also designated NGC 5128) is a powerful radio and X-ray source.

as an elliptical galaxy crossed by a prominent dust lane. It appears to be a massive elliptical galaxy in collision with a smaller spiral with a high content of dust.

central meridian (c.m.) The imaginary line joining the north and south poles of a planet or the Sun, bisecting its disk. It is used by observers as a reference line to estimate the longitude of features as the body rotates.

centre of mass The point in a body, or a system of bodies, at which the total mass of the body or system may be regarded as concentrated. It is the point from which the gravitational attraction of the system as a whole appears to act. SEE ALSO **barycentre**.

Cepheid variable One of an important class of variable stars that pulsate in a regular manner, accompanied by changes in luminosity. They take their name from the first of the type to be discovered, Delta Cephei, whose variability was noted by John **Goodricke** in 1784. Cepheids periodically expand and contract, changing in size by as much as 30% in each cycle. The diagram shows the pulsations and light curve of Delta Cephei itself. A typical Cepheid variable has a surface temperature that varies between 6000 and 7500 K and a spectral type that ranges from G2 at minimum to F2 at maximum. The average luminosity is 10,000 times that of the Sun.

Cepheids became important in cosmology in 1912, when Henrietta **Leavitt** discovered a simple relationship between the period of light variation and the absolute magnitude of a Cepheid. This relationship, the **period–luminosity** law, enables the distances of stars to be ascertained, not only in our Galaxy but also in other galaxies.

There are two classes of Cepheid: type I, or *classical Cepheids*, are younger and more massive than type II, or **W Virginis stars**. Both types follow a period–luminosity relationship, but their curves are different. The classical Cepheids, such as Delta Cephei itself, are yellow supergiants

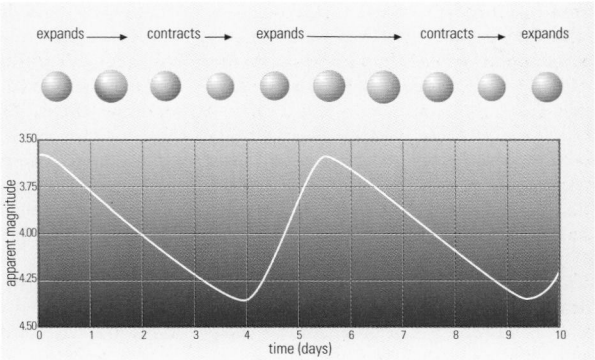

► **Cepheid variable** Their extremely regular light variations make Cepheids valuable standard candles: the longer the period from one maximum to the next, the greater the star's intrinsic luminosity. As shown, the typical Cepheid light curve shows a rapid rise to peak brightness, followed by a slower decline to minimum.

of Population I, while the second type are older, Population II stars and are found in globular clusters and the centre of the Galaxy (SEE **stellar populations**).

In using Cepheids to determine distances it is necessary to know which type is being observed. Classical Cepheids are one to two magnitudes brighter than type II Cepheids of the same period. The periods of Cepheids are in the range 1–50 days; classical Cepheids have a somewhat shorter period on average than those of type II.

Cepheus A constellation in the far north, adjoining Ursa Major, but not very distinctive. In mythology it represents the husband of Cassiopeia and the father of Andromeda. Alpha (Alderamin) is magnitude 2.4. Beta (Alfirk), magnitude 3.2, is variable over a small range, and is a prototype of the class known as either *Beta Cephei stars* or *Beta Canis Majoris stars*. Delta is the prototype **Cepheid variable**, with a range from magnitude 3.5 to 4.4 and a period of 5.37 days; it has a wide optical companion of magnitude 7.5. Mu, which William **Herschel** named the Garnet Star, is a semiregular red supergiant with a range of 3.4 to 5.1 and a period of about 2 years. T Cephei is a Mira variable (5.2 to 11.3, 388 days); VV is a giant eclipsing variable, 4.8 to 5.4, period 7430 days.

Ceres The first **asteroid** to be discovered, by Giuseppe **Piazzi** on 1 January 1801. Its diameter of 934 km (580 miles) makes it the largest asteroid. It orbits in the main asteroid belt, at an average distance from the Sun of 2.8 AU (414 million km; 257 million miles), the distance of the 'missing' planet predicted by **Bode's law**.

Cerro Tololo Inter-American Observatory An observatory on Cerro Tololo peak near La Serena, Chile, part of the US National Optical Astronomy Observatories. Its largest telescope is the Victor M. Blanco Telescope, a 4-m (158-inch) reflector, the twin of that at Kitt Peak. Other instruments include the 0.61-m (24-inch) Curtis Schmidt Telescope of the University of Michigan.

Cetus A large constellation, crossed by the celestial equator, but not very conspicuous; it represents the sea monster of the Andromeda legend. Its brightest star is Beta, magnitude 2.0. The most famous star in the constellation is the prototype long-period variable, **Mira**. Tau Ceti, magnitude 3.5, 11.9 light years away, is one of the closest Sun-like stars, being a main-sequence star of type G8. UV Ceti is a **flare star**; it is a pair of 12th-magnitude red dwarfs 8.7 light years away and exhibits large flares every few hours.

Chamaeleon A small constellation near the south pole, representing a chameleon. Its brightest stars are only of magnitude 4.1.

Chandrasekhar, Subrahmanyan (1910–95) Indian-American astrophysicist. He made significant contributions to the theories of stellar evolution, stellar atmospheres and relativity. His theory of white dwarf stars suggests there is a limit to their mass (SEE **Chandrasekhar limit**).

Chandrasekhar limit The maximum possible mass for a **white dwarf** star, first calculated in 1931 by Subrahmanyan **Chandrasekhar**. The value is about 1.4 solar masses, although more recent computations suggest that a higher value is possible for a rapidly rotating white dwarf. If a star has a mass which exceeds the Chandrasekhar limit it will collapse to become a **neutron star** or possibly a **black hole**.

Chandra X-ray Observatory (CXO) A NASA satellite for studying the sky at X-ray wavelengths, formerly known as the Advanced X-ray Astrophysics Facility, launched on 23 July 1999. It carries the highest-resolution **X-ray telescope** ever constructed. The main instruments are a CCD imaging spectrometer and a high-resolution camera. It is named in honour of Subrahmanyan **Chandrasekhar**.

charge-coupled device SEE **CCD**.

Charon The **satellite** of **Pluto**, discovered in 1978 by James Christy. Its diameter of 1186 km (737 miles) makes it the largest satellite in relation to its primary in the Solar System. Its mass is about 12% of Pluto's, and its density is about that of Pluto, 2.0 g/cm^3. It is unique among satellites in having a **synchronous orbit** with a period (6.4 days) matching the rotation period of its primary. From Charon's spectrum, it seems to have a greyish surface of water ice and some ammonia ice.

Chiron A small body in the outer Solar System, originally designated as an asteroid (no. 2060) following its discovery in 1977 by Charles Kowal. However, it has an eccentric, cometary orbit (period 50.7 years) which takes it from within Saturn's orbit out nearly as far as Uranus. In 1989, seven years before reaching perihelion, it was found to have a comet-like **coma** and was given a periodic comet designation (95P/Chiron). It has a diameter of roughly 200 km (125 miles), much larger than any 'ordinary' comet nucleus. It is now classified as a **Centaur**.

chondrite A type of **stony meteorite**, consisting of an agglomeration of *chondrules* – millimetre-sized globules believed to have remained unchanged since they condensed out of the nebula from which the Sun and the Solar System formed. They are therefore more ancient than **achondrites**. Chondrites consist mainly of iron, calcium, aluminium, magnesium and sodium silicates. They are classified into various types, including **carbonaceous chondrites**.

chromatic aberration An **aberration** (2) in lenses that results from the unequal refraction of the different colours that make up white light. Each colour is brought to a focus at a slightly different distance from the lens, causing the image of an object to be fringed with prismatic colours. Chromatic aberration is reduced by the use of an **achromatic lens**.

chromosphere ('sphere of colour') The layer of the Sun's atmosphere between the photosphere and the corona. The chromosphere is normally invisible because of the glare of the photosphere shining through it, but it is briefly visible near the beginning and end of a total solar eclipse as a spiky red rim around the Moon's disk. At other times it can be studied by spectroscopy or with a **spectroheliograph**. The chromosphere is about 10,000 km (6000 miles) thick, with a temperature of 10,000–25,000 K. **Filaments**, **flocculi**, **plages** and **prominences** originate in the chromosphere. SEE ALSO **supergranulation**.

Circinus A small constellation near Alpha and Beta Centauri, representing a pair of compasses. Its only fairly bright star is Alpha, magnitude 3.2.

circumpolar stars Stars which from a given place of observation never dip below the horizon. Their polar distance (90° minus their declination) is less than the observer's latitude.

civil twilight SEE **twilight**.

Clementine A US Department of Defense/NASA space probe launched on 25 January 1994 which from lunar orbit imaged the entire surface of the Moon in several wavelength bands. This yielded the

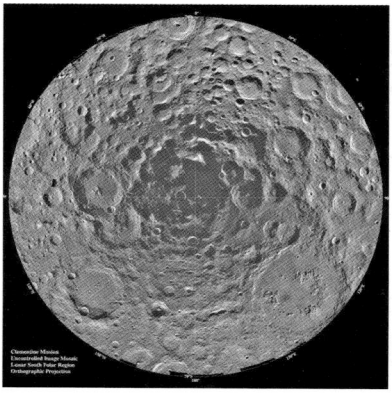

▲ **Clementine** The Moon's south pole is seen in this image obtained by the orbiting Clementine spacecraft in 1994. This picture has been compiled from some 1500 individual images.

first complete mapping of the Moon at visual wavelengths and a detailed geological survey. Clementine returned the first tentative evidence that the floors of some polar craters, permanently in shadow, contain deposits of water ice.

closed universe In cosmology, a mathematical model that describes the behaviour of a universe in which there is enough matter to halt the expansion triggered in the **big bang**. The universe collapses back on itself, ending in an event that is the opposite of the big bang, sometimes called the *big crunch*. In a closed universe, the **curvature of space** is positive, the **deceleration parameter** is greater than 0.5, and the **density parameter** is greater than 1 (i.e. the mean density of the Universe is greater than the **critical density**). COMPARE **flat universe**, **open universe**.

cluster of galaxies A group of galaxies, which can range in size from a few dozen members, as in our own **Local Group,** to a rich cluster containing thousands of galaxies. Poor clusters are usually irregular in shape, but rich clusters tend to be spherical and condensed at the centre, often with a giant elliptical galaxy or a cD galaxy dominating the central region. cD galaxies are thought to have grown by *cannibalism*, their overwhelming gravity dragging in smaller galaxies. The nearest rich clusters of galaxies are the **Virgo Cluster**, 45 million light years away, and the Coma Cluster, 300 million light years away. Clusters are grouped into *superclusters* up to about 100 million light years across. The Local Supercluster, of which the Local Group is part, is centred on the Virgo Cluster.

CMB ABBREVIATION FOR **cosmic microwave background**.

CME ABBREVIATION FOR **coronal mass ejection**.

Coalsack A prominent dark nebula in the constellation Crux, the Southern Cross. It is seen in silhouette against the background stars of the Milky Way.

Coathanger A bright group of ten stars in the constellation Vulpecula. Six of the stars are in an east–west line extending for an angular distance of 1.5°, while the other four form a 'hook' to the south. Catalogued as Collinder 399, the Coathanger is also known as *Brocchi's Cluster*; it is now known to be an **asterism** and not a genuine cluster, the component stars lying at various distances between 1140 and 2180 light years.

COBE ABBREVIATION FOR **Cosmic Background Explorer**.

coelostat Two mirrors arranged so that slowly rotating one of them counteracts the apparent rotation of the sky, enabling light from a celestial object to be directed along a fixed path. The first mirror is mounted equatorially and is clock-driven at half the rate of the Earth's rotation. The beam of light it receives is reflected onto a second (fixed) mirror, which in turn reflects the beam into the instrument being used for the observation. A coelostat is used with heavy or fixed equipment, for example a spectroheliograph. It is an improved form of **heliostat** and **siderostat**. SEE ALSO **coudé**.

collimation The process of aligning the optical components of a telescope so as to bring the rays of light from the object being observed to a focus at the correct position. A refracting telescope requires only occasional adjustment of the object glass in its cell. Collimation is more of a problem with reflectors because the primary and secondary mirrors both need to be adjusted, and the secondary can become misaligned quite easily. An incorrectly collimated telescope can suffer from **coma** (1).

colour index The difference in brightness of a star at two different wavelengths, used as a measure of the star's colour and hence its temperature. The brightness of the star is measured through coloured filters, usually at blue (B) and yellow (V, for visual) wavelengths, and the difference between the two readings is the B−V colour index. The B−V colour index for cool red stars is positive (up to about +2.0), while for hot blue stars it is negative (about −0.5 maximum). A white star of spectral type A0 such as Vega has a B−V colour index of exactly 0.0. A measurement at U (ultraviolet) wavelengths may also be taken, and a U−B colour index derived. Other measurements in the red and infrared can also be used.

The colour index of light from a star may change as the light passes through dust in interstellar space. Dust scatters blue light but allows red light to pass more freely, thereby reddening the star. This effect must be taken into account when determining the colour index of a star. SEE ALSO **UBV system.**

Columba A southern constellation adjoining Canis Major and Carina, representing a dove. Alpha (Phakt) is magnitude 2.6, but the constellation contains little of interest.

colure A **great circle** on the celestial sphere that passes through both celestial poles. SEE ALSO **equinoctial colure, solstitial colure**.

coma (1) A defect in an optical system which results in the image of a point appearing as a blurred pear-shaped patch with a flared appearance resembling a comet. It is caused by incident light striking a lens obliquely. The defect increases with distance from the centre of the field of view.

coma (2) The spherical cloud of gas and dust surrounding the nucleus of an active **comet**.

Coma Berenices A northern constellation adjoining Boötes, representing the hair of Queen Berenice of Egypt. Its brightest stars, of 4th and 5th magnitude, form a very wide star cluster. It is rich in faint galaxies; some are part of the **Virgo Cluster**, but others are part of the separate and more distant *Coma Cluster*, 330 million light years away.

Coma–Virgo Cluster SEE **Virgo Cluster**.

comes (plural **comites**) The fainter companion in a **binary star**.

comet A small, icy Solar System body in an independent, eccentric orbit around the Sun. The solid *nucleus* of a comet is small – that of **Halley's Comet**, for example, measures just 16 × 8 km (10 × 5 miles). The nucleus is made up of ices, largely water ice (plus some methane, carbon dioxide and carbon monoxide, with traces of organic substances), in which are embedded rock and dust particles, and is covered by a thin dark crust. As the comet

SOME PERIODIC COMETS AND ASSOCIATED METEOR SHOWERS

Name	Discovery	Orbital period (years)	Perihelion distance (million km)	Eccentricity	Inclination (°)	Shower(s)
Encke	1786, 1818	3.28	50	0.85	12	Taurids
Pons–Winnecke	1819, 1858	6.37	188	0.64	22	Pons–Winneckids
Giacobini–Zinner	1900, 1913	6.61	155	0.71	32	Draconids (Giacobinids)
Tuttle	1790, 1858	13.5	149	0.82	55	Ursids
Tempel–Tuttle	1865	32.9	147	0.90	163 *R*	Leonids
Halley	240 BC	76.0	88	0.97	162 *R*	Eta Aquarids, Orionids
Swift–Tuttle	1737, 1862	135*	143	0.96	113 *R*	Perseids
Thatcher (1861 G1)	1861	415*	138*	0.98	80	Lyrids

** approximate values* *R indicates retrograde orbit*

approaches the Sun and gets warmer, sublimation begins, and jets of gas and dust escape through the crust to form the luminous, spherical *coma*. The coma can be huge, a million kilometres or so across, but extremely tenuous. Closer in, radiation pressure from the Sun and the solar wind may send dust and gas streaming away as a *tail*, tens of millions of kilometres long. The *gas tail* or *ion tail* is straight and long, up to 10 million km (6 million miles), while the *dust tail* is shorter, broader and curved.

Comets are thought to originate in the **Oort Cloud**, where there may be as many as 6 trillion (6×10^{12}) of them, and in the **Kuiper Belt**. They are believed to be **planetesimals** left over from the formation of the Solar System. A comet may be gravitationally perturbed (by a passing star, for example) towards the Sun, swinging past the Sun on a parabolic orbit to return to the outer Solar System, or entering an elliptical orbit as a *long-period comet*, with a period of thousands or even millions of years (SEE **parabola**). Further perturbation by one of the giant planets may drive it into a smaller orbit as a *short-period comet*, with a period of 200 years or less. Further such close encounters can subsequently modify the orbit: some comets may even be ejected from the Solar System on hyperbolic trajectories.

With their loss of material at every approach to the Sun, periodic comets have a limited lifetime. The debris a comet leaves in its wake spreads around its orbit to form a **meteor stream**. Some comets, once they have expended all their gas and dust, may be strong enough to hold together and continue to orbit the Sun – the **Apollo asteroids**, **Amor asteroids** and **Aten asteroids** may be extinct comets. Others, among them Comets **Biela**, **West** and **Shoemaker–Levy 9**, have been observed to break up. Some objects first classed as asteroids have later behaved as comets – **Chiron**, for example, developed a coma.

Over a thousand comets have had their orbits calculated in detail, of which some 150 are short-period comets. Several new comets are discovered every year. After its orbit has been confirmed, it is customary for a new comet to be given the name of its discoverer. Donati's comet (or, more formally, Comet Donati), for example, was discovered by Giovanni Donati in 1858. A few, such as Halley's Comet and Encke's Comet, are named after the people who first calculated their orbit. In the case of co-discoveries, up to three names (since 1995, only two) may be given, as in Comet IRAS–Araki–Alcock (discovered independently by two amateur astronomers and the IRAS satellite). Formally, periodic comets are also given the prefix 'P' plus a permanent number once the orbit is established, as in 1P/Halley. Non-periodic comets are similarly assigned the prefix 'C'.

cometary globule A reflection nebula, which can superficially resemble a comet, associated with a young star such as a **T Tauri star**.

commensurable Describing two numbers with a common factor. If the orbital periods of two planets around the Sun, or of two satellites round a planet, can be expressed as a ratio of small whole numbers, then these periods are commensurable. Because of tidal interactions (SEE **tides**) between bodies of similar size, stable orbits tend to have (nearly) commensurable periods. Examples are the orbital periods of Saturn and Jupiter, which are in a ratio of nearly 5:2, and the periods of Jupiter's **Galilean satellites**. Orbital **resonances** may prevent bodies which are very different in size from having commensurable orbital periods; this explains the **Kirkwood gaps** in the main asteroid belt and the gaps in Saturn's ring system.

comparator A device with which two images of the same starfield obtained at different times can be compared. If any object has changed its position or varied in brightness during the interval between the times when the two images were taken, the comparator will show it up. In the *blink comparator* the two images are rapidly alternated in the field of view. The

► conjunction
Superior planets come to conjunction with the Sun when on the far side of it, as seen from Earth (and are therefore lost from view). The inferior planets, Mercury and Venus, can undergo conjunction at two stages in their orbit. At superior conjunction, they lie on the far side of the Sun from Earth, while at inferior conjunction they are between the Sun and the Earth. Under certain circumstances, Mercury and Venus can transit across the Sun's disk at inferior conjunction.

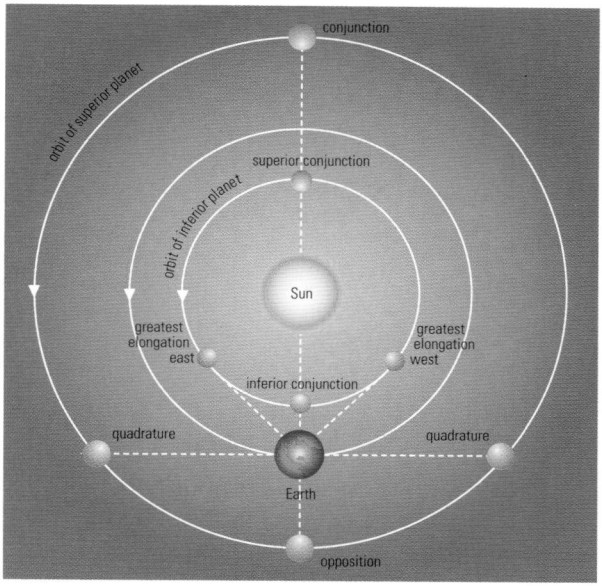

stereo comparator is a simple device for viewing the two images simultaneously with binocular vision; another arrangement uses two slide projectors and a rotating shutter. Hunters of novae and asteroids use comparators. Variable stars reveal themselves as a pulsation, because they give images of different sizes according to their brightness.

Compton Gamma Ray Observatory (GRO) A NASA satellite for studying the sky at gamma-ray wavelengths, launched on 5 April 1991. In its nine years of operation to June 2000, GRO increased the number of known gamma-ray sources tenfold, and recorded 2500 **gamma-ray bursts**.

Cone Nebula A V-shaped region of dark nebulosity seen in silhouette against brighter material in the constellation Monoceros. Visually elusive, but prominent in long-exposure images, this is part of a star-forming region, including the open cluster NGC 2264, at a distance of 2400 light years.

conjunction The alignment of the Earth with two other bodies in the Solar System so that from the Earth the other two appear in (nearly) the same position. The term is used mainly for alignments with the major planets (SEE the diagram). A **superior planet** is at conjunction when on the opposite side of the Sun from the Earth. An **inferior planet** is at *inferior conjunction* when it lies between the Earth and the Sun, and at *superior conjunction* when it is on the far side of the Sun. Because planetary orbits are inclined at various angles to the plane of the ecliptic, alignments at conjunction are rarely exact, and conjunction is defined to occur when the two bodies have the same celestial longitude as seen from the Earth.

Two or more planets – including in this case the Moon – are also said to be in conjunction when they are close together in the sky.

constellation A region of the sky containing an arbitrary grouping of stars, forming an imaginary figure. The stars in a constellation lie at very different distances

THE CONSTELLATIONS

Name	Genitive	Abbreviation	Order of size
Andromeda	Andromedae	And	19
Antlia	Antliae	Ant	62
Apus	Apodis	Aps	67
Aquarius	Aquarii	Aqr	10
Aquila	Aquilae	Aql	22
Ara	Arae	Ara	63
Aries	Arietis	Ari	39
Auriga	Aurigae	Aur	21
Boötes	Boötis	Boo	13
Caelum	Caeli	Cae	81
Camelopardalis	Camelopardalis	Cam	18
Cancer	Cancri	Cnc	31
Canes Venatici	Canum Venaticorum	CVn	38
Canis Major	Canis Majoris	CMa	43
Canis Minor	Canis Minoris	CMi	71
Capricornus	Capricorni	Cap	40
Carina	Carinae	Car	34
Cassiopeia	Cassiopeiae	Cas	25
Centaurus	Centauri	Cen	9
Cepheus	Cephei	Cep	27
Cetus	Ceti	Cet	4
Chamaeleon	Chamaeleontis	Cha	79
Circinus	Circini	Cir	85
Columba	Columbae	Col	54
Coma Berenices	Comae Berenices	Com	42
Corona Australis	Coronae Australis	CrA	80
Corona Borealis	Coronae Borealis	CrB	73
Corvus	Corvi	Crv	70
Crater	Crateris	Crt	53
Crux	Crucis	Cru	88
Cygnus	Cygni	Cyg	16
Delphinus	Delphini	Del	69
Dorado	Doradus	Dor	72
Draco	Draconis	Dra	8
Equuleus	Equulei	Equ	87
Eridanus	Eridani	Eri	6
Fornax	Fornacis	For	41
Gemini	Geminorum	Gem	30
Grus	Gruis	Gru	45
Hercules	Herculis	Her	5
Horologium	Horologii	Hor	58
Hydra	Hydrae	Hya	1
Hydrus	Hydri	Hyi	61
Indus	Indi	Ind	49

THE CONSTELLATIONS (continued)

Name	Genitive	Abbreviation	Order of size
Lacerta	Lacertae	Lac	68
Leo	Leonis	Leo	12
Leo Minor	Leonis Minoris	LMi	64
Lepus	Leporis	Lep	51
Libra	Librae	Lib	29
Lupus	Lupi	Lup	46
Lynx	Lyncis	Lyn	28
Lyra	Lyrae	Lyr	52
Mensa	Mensae	Men	75
Microscopium	Microscopii	Mic	66
Monoceros	Monocerotis	Mon	35
Musca	Muscae	Mus	77
Norma	Normae	Nor	74
Octans	Octantis	Oct	50
Ophiuchus	Ophiuchi	Oph	11
Orion	Orionis	Ori	26
Pavo	Pavonis	Pav	44
Pegasus	Pegasi	Peg	7
Perseus	Persei	Per	24
Phoenix	Phoenicis	Phe	37
Pictor	Pictoris	Pic	59
Pisces	Piscium	Psc	14
Piscis Austrinus	Piscis Austrini	PsA	60
Puppis	Puppis	Pup	20
Pyxis	Pyxidis	Pyx	65
Reticulum	Reticuli	Ret	82
Sagitta	Sagittae	Sge	86
Sagittarius	Sagittarii	Sgr	15
Scorpius	Scorpii	Sco	33
Sculptor	Sculptoris	Scl	36
Scutum	Scuti	Sct	84
Serpens	Serpentis	Ser	23
Sextans	Sextantis	Sex	47
Taurus	Tauri	Tau	17
Telescopium	Telescopii	Tel	57
Triangulum	Trianguli	Tri	78
Triangulum Australe	Trianguli Australis	TrA	83
Tucana	Tucanae	Tuc	48
Ursa Major	Ursae Majoris	UMa	3
Ursa Minor	Ursae Minoris	UMi	56
Vela	Velorum	Vel	32
Virgo	Virginis	Vir	2
Volans	Volantis	Vol	76
Vulpecula	Vulpeculae	Vul	55

from us, so the groupings have no physical significance. The original number of 48 constellations given in Ptolemy's ***Almagest*** has grown to 88 (SEE pages 48–9), which were assigned boundaries on the celestial sphere by the International Astronomical Union in 1930. Constellations are very unequal in size and prominence: they range from the vast Hydra down to the tiny but brilliant Crux. Many of the names are drawn from ancient mythology, while more recent mapping of the southern-hemisphere skies introduced some modern-sounding names such as the Octant and the Telescope. Prominent stars within a constellation are designated by Greek letters followed by the name of the constellation in the genitive (possessive) case, as with, for example, Alpha Lyrae; these are often abbreviated in the form α Lyr.

continuous spectrum The unbroken sequence of colours, merging one into the other, produced when white light is decomposed by refraction through a prism. An incandescent solid, liquid or dense gas emits a continuous spectrum. The spectrum of a star consists of a continuous spectrum crossed by **absorption lines**. SEE **emission spectrum**.

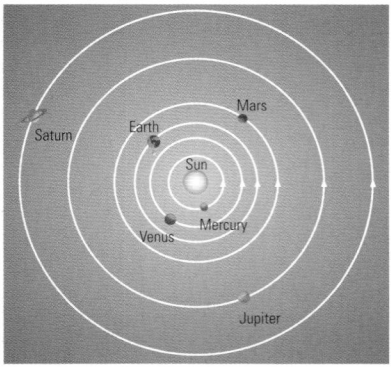

▲ **Copernicus, Nicholas** Copernicus's model of the Solar System put the Sun at the centre, with Earth and the other planets in orbit about it. Replacing the earlier geocentric system, the Copernican system removed Earth from any special position in the Solar System.

continuum SEE **spacetime**.

co-orbital Describing a satellite orbiting a planet at the same average distance as another satellite. Although the distances – and hence the periods – are the same, the orbital inclinations and eccentricities may differ slightly. An example is **Helene**, a small satellite of Saturn with a very similar orbit to the larger satellite Dione.

Coordinated Universal Time SEE **Universal Time**.

coordinates A pair of values which, in conjunction with a reference plane that serves as a zero, are used to define the position of a point or body on the **celestial sphere**. Several systems of coordinates are used in astronomy.

The most commonly used is the *equatorial system*, in which the reference plane is that of the celestial equator, and the coordinates are usually **right ascension** and **declination** (although **hour angle** and **polar distance** can be used alternatively).

In the *horizontal system* the reference plane is that of the observer's horizon, and the coordinates are **azimuth** and **altitude**.

In the *ecliptic system* the reference plane is that of the ecliptic, and the coordinates are celestial latitude and longitude. *Celestial latitude* is the angular distance between the object and the ecliptic, measured at right angles to the ecliptic. It is measured from 0° to 90°, positively towards the north ecliptic pole and negatively towards the south ecliptic pole. *Celestial longitude* is given by the angle between the object and the vernal equinox, measured along the ecliptic from 0 to 360° eastwards from the vernal equinox. Celestial latitude and longitude are not measured with reference to the celestial equator; they are sometimes more correctly called *ecliptic latitude* and *ecliptic longitude*.

For specifying positions with respect to the Galaxy, the system of *galactic coordinates* is used, in which the plane of reference is the galactic plane (the plane of the Milky Way). Its coordinates are galactic latitude and longitude. *Galactic latitude* is measured from 0° to 90° from the galactic equator, positively towards the north galactic pole. *Galactic longitude* is measured from 0° to 360° eastwards along the galactic equator. As defined by the IAU, the position of zero galactic longitude is RA 17h 46m, dec. −28° 56′ (epoch 2000.0).

Copernicus, Nicholas (1473–1543) Polish churchman and astronomer. As canon at Frauenburg Cathedral and a local administrator, he pursued astronomy as a side-interest. The accepted view in his day was the *geocentric* (Earth-centred) universe as expounded by **Ptolemy** nearly 1500 years before, with its complicated system of orbits. Through his study of planetary motions, Copernicus developed a *heliocentric* (Sun-centred) theory of the universe. In this *Copernican system*, as it is called, the planets' motions in the sky were explained by having them orbit the Sun. The motion of the sky was simply a result of the Earth turning on its axis, and the stars remained fixed as the Earth orbited the Sun because they were so very far away.

An account of his work, *De revolutionibus orbium coelestium*, was published in 1543. Because the new system was inferior at calculating planetary positions, most astronomers continued to believe in the Ptolemaic system. It was not until the discoveries by **Galileo** and Johann **Kepler** that the superiority of the Copernican system became apparent and its reality began to take hold.

Copernicus A large lunar crater, 93 km (58 miles) in diameter, in Oceanus Procellarum. It is one of the most conspicuous craters, and is surrounded by a system of rays stretching for well over 500 km (300 miles).

Cordelia The innermost small **satellite** of Uranus, discovered in 1986 during the Voyager 2 mission. Cordelia and Ophelia act as **shepherd moons** to the planet's Epsilon Ring.

Coriolis force A force that appears to act on a body (such as a mass of air) moving freely across the Earth so as to deflect it from a straight path. The deflection is to the right in the northern hemisphere, to the left in the southern. The effect is a consequence of us observing the motion from within a rotating frame of reference – the spinning Earth. In 1735 the English meteorologist George Hadley (1685–1768) recognized the effect of the Earth's rotation on the movement of air currents; a general deduction of the displacing force of the Earth's rotation was made by Gaspard de Coriolis (1792–1843). The Coriolis force accounts for the circulation of air around cyclones on the Earth, and is a driving force for the meteorology of the Earth and other planets with substantial atmospheres.

▲ **corona** Loops in the Sun's inner corona were imaged by the TRACE spacecraft in 1998. These loops are shaped by magnetic fields.

corona The outermost layer of the Sun's atmosphere, extending for many millions of kilometres into space. It springs into view as a white halo during a total solar eclipse; at other times it can be observed in visible light only by using a special instrument called a **coronagraph**. The corona emits strongly in the extreme ultraviolet and X-ray regions of the electromagnetic spectrum, and has been studied by spacecraft such as SOHO and the **Transition Region and Coronal Explorer**. Its overall shape changes during the 11-year solar cycle, from regular and symmetrical at solar minimum to uneven with long streamers at solar maximum. The corona has a temperature of 1–2 million degrees K. When lines of the Sun's magnetic field connect, large currents flow and heat the corona to this high temperature.

The corona may be divided into the inner *K corona*, which shines by light from the photosphere scattered by high-energy electrons, and the outer *F corona*, which shines by light scattered by lower-energy dust particles. The K corona has a **continuous spectrum** (the 'K' standing for the German word *Kontinuum*), on which are superimposed the **Fraunhofer lines** of the F corona. There are also emission lines in the spectrum contributed by the *E corona*, consisting of lower-energy metal ions such as calcium which are present out to about two solar radii. SEE ALSO **coronal hole**, **coronal mass ejection**, **flare**, **Sun**.

Corona Australis A small southern constellation representing a crown or wreath. It contains no star above fourth magnitude, but its arc of stars makes it easy to identify. It is sometimes called Corona Austrinus.

Corona Borealis A northern constellation, representing the wedding crown of Ariadne, adjoining Boötes. Alpha (Gemma or Alphecca), magnitude 2.2, is the only bright star, but the curve of five stars makes the constellation easy to find. In the 'bowl' of the crown is the famous variable star **R Coronae Borealis**, the prototype of a group of stars that fade suddenly and unpredictably. T Coronae Borealis is a recurrent nova, known as the Blaze Star, which erupted in 1866 and 1946.

coronagraph (sometimes spelled **coronograph**) An instrument, used in conjunction with a telescope, which allows the Sun's corona and prominences to be viewed by producing an 'artificial eclipse', masking the light from the Sun's disk with a circular obstruction. Before the invention of the coronagraph, such observations were possible only during the brief duration of a total eclipse.

coronal hole A relatively cool region of the solar **corona**. Coronal holes are associated with weak regions of the Sun's magnetic field, and are sources of high-speed streams in the **solar wind**.

coronal mass ejection (CME) The eruption into interplanetary space of millions of tonnes of plasma from the Sun's corona, triggered by disturbances in the Sun's magnetic field. CMEs are more frequent at solar maximum (several a day) than at solar minimum (about one a week). They are often associated with eruptive **prominences** in the chromosphere and **flares** in the inner corona above active sunspot groups. CMEs commonly travel at hundreds of kilometres per second, and produce shock waves in the slower **solar wind**. The arrival at Earth of energetic particles and strong magnetic fields associated with CMEs can cause *geomagnetic storms*, producing **aurorae** and interfering with radio communications.

Corvus A constellation adjoining Hydra and Crater, representing a crow. Its four brightest stars, of third magnitude, form a quadrilateral which is easy to find. Corvus contains the Antennae, a pair of 10th-magnitude interacting galaxies.

cosmic abundance The relative proportions of the various chemical elements (SEE **element, chemical**) in the Universe as a whole. It has been found from spectroscopic studies of the atmospheres of the Sun and stars that (with the exception of hydrogen and helium) the cosmic abundance is much the same as in the Sun. Stars are roughly two-thirds hydrogen and one-third helium, with other elements – chiefly carbon and oxygen – accounting for only 2%. The chemical make-up of a particular star depends on its age (SEE **stellar evolution**). Some rare types of star, for example **Wolf–Rayet stars**, show unusual abundances of certain elements; others, such as *technetium stars*, contain elements it is difficult to explain by currently accepted theories of how stars evolve. SEE ALSO **interstellar matter**.

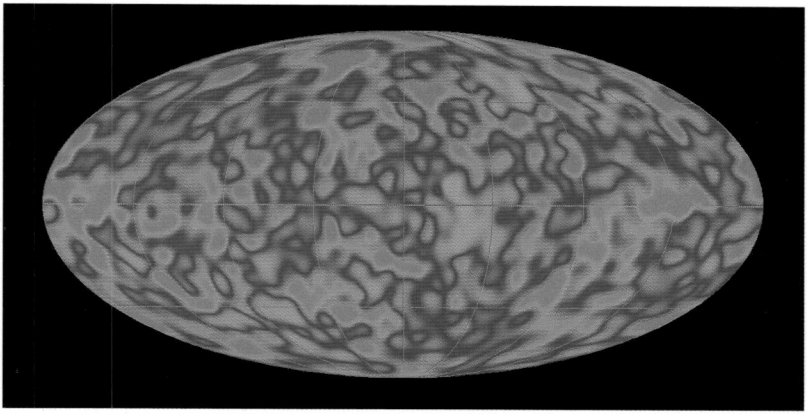

▲ **cosmic microwave background** This all-sky temperature map from COBE, from which the effects due to the motion of the Earth have been subtracted, shows 'ripples' in the cosmic microwave background. Although the extent of these variations is tiny – 30 millionths of a degree – they are believed to have been sufficient to trigger formation of the first galaxies after the big bang.

Cosmic Background Explorer (COBE) A NASA satellite launched on 18 November 1989 to study the **cosmic microwave background** radiation. It determined that the microwave background has a **black body radiation** spectrum with a temperature of 2.73 K. In 1992 it detected slightly warmer and cooler spots in the background radiation, differing by about 30 millionths of a degree from the average; these *ripples* are attributed to density fluctuations in the early Universe that marked the first stage in the formation of galaxies.

cosmic microwave background (CMB) Weak electromagnetic radiation from the Universe, first detected in 1965 by Arno **Penzias** and Robert **Wilson** of the Bell Telephone Laboratories. The microwave background is **black-body radiation** at a temperature of 2.73 K and has an almost equal intensity in all directions in space. It is considered to be the remnant of the radiation from the hot, early Universe following the **big bang**. The background radiation increased its wavelength as the Universe expanded until today it peaks in the millimetre region of the spectrum. It is called the microwave background because it was first detected in the form of microwaves. Penzias and Wilson won the 1978 Nobel Prize for Physics for their discovery. Observations from the **Cosmic Background Explorer** and **Wilkinson Microwave Anisotropy Probe**, as well as from ground-based and balloon-based instruments, have shown that the background varies on all angular scales by a few thousandths of 1%, and this requires the total average density of the Universe to be within 2% of the **critical density**.

cosmic rays High-energy atomic particles, moving at speeds approaching the speed of light, which enter the Earth's atmosphere from space. Cosmic rays (called 'rays' because they were originally thought to be gamma-rays) consist mainly of the nuclei of atoms, mostly hydrogen, stripped of their electrons. Heavier nuclei are present – practically every known element has been detected – but in exceptionally small numbers. While travelling through space they are called *primary cosmic rays*. When they enter the atmosphere they cause disintegration of the atoms they encounter and produce various atomic and subatomic particles, called *secondary cosmic rays*, which may be detected at ground level. The highest-energy cosmic rays come from X-ray sources such as the centre of the Galaxy and the Virgo Cluster of galaxies. Cosmic rays of medium energy are now known to originate in supernovae; low-energy cosmic rays are ejected from the Sun by solar **flares**.

cosmogony The study of how the Universe and objects within it came into being. **Cosmology**, by contrast, aims to understand the structure, history, future and governing laws of the Universe as it now exists. Mythical cosmogony covers non-scientific descriptions of the world's creation, ranging from myths of antiquity to accounts from various belief systems. Scientific cosmogony covers theories of the origin of the Universe (SEE **big bang**, **cosmology**) and of the Solar System.

For the Solar System, the scenario now accepted by most astronomers is a development of the *nebular hypothesis* first proposed in the 18th century by Immanuel Kant and Pierre Simon de **Laplace**. As a **molecular cloud** collapsed under its own gravity, a central condensation grew into a **protostar** which later became the Sun. Around the protostar, material from the cloud spread into a broad disk (SEE **protoplanetary disk**). Small grains of material stuck to one another by **accretion**, gradually forming **planetesimals**. (**Comets** and objects in the **Kuiper Belt** are believed to be remnant populations of these planetesimals.) The planetesimals collided with one another and grew into larger bodies, **protoplanets**, which swept up most of the remaining bits and pieces, as evidenced by **impact features** visible on many planetary bodies today. Gravitational perturbations between the resulting planets then moved them into their present orbits.

The different chemical compositions of the planets thus arose as a result of substances of different chemical composition condensing out at different distances from the Sun (SEE **volatile**), and by the proto-Jovian planets becoming large enough for their gravitational fields to scoop up the still gaseous hydrogen and helium. By this theory, planets are a normal consequence of star formation. Strong observational support for these ideas comes from the large number of extrasolar planets that have now been found and the detection of equatorial dust disks around many stars.

Earlier 'catastrophe' theories had explained the formation of planets by a chance event, such as material being torn from the Sun by the close approach of a passing star. However, planetary systems would be extremely rare if they formed in this manner.

cosmological constant A constant arbitrarily added by Albert Einstein to the relativistic equations describing the behaviour of the universe so that they would have a static solution. When the Universe was shown to be expanding, Einstein regretted using the cosmological constant, but recent observations of the **accelerating Universe** suggest that it may be real. SEE ALSO **dark energy**, **inflation**.

cosmological principle The principle that the Universe is both *homogeneous* and *isotropic*: in other words, that it is the same (on a sufficiently large scale) in all directions. The principle was assumed to be true by Einstein and others as it helped to solve the equations of general relativity. The discoveries that the **cosmic microwave background** is indeed isotropic down to about the 0.001% level, and that the large-scale distribution of galaxies in the Universe appears to be homogenous, show that the principle is real.

cosmological redshift SEE **redshift**.

cosmology The study of the structure, history, future and governing laws of the Universe on the largest scale. The starting point for modern cosmology was the discovery in the 1920s by Edwin **Hubble** of the large redshifts of galaxies, interpreted as evidence that the Universe is expanding. The rate of expansion is given by the **Hubble constant**. Other observational evidence shows that the Universe appears much the same in all directions (*isotropy*) and at all distances (*homogeneity*).

A *cosmological model* is a description of the Universe in terms of mathematical equations. These models aim to predict how the Universe behaves, and are tested by comparing them with observations. Cosmological models have included the **steady-state theory**, in which the Universe is not only the same in all places but also at all times. It therefore had no beginning, will have no end, and never changes at all when viewed on the large scale. This theory required matter to be created as the Universe expanded in order that the overall density of galaxies should not decrease. For this reason it is also referred to as the *continuous creation* model. On the other hand, according to the **big bang** theory, the Universe was created in a single instant about 13.7 billion years ago, and has been expanding ever since. In the future it may continue to expand or possibly collapse back on itself, depending on the total amount of matter and energy in the Universe – on whether it is above or below its **critical density**. An important consideration is the question of the **missing mass**: the amount of ordinary (or luminous) matter we see in the Universe is less than 5% of the amount actually present, as inferred from the first observations by the **Wilkinson microwave anisotropy probe**.

Cosmologists have made several important advances in discriminating between cosmological models. The discovery of the **cosmic microwave background** radiation provided strong evidence against the steady-state theory. This background radiation is very uniform in all directions, which has led cosmologists to postulate that there was a very rapid period of expansion, known as **inflation**, shortly after the big bang. The abundances of helium and other light elements also support the big bang model since these would have been formed naturally from protons and neutrons in the early stages of the explosion. The Hubble constant is currently thought to have a value of around 70 km/s/megaparsec. The **deceleration parameter** – which determines whether the Universe will expand for ever or collapse back on itself – has also been estimated, and seems to have a value close to 0.5, the value for a Universe which will just continue to expand for ever.

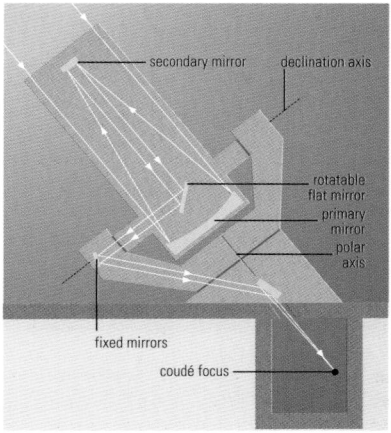

▲ **coudé** In this optical configuration, mirrors direct light from the telescope along the polar axis to a fixed observing position. This arrangement is advantageous for observations that require use of heavy or bulky detectors.

coudé An arrangement of auxiliary mirrors used with a telescope on an equatorial mount to direct the light path down the polar axis to a fixed focal point. The coudé system is used in conjunction with large, immovable instruments, such as spectrographs, which are too massive to be mounted on the telescope itself. (The word comes from the French *coudé*, meaning bent like an elbow, and is not the name of a person, as is sometimes thought.) SEE ALSO **Nasmyth focus**.

Crab Nebula The nebula M1 (NGC 1952), about 6500 light years away in Taurus, the remnant of a supernova that was noted by Chinese astronomers in July 1054, when it shone as brightly as Venus and was visible even in daylight. The nebula, now about 8th magnitude, was discovered in 1731 by the English astronomer John Bevis and independently by Charles **Messier** in 1758. It gained its popular name of the

▲ **Crab Nebula** A supernova remnant, the Crab Nebula is an expanding gas cloud with filamentary structure and a pulsar at its heart.

Crab Nebula after its crab-like appearance in a sketch made by Lord **Rosse**.

The Crab Nebula is an intense source of radio emission, known as *Taurus A*, and is also a source of X-rays (Taurus X-1). Gas ejected in the supernova explosion forms filaments that appear red on colour images, surrounding an ionized gas that emits a yellow glow. The glow is **synchrotron radiation**, caused by electrons spiralling in an intense magnetic field. This radiation extends from the radio domain into the visible, and the Crab Nebula is the only synchrotron nebula that can be seen in a small telescope. Inside the nebula lies the *Crab Pulsar*, the remaining core of the supernova of 1054. It spins 30 times a second, emitting electrons to replenish the synchrotron radiation. The Crab Pulsar also flashes at visible wavelengths. SEE **pulsar**.

crater A circular formation, found on many bodies in the Solar System – planets, satellites and asteroids. Nearly all craters are **impact features**, although on some bodies there are volcanic craters. A typical impact crater has a raised rim, or wall, and its floor is below the level of the surrounding terrain. Some have a central peak. Impact craters range in size from the microscopic (as found on lunar rocks returned by the Apollo missions) to features a third of the diameter of the body (as on Saturn's moon **Mimas**). Heavily cratered bodies such as **Callisto** are geologically inactive and preserve a record of the bombardment that took place in the early Solar System, when there were still plenty of free **planetesimals**. Less cratered bodies are usually those on which geological activity has led to some degree of resurfacing, obliterating older impact sites.

Crater A small constellation adjoining Hydra and Corvus, representing a cup or chalice. Its brightest stars are of fourth magnitude and it contains no objects of note.

Crepe Ring A popular name for Ring C of **Saturn**. SEE ALSO **ring, planetary**.

Cressida An inner **satellite** of Uranus, discovered in 1986 during the Voyager 2 mission.

critical density The average density of matter in the Universe that would ensure that it would just barely continue expanding endlessly, without ever collapsing back on itself. For a value of the **Hubble constant** of 71 km/s/megaparsec, the critical density is about 10^{-26} kg/m^3, or 6 hydrogen atoms per cubic metre. The observed density of 'normal' or luminous matter in the Universe is about 5% of the critical value. However, the first results from the **Wilkinson microwave anisotropy probe** place the total density of the universe to within $\pm 2\%$ of the critical density. The deficiency is thought to be accounted for by **dark matter** (23%) and **dark energy** (72%). The ratio of the observed and critical densities is the *density parameter*, Ω. Hence, if the Universe has the critical density, then $\Omega = 1$. Ω is twice the value of q_0, the **deceleration parameter**. SEE ALSO **missing mass**.

Crux The smallest constellation in the sky, popularly known as the Southern Cross. It is almost surrounded by Centaurus. Its brightest star, Alpha (Acrux), is of magnitude 0.77 and lies about 321 light years away. It is actually a visual double, magnitudes 1.3 and 1.7, both type B. With

Beta (Mimosa, or Becrux), magnitude 1.2, Gamma (Gacrux, 1.6) and Delta (2.8), it makes a pattern that is more like a kite than a cross; the symmetry is disturbed further by Epsilon (3.6), the only other star above magnitude 4. Interesting objects include the **Jewel Box** cluster and the **Coalsack**, a dark nebula.

C star SEE **carbon star**.

culmination The passage of a celestial body across the observer's meridian. At culmination, an object reaches the greatest altitude above, or least altitude below, the observer's horizon. For circumpolar stars both culminations are observable. *Lower culmination* occurs when a star transits between the pole and horizon; the star's **hour angle** is then exactly 12h. *Upper culmination* occurs when a star transits between the pole and zenith; the star's hour angle is then exactly 0h.

curvature of space The distortion of **spacetime** in the neighbourhood of matter, one of the consequences of **Einstein**'s General Theory of Relativity. This curvature makes rays of light and particles of matter follow curved paths called *geodesics* which, for the planets, for example, are their orbits around the Sun. If the Universe contains enough mass it will have enough curvature to be a **closed universe**; if not, it will be an **open universe** or a **flat universe** (SEE **cosmology**).

cusp One of the points or horns of the crescent Moon, or of an inferior planet seen at crescent phase.

cusp caps The bright areas at the **cusps** of the crescent Venus that are sometimes reported by observers.

CXO ABBREVIATION FOR **Chandra X-ray Observatory**.

Cygnus One of the most distinctive constellations, sometimes called the Northern Cross. Its brightest star is first-magnitude **Deneb**. Beta is Albireo, a beautiful coloured double with a giant primary of type K3, magnitude 3.1, and a main-sequence companion of type B9, magnitude 5.1, separation 34 arc seconds. Chi has the largest visual range of any Mira variable, 3.3 to 14.2, period about 400 days. P Cygni is a B-type supergiant variable that has ranged between magnitudes 3 and 6 in the past, because of mass loss, and is now around 5th magnitude. **61 Cygni** is a binary pair of orange dwarf stars, magnitudes 5.2 and 6.0. The constellation also contains the radio galaxy **Cygnus A**, the probable black hole **Cygnus X-1**, the **North America Nebula** and the **Cygnus Loop** supernova remnant.

Cygnus A The strongest source of radio emission outside our Galaxy, situated in the constellation Cygnus. It is thought to be two galaxies in collision. Cygnus A is also a source of X-ray emission.

Cygnus Loop A **supernova remnant** consisting of a vast loop of gas ejected from a star that exploded about 30,000 years ago. It lies about 2500 light years away. The Loop is about 3° in diameter and is expanding at about 100 km/s. Different parts of it bear the designations NGC 6960, 6979, 6992 and 6995. The brightest part, NGC 6992, is known as the Veil Nebula.

Cygnus Rift A prominent dark nebula that splits the Milky Way in the constellation Cygnus, also called the *Northern Coalsack*; its extension through Aquila to Ophiuchus is known as the *Great Rift*. The rift is a string of **molecular clouds** lying in the **Orion Arm**, about 2200 light years away.

Cygnus X-1 A strong source of X-ray emission in the constellation Cygnus, believed to contain a **black hole**. At the position of the source lies a spectroscopic binary star, HDE 226868, with an orbital period of 5.6 days. The visible star is a blue supergiant. Its invisible companion is calculated to be around 10 solar masses or more, far too great for a white dwarf or neutron star, so a black hole seems the only possibility.

D

Danjon scale A rough measure of how bright the Moon appears in a **lunar eclipse**, devised by the French astronomer André Danjon (1890–1967). Eclipses are graded from very dark (0) to bright (4), the former occurring when Earth's atmosphere is more opaque, and therefore refracts less sunlight to the eclipsed Moon, as a result of clouds or volcanic dust.

dark adaptation (dark adaption) The process by which the human eye adjusts from vision under conditions of high illumination to night vision. Dark adaptation takes 20 minutes or more to develop fully, during which time the pupil widens and different receptors come into play. It is essential for observers leaving a brightly lit room to allow their eyes to adapt to the dark before they begin observing. Dark adaptation is instantly destroyed by any bright light, so observers use dim red lights when they need to consult star charts or record observations.

dark energy A hypothetical form of energy that may pervade the whole Universe and leads to a repulsive force between distant objects. It was postulated in the 1990s to explain the **accelerating universe**, and its effects may be modelled mathematically using the **cosmological constant**.

The nature of dark energy is unknown, although vacuum energy – a consequence of quantum mechanics – has been suggested, even though this seems to be too large by a huge factor (10^{120}). An alternative to dark energy is *quintessence* (a fifth 'essence', or force) that generates a negative pressure. It may be possible to distinguish between different forms of dark energy and quintessence when more accurate measurements of the changes in the rate of expansion of the Universe with time become available. If dark energy exists, then it may make up over 70% of the Universe: early results from the **Wilkinson microwave anisotropy probe** show that the mean density of the universe is within 2% of the **critical density**, yet ordinary and **dark matter** amount to only about 30% of that value.

dark matter Unseen matter, inferred to exist in galactic halos and in the space between galaxies, that is thought to make up about 23% of the mass of the Universe. Whatever form it takes, it must not be able to undergo **nucleosynthesis** reactions, otherwise elements heavier than helium would have been produced in the early stages of the **big bang**. Thus it could consist of objects such as black holes, **WIMPS** or other unknown atomic particles which are termed *cold dark matter*; or it could be in the form of a 'sea' of fast-moving neutrinos, known as *hot dark matter*; or it could be a mixture of both. SEE ALSO **MACHO**, **missing mass**.

dark nebula A cloud of gas and dust that is not illuminated, and can thus be seen only in silhouette against stars or bright nebulae beyond. They range in size from minute **Bok globules** to the naked-eye clouds of the **Coalsack** nebula in Crux and the gigantic Rho Ophiuchi Dark Cloud, which covers about 1000 square degrees (2%) of the sky. Dust comprises only about 0.1% of the mass of a dark nebula, but is believed to play an important part in the formation of molecules in space, since the surface of the dust particles provides a site for atoms to adhere and combine into molecules. The interiors of these molecular clouds are very cold, typically only 10 K, which allows them to collapse under their own gravity into stars.

David Dunlap Observatory The observatory of the University of Toronto, Canada. Its principal instrument is a 1.88-m (74-inch) reflector, opened in 1935.

Dawes limit SEE **resolving power**.

day The time taken by the Earth to rotate once on its axis. There are various definitions. An *apparent solar day* is the interval between two successive transits of the Sun across the meridian. This interval is variable because the Sun's apparent motion over the course of the year is not uniform (SEE **apparent solar time**). For convenience, an imaginary *mean Sun* is defined which moves along the celestial equator at a constant rate, giving the *mean solar day*. A *sidereal day* is defined as the interval between two successive transits of the spring **equinox** across the meridian. This day is about four minutes shorter than the mean solar day owing to the Sun's apparent eastward movement of about 1° each day. A mean solar day is 24h 03m 56.555s of mean sidereal time. A mean sidereal day is 23h 56m 04.091s of mean solar time. SEE ALSO **timescale**.

deceleration parameter (symbol q_0) A figure describing the rate at which the expansion of the Universe is slowing down. If q_0 is greater than 0.5, the expansion will eventually stop and reverse. If q_0 is less than 0.5, the Universe will expand for ever. Theoretical considerations suggest that the true value of q_0 should be exactly 0.5, although in an **accelerating universe** it can be negative. SEE ALSO **critical density**, **missing mass**.

declination (dec., symbol δ) The angular distance of a celestial object north or south of the celestial equator. It is reckoned positively from 0 to 90° from the equator to the north celestial pole, and negatively from 0 to 90° from the equator to the south celestial pole. SEE ALSO **celestial sphere**.

decoupling era The time, about 300,000 years after the big bang, by which the Universe had cooled sufficiently for the first atoms to form; also known as the *recombination era*. Before decoupling, particles and photons underwent constant interactions (coupling); afterwards, the Universe was effectively transparent and the **cosmic microwave background** originated from the radiation that was released then.

deep sky The Universe beyond the Solar System. Double stars, star clusters, nebulae and galaxies are known as *deep-sky objects*; the term is not usually applied to individual stars.

Deep Space 1 The first space probe in NASA's New Millennium series, launched on 24 October 1998, intended to test advanced technologies, including an ion propulsion system. It rendezvoused with asteroid 9969 Braille in July 1999 and with Comet Borrelly in September 2001.

deferent SEE **Ptolemy**.

► **Deep Space 1** Test firing of the xenon ion engine of Deep Space 1. Launched in October 1998, the spacecraft had a highly successful encounter with Comet Borrelly in September 2001.

degenerate matter A state of matter existing in stars in the final stage of their evolution (SEE **stellar evolution**), when they have ceased producing energy at their cores. Atomic nuclei and electrons are packed closely together at ultra-high densities, and the laws of classical physics no longer apply. Pressure ceases to be dependent on temperature and is a function only of density.

Deimos The smaller of Mars's two **satellites**, discovered in 1877 by Asaph Hall. It is a dark, irregular body, measuring about 15 × 12 × 10 km (9 × 7.5 × 6 miles), and may well be a captured asteroid.

Delphinus A small but distinctive constellation near Aquila, representing a dolphin. Its leading stars, of fourth magnitude, make up what looks like a widespread cluster.

Delta Aquarids A **meteor shower** which occurs between 15 July and 20 August. It has a double radiant in Aquarius, the more active radiant lying near the star Delta Aquarii, and peaks at about **ZHR** = 25 close to 28 July.

Delta Scuti star A short-period (0.01–0.2 day) pulsating variable of spectral type between A0 and F5. Delta Scuti stars lie at the lower end of the **instability strip**, either on the main sequence or nearby among the subgiants and giants, and are Population I stars (SEE **stellar populations**). They have very small amplitudes (0.003–0.9 mag.), frequently with multiple periodicities. Their pulsations are driven by hydrogen-burning instabilities. They are related to the *AI Velorum stars*, which have greater amplitudes (0.3–1.2 mag.) and higher luminosities, and to the *SX Phoenicis stars*, which are subdwarfs belonging to Population II and have multiple periods and amplitudes of about 0.7 mag.

Deneb The star Alpha Cygni, magnitude 1.25, a supergiant of type A2, luminosity half a million times the Sun's, distance about 3230 light years. It is the brightest star in the constellation Cygnus.

density parameter (Ω) The ratio of the actual total density of the Universe to its **critical density** – the density it would have if it contained just enough mass to halt expansion after an infinite time. The density parameter therefore determines whether the Universe will continue to expand for ever, or whether its expansion will halt and it will ultimately collapse back on itself. If Ω is greater than 1 the Universe will collapse; if it is less than 1 the Universe will continue to expand. $\Omega = 1$ corresponds to the critical density. In cosmological models (SEE **cosmology**), it is the parameter that distinguishes between **open** and **closed universes**

descending node The point at which an orbit crosses from north to south of a reference plane such as the celestial equator or ecliptic. SEE **node**.

Desdemona One of the small inner **satellites** of Uranus discovered in 1986 during the Voyager 2 mission.

de Sitter, Willem (1872–1934) Dutch cosmologist. His hypothetical *de Sitter Universe*, derived from Einstein's general theory of relativity, contained no mass, but provided the first theoretical indication of an expanding Universe.

Despina One of the small inner **satellites** of Neptune discovered in 1989 during the Voyager 2 mission. It appears to be the inner **shepherd moon** to the planet's Le Verrier Ring.

diagonal (1) An optical device, used in conjunction with a telescope's eyepiece, for deflecting the light path through 90° in order to make it easier for the observer to view the image when the telescope is at an awkward angle. This can be a problem with amateur refracting telescopes. In this device, also called a *star diagonal*, either a mirror or a right-angled prism is set at 45° to the light path.

diagonal (2) Another name for the flat secondary mirror in a **Newtonian telescope**.

diamond-ring effect A short-lived phenomenon seen during a total solar eclipse, just before or after totality, when just the bright central part of the edge of the Sun's disk is visible, shining through a depression on the Moon's limb as an intense point of light, together with the faint inner corona, giving the appearance of a diamond ring.

dichotomy The phase of Mercury or Venus when it appears exactly half-illuminated by the Sun, and the **terminator** is a straight line. The term is also applied, less commonly, to the half Moon.

differential rotation The rotation of different parts of a non-solid body at different rates. For example, the deep atmospheres of the giant planets rotate slightly faster at the equator than at the poles. The Sun shows a similar differential rotation, but rather more pronounced.

differentiation The process by which a planetary body evolves a layered structure. In a body which has grown to a sufficient size by **accretion**, energy supplied by gravitational compression and radioactive decay, supplemented by the kinetic energy of impacting bodies, will cause the interior to melt. Under the action of gravity, denser materials will then sink towards the centre, where they will eventually form the *core*, and less dense materials will rise towards the surface, where they will form the *mantle* and the *crust*. Differentiation has given the terrestrial planets, for example, cores of nickel–iron and outer layers of a predominantly rocky composition.

diffraction The slight sideways spreading of a beam of light as it passes by a sharp edge or through a narrow slit. Starlight passing through a telescope is diffracted by the edges of various components so that, instead of coming to a sharply focused point, it appears as a small disk of light surrounded by concentric *diffraction rings* of light, and radial *diffraction spikes*. The bright central disk is called the **Airy disk**.

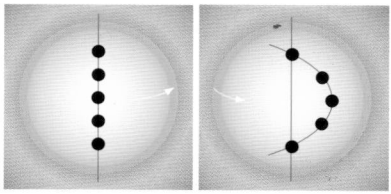

▲ **differential rotation** In a fluid body, such as a star or gas giant planet, the equatorial regions rotate more rapidly than the poles. As shown, a consequence of this is that a set of points lined up on the central meridian will become spread out in longitude over the course of a rotation. Points close to the equator will return to the central meridian earlier than those near the poles.

diffraction grating A surface on which a very large number of equidistant parallel lines have been ruled very close together. Light striking a grating is dispersed by **diffraction** into a spectrum. There are metal gratings that function by reflection, and glass gratings that function by transmission. The lines are ruled at, typically, 100 to 1000 per millimetre. Diffraction gratings produce very high-quality spectra, and are used in astronomical **spectrographs**.

diffuse nebula A luminous cloud of gas in space. The term 'diffuse' refers to the fact that through a telescope diffuse nebulae cannot be resolved into individual stars, unlike star clusters and galaxies. They come in two varieties: **emission nebulae**, also known as **H II regions**, which shine by fluorescence; and **reflection nebulae**, which shine by starlight reflected off dust particles.

Dione A medium-sized **satellite** of Saturn, diameter 1120 km (695 miles), discovered in 1684 by Giovanni Cassini. It has the second-highest density, at 1.44 g/cm^3, of Saturn's main satellites. The surface is bright and icy, but one hemisphere is darker than the other. There are three main types of terrain: cratered terrain, cratered

▲ **Dione** This Voyager image clearly shows Dione's uneven surface. Central peaks can be seen in the larger craters.

plains and smooth plains. Evidence of past geological activity includes a number of **tectonic** features such as Palatine Chasma – a trough over 600 km (nearly 400 miles) long and up to 8 km (5 miles) wide. Dione shares its orbit with a small **co-orbital** satellite, Helene.

direct (prograde) (1) Describing orbital or rotational motion in the same direction as the Earth's motion: anticlockwise as seen from above the Sun's north pole. The orbital motion of all planets and most satellites is direct. Orbital or rotational motion is direct if the orbital or axial inclination is less than 90°. COMPARE **retrograde** (1).

direct (prograde) (2) Describing the regular, west-to-east motion of Solar System bodies as seen from Earth relative to background stars. COMPARE **retrograde** (2).

disconnection event The shearing of a comet's ion tail in response to changes in the local solar wind. Disconnection is usually followed by the development of a new ion tail in a slightly different orientation.

Discovery A series of NASA space probes built on a 'faster, better, cheaper' principle: low-cost missions that can be designed and launched quickly. The first was the **Near Shoemaker** probe; another was **Mars Pathfinder**.

disk galaxy A name given to any type of **galaxy** in the form of a thin disk of stars which are in roughly circular orbits around the centre; the brightness of the galaxy tails off with distance from the centre. Disk galaxies include spiral galaxies and lenticular galaxies.

dispersion The separation of white light into its constituent colours by a lens, prism or diffraction grating. This happens because the different wavelengths of light are refracted more at the red end than at the violet end of the spectrum. Dispersion has to be countered in telescope lenses to avoid **chromatic aberration**, but it is exploited in astronomical instruments such as the **spectrograph** to study the spectra of stars.

distance modulus The difference between the apparent magnitude (m) and the absolute magnitude (M) of a star. It is a measure of the star's distance (r) in parsecs:

$$m - M = 5(\log r) - 5$$

diurnal motion The apparent daily motion of celestial bodies across the sky from east to west. It is a result of the Earth's axial rotation.

diurnal parallax SEE **parallax**.

Dobsonian telescope A Newtonian reflecting telescope on a simple form of **altazimuth** mount. The use of lightweight materials makes it possible to build a low-cost, large-aperture telescope that is portable, making the Dobsonian increasingly popular with amateur observers. A typical Dobsonian is mounted at its base, on a 'rocker-box' assembly which rotates on a horizontal baseplate. It is named after its inventor, the American amateur John Dobson (1915–).

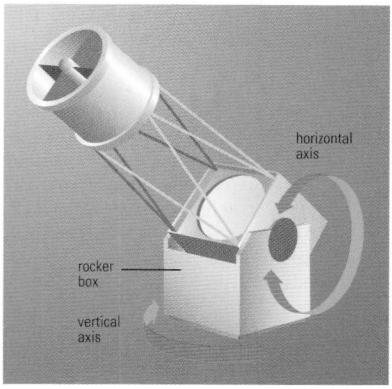

▲ **Dobsonian telescope** This simple altazimuth mounting has enabled many amateur astronomers to construct large-aperture instruments comparatively cheaply. The Dobsonian offers considerable advantages of portability.

Dollond, John (1706–61) English optician. Formerly a silk-weaver, in 1752 he joined the optician's business founded by his son, **Peter Dollond** (1730–1820). He found that the deviation of light rays could be achieved without **dispersion**, and combined crown and flint glass to make **achromatic lenses**. These he used in commercially successful refracting telescopes, which were further refined by Peter. John also invented the **heliometer**, a refractor modified so as to measure small angular distances. Peter Dollond was the first to make an achromatic triplet lens.

Dominion Astrophysical Observatory An observatory near Victoria, BC, Canada, opened in 1917. It has 1.83-m (72-inch) and 1.22-m (48-inch) reflecting telescopes. The Dominion Radio Astrophysical Observatory, at Penticton, BC, has a 26-m (85-ft) radio dish.

Doppler effect The apparent increase in frequency (and decrease in wavelength) of radiation from a source moving towards the observer, or the similar decrease in frequency (increase in wavelength) of radiation from a source moving away. In astronomy, its importance is that it explains the displacement of spectral lines. If a celestial object is moving towards the Earth, its light will be shifted to shorter (bluer) wavelengths. But if is moving away, then the shift will be towards longer (redder) wavelengths – a **redshift**.

This *Doppler shift*, as it is called, makes it possible to detect and measure relative motion in the line of sight, in other words **radial velocity**, and rotation of celestial objects. The redshifts of distant galaxies provided the first evidence that the Universe is expanding. Although galaxies and quasars do exhibit their own individual motions, the redshift is predominantly caused by the expansion of space itself. The effect is named after Austrian physicist Christian Doppler (1803–53), who published the first detailed discussion of it.

Dorado A southern constellation, representing a goldfish. It contains the main part of the Large **Magellanic Cloud**, including the **Tarantula Nebula** and the highly luminous eruptive variable star S Doradus. Alpha, its brightest star, is magnitude 3.3.

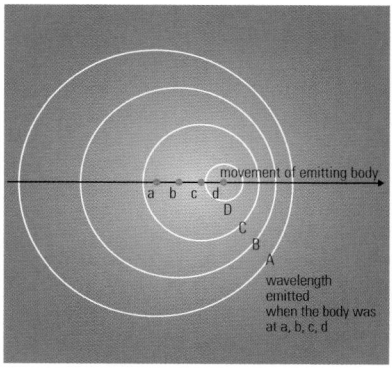

▲ **Doppler effect** Electromagnetic waves emitted from a moving point source appear compressed ahead of, and more spread out away from, the direction of motion. The result is that, for light, the wavelength is shortened (blueshift) ahead of a source approaching the observer and lengthened (redshifted) behind a receding source such as a distant galaxy.

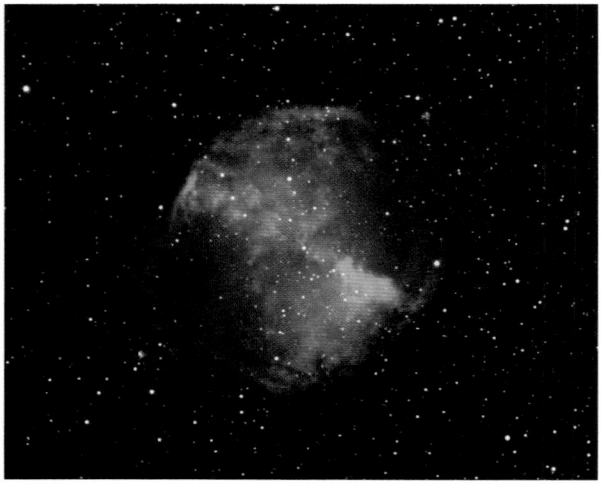

◀ **Dumbbell Nebula**
The brightest of the planetary nebulae, the Dumbbell takes its name from the two prominent lobes of material to either side of the central, illuminating star.

Double Cluster Two large star clusters in the constellation Perseus, visible to the naked eye. They bear the designations NGC 869 and NGC 884, and are also known as h and χ (chi) Persei. They lie about 7200 and 7500 light years away.

double star Two stars that appear close together in the sky. There are two types of double star. If the stars are genuinely close together in space and are connected by gravity they are known as a **binary star**. Two stars which are quite distant from each other, but appear close together as a result of chance alignment, are known as an **optical double**.

doublet A lens with two components, which are either cemented together or separated by an air-gap. This construction is used to reduce **chromatic aberration**.

Draco A long, winding northern constellation representing the dragon slain by Hercules, extending between Ursa Major and Ursa Minor, with the dragon's head near the star Vega. Gamma (Eltanin), magnitude 2.2, is its brightest star. Alpha (Thuban), magnitude 3.7, was the north pole star in ancient times. Draco is a constellation largely devoid of interest, but one of the stars in the head, Nu, is a double with a separation of 62 arc seconds; each component is of magnitude 4.9.

draconic month The time taken for one revolution of the Moon around the Earth relative to its ascending **node**. It is equal to 27.21222 days of **mean solar time**.

Draconids SEE **Giacobinids**.

Draper, Henry (1837–82) American pioneering spectroscopist. In 1872 he obtained the spectrum of Vega, the first time this had been done for a star other than the Sun. After his death, his widow made available the funds that made possible the **Henry Draper Catalogue**. His father, the English scientist **John William Draper** (1811–82), was a pioneering astrophotographer who took the first images of the Moon and the Sun's infrared spectrum.

Dreyer, Johan Ludvig Emil (1852–1926) Danish astronomer, who from 1874 lived and worked in Ireland, initially at Lord **Rosse**'s observatory. He compiled the *New General Catalogue of*

Nebulae and Clusters of Stars (SEE **NGC**) and its two supplementary *Index Catalogues*.

drive A means of moving a telescope to compensate exactly for the rotation of the Earth, so keeping the object being observed in the field of view. **Equatorial telescopes** are driven about their polar axis to track objects being observed. Computer control allows professional **altaz-imuth** mounted instruments to be driven simultaneously about their vertical and horizontal axes.

Dumbbell Nebula A large planetary nebula, also designated M27, about 1250 light years away in the constellation Vulpecula. It was named by Lord **Rosse** for its hourglass shape. The Dumbbell is visible in binoculars.

dwarf galaxy A galaxy that is much smaller than a normal galaxy and is of low luminosity. Such galaxies are usually elliptical or irregular. There are several dwarf ellipticals in the **Local Group** of galaxies.

dwarf nova A member of a class of irregular variable stars whose light curves resemble those of novae. Their luminosity stays the same for long periods, then rapidly increases, and finally returns slowly to normal. **U Geminorum stars** and **Z Camelopardalis stars** are the main types. Dwarf novae are close binary stars in which one component is a white dwarf and the other is a subgiant or dwarf.

dwarf star The commonest type of star in the Galaxy, constituting 90% of its stars and 60% of its mass. Dwarfs are also known as **main-sequence** stars, from their position on the **Hertzsprung–Russell diagram**. The term 'dwarf' refers to luminosity rather than size, so dwarf stars should be thought of as normal rather than diminutive. The Sun is a typical dwarf, roughly midway in the range of properties of dwarf stars.

The hottest and most massive dwarfs (spectral types O and B) have relatively short lifetimes, a few hundred million years or less, so they are rare in the Galaxy. The lifetimes of the lowest-mass dwarfs (type K and M) are so long that none have evolved away from the main sequence since the Galaxy formed. Stars below 0.8 solar masses are known as **red dwarfs**, and those below 0.08 solar masses as **brown dwarfs**. SEE ALSO **stellar evolution, white dwarf**.

dynamical parallax The distance to a **binary star** determined from a relationship between the known masses of the components, the size of the orbit and the period of revolution.

E

Eagle Nebula A bright nebula surrounding a cluster of new-born stars, collectively designated M16 or NGC 6611, in the constellation Serpens. In the nebula (known also as IC 4703) are dark columns of gas and dust in which stars are being born. The nebula and stars are about 6500 light years away.

early-type star A hot star of spectral type O, B or A. The name was given when it was thought that the sequence of spectral types was an evolutionary one, hot stars being the youngest and cool stars the oldest. Although that is now known not to be so, the term has remained in use.

▲ **Eagle Nebula** One of the Eagle Nebula's dust pillars is shown in this Hubble Space Telescope image. These dark fingers of nebulosity are associated with star formation.

Earth The third major planet from the Sun, and the largest of the four inner, or terrestrial planets. Seventy per cent of the surface is covered by water, and it is this and the Earth's average surface temperature of 13°C (286 K) that make it suitable for life. The continental land masses make up the other 30%. Our planet has one natural satellite, the **Moon**. The main data for the Earth are given in the first table.

Monitoring the propagation of seismic waves from earthquakes has revealed the Earth's internal structure. Like all the terrestrial planets, the Earth has a dense *core* rich in iron and nickel, surrounded by a *mantle* consisting of silicate rocks. The thin, outermost layer of lighter rock is called the *crust*. Continental crust can be as much as 50 km (30 miles) deep, but oceanic crust has an average depth of only 10 km (6 miles). The mantle extends to a depth of 2890 km (1795 miles). The outer part of the core, down to 5150 km (3200 miles), is molten, but the inner core, 2460 km (1525 miles) in diameter, is solid because of the greater pressure there. The temperature and pressure at the Earth's centre are estimated to be about 4000°C and 4 million bars (i.e. 4 million times the pressure at the surface). The solid inner core rotates two-thirds of a second faster than the layers outside it, and this, together with currents in the molten outer core, gives rise to the Earth's magnetic field.

The crust and the uppermost layer of the mantle together form the *lithosphere*, which consists of eight major and over twenty minor tightly fitting slabs called *plates* that move slowly with respect to one another, and are supported on a semi-molten layer of mantle called the *asthenosphere*. Plates grow at *mid-oceanic ridges*, boundaries between adjoining plates where molten rock rises and solidifies into new oceanic crust, pushing the plates apart; an example is the Mid-Atlantic Ridge. Where two plates meet, one may be forced below the other, descending into the mantle and melting. This process of *subduction* produces earthquakes and volcanoes at the edge of the plate above. Where plates collide and no subduction occurs, they crumple up to form folds that develop into mountain ranges. All these interactions are known collectively as *plate tectonics*.

Because of tectonics, most of the Earth's surface is young compared with the planet's age of 4.6 billion years. The oldest rocks belong to the *Precambrian shields*, areas over 600 million years old, as in parts of Canada; some individual rocks have been dated to more than 4 billion years ago.

The Earth's atmosphere consists of 78% (by volume) nitrogen, 21% oxygen, 0.9% argon, and 0.03% carbon dioxide, other gases such as neon making up the remaining 0.07%. In addition there can be up to 3% water vapour, depending on geographical location and weather conditions. The atmosphere produces a **greenhouse effect**, raising the surface temperature by 35 K.

The atmosphere is divided into a number of layers according to the way in which its temperature varies with altitude (SEE the second table). The *troposphere* contains most of the atmosphere, and is where weather systems operate. The *stratosphere* contains the *ozone layer*, which absorbs high-energy ultraviolet radiation from the Sun. **Meteors** occur just above the mesosphere, and **aurorae** in the thermosphere. The thermosphere is extremely rarefied, and its high temperature indicates the high kinetic energy of its molecules, rather than its heat content. Ionized atoms and molecules in the mesosphere and thermosphere constitute the **ionosphere**. Above the thermosphere is the exosphere, which contains the Earth's **magnetosphere** and **Van Allen Belts**, and merges into interplanetary space.

EARTH: DATA

Globe	
Diameter (equatorial)	12,756 km
Diameter (polar)	12,714 km
Density	5.52 g/cm^3
Mass	5.974×10^{24} kg
Volume	1.083×10^{12} km^3
Sidereal period of axial rotation	23h 56m 04s
Escape velocity	11.2 km/s
Albedo	0.37
Inclination of equator to orbit	23° 26′
Surface temperature (average)	290 K
Orbit	
Semimajor axis	1 AU = 149.6×10^6 km
Eccentricity	0.0167
Inclination to ecliptic	0° (by definition)
Sidereal period of revolution	365.256d
Mean orbital velocity	29.8 km/s
Satellites	1

earthgrazer An alternative name for a **near-Earth asteroid**.

earthshine Sunlight reflected from the Earth onto the part of the Moon's nearside that is in the Sun's shadow. It is easiest to see when the Moon is a thin crescent, when it appears as a pale greyish light over the remainder of the lunar disk.

EARTH: LAYERS OF THE EARTH'S ATMOSPHERE

Layer	Upper boundary	Altitude (km)	Temperature (°C)	Pressure (mbar)
Troposphere	Tropopause	20*	−80*	250
Stratosphere	Stratopause	50	0	0.9
Mesosphere	Mesopause	80	−90	0.007
Thermosphere	—	500	1500	10^{-9}

The altitude, temperature and pressure are values at the upper boundary.
** Above the equator. At the poles the altitude of the tropopause is 10 km (6 miles).*

eccentric (1) Describing an orbit with a high **eccentricity**.

eccentric (2) SEE **Ptolemy**.

eccentric anomaly SEE **anomaly**.

eccentricity (symbol *e*) One of the elements of an **orbit**. It indicates how much an elliptical orbit departs from a circle. The eccentricity is found by dividing the distance between the two foci of the **ellipse** by the length of its major axis. A circle has an eccentricity of 0, and a parabola has an eccentricity of 1.

echelle grating A **diffraction grating** designed to yield spectra of high **dispersion** and detail. Its lines are ruled farther apart than on an ordinary grating, and are shaped so as to produce high resolution over a narrow band of wavelengths at angles of illumination greater than 45°. Such a grating is used in an *echelle spectrograph*.

eclipse The partial or total obscuration of the light from a celestial body as it passes through the shadow cast by another body. A body may be eclipsed by the passage of another body between it and the observer, as in a **solar eclipse**, or by the intervention of another body between it and the source of the light it reflects, as in a **lunar eclipse**. The eclipse of a star by the Moon or by a planet or other Solar System body is called an **occultation**. SEE ALSO **annular eclipse**, **transit**.

eclipse year The interval between successive passages of the Sun through a given node of the Moon's orbit, equal to 346.62003 days of mean solar time. It differs significantly from other types of year because the nodes of the Moon's orbit move westward by over 19° per year (SEE **regression of the nodes**). Nineteen eclipse years are equal to 6585.78 days, almost exactly the same length of time as a **saros**.

eclipsing binary A type of **binary star** in which the orbital plane of the two stars is viewed almost edge-on, leading to mutual eclipses and consequent variations in the two stars' combined light output. Two eclipses may be expected during an orbital cycle, one usually causing a bigger drop in light than the other since the two stars are rarely equal in size or brightness. The deeper minimum is called the primary minimum, and the shallower minimum is the secondary minimum. If the stars are very unequal in brightness the secondary minimum may hardly be noticeable.

Eclipsing binaries are classified into three types on the basis of their light curves: **Algol** type (EA), **Beta Lyrae stars** (EB) and **W Ursae Majoris stars** (EW). The EA types have a constant light level outside eclipses, but EB and EW types show continuous variability. This is because one or both stars in the binary are elliptical in shape, having expanded to fill their **Roche lobe**. (If only one star has filled its lobe, the system is called *semidetached*; a system

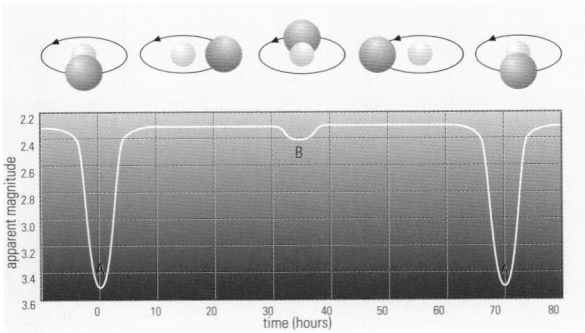

◀ eclipsing binary
The light curve (bottom) of an Algol-type binary can be accounted for by partial eclipses of each star by the other (top). At the deep primary minimum, the brighter, smaller star is largely covered by the fainter. At the secondary minimum, the brighter star passes in front of the fainter.

in which neither lobe is filled is known as *detached*.) The EW stars show more or less equal depths of primary and secondary minima. In such systems, both stars are thought to be filling their Roche lobes and are thus touching each other. They are therefore known as *contact binaries*, and in most systems the two stellar cores are surrounded by a common convective envelope.

ecliptic The apparent yearly path of the Sun against the background stars. It intersects the celestial equator at the two equinoxes. The ecliptic is really the projection of the Earth's orbit around the Sun onto the celestial sphere. Because of the tilt of the Earth's axis the ecliptic is inclined to the celestial equator by an angle of about $23\frac{1}{2}°$; this tilt is known as the *obliquity of the ecliptic*. The *ecliptic poles* are the points 90° from the ecliptic, and lie in the constellations Draco and Dorado.

ecliptic coordinates SEE **coordinates**.

ecliptic limits The greatest angular distance the Sun or Moon can be from the Moon's nodes for which a **solar eclipse** or **lunar eclipse** can take place. The limits are 37° for the Sun at a solar eclipse, and 24° for the Moon at a lunar eclipse.

E corona SEE **corona**.

ecosphere The shell-shaped region of space surrounding the Sun or another star within which the temperature is such that any suitable planet would be capable of sustaining life. The dimensions of the ecosphere vary greatly with the type of star. The ecosphere of a small dwarf star would be narrow and lie close in, whereas that of a luminous giant would be broad and a considerable distance from the star.

Eddington, Arthur Stanley (1882–1944) English astronomer and physicist. He pioneered the use of atomic theory to study the internal constitution of stars, and explained the role of radiation pressure in preventing stars from collapsing under gravity. Among his many discoveries were the mass–luminosity relationship, and the fact that white dwarfs contain degenerate matter. He supported and popularized Einstein's theory of relativity, and in 1919 he obtained experimental proof of the theory's prediction that gravity bends light by measuring star positions close to the Sun during a solar eclipse.

Eddington limit The maximum ratio of luminosity to mass that a star can have before radiation pressure overcomes the gravitational force holding it together, and the star consequently blows apart. It gives a theoretical limit to stellar mass of between about 50 and 100 solar masses.

Edgeworth–Kuiper Belt SEE **Kuiper Belt**.

effective temperature (symbol T_{eff}) The temperature a star would have if its output were **black-body radiation** at the same energy and at the same wavelengths as the star. As stellar spectra approximate closely to black-body spectra, a star's effective temperature is a good approximation to its actual surface temperature.

Effelsberg Radio Telescope A radio telescope near Bonn, Germany, operated by the Max-Planck-Institut für Radioastronomie, completed in 1970. With a diameter of 100 m (328 ft), it is the world's largest fully steerable radio dish. It is used either alone or as part of a **very long baseline interferometry** network.

Einstein, Albert (1879–1955) German-Swiss-American theoretical physicist. His work had an enormous impact on 20th-century science. For astronomy, and particular in cosmology, his theories of relativity had profound consequences. The Special Theory of Relativity (1905) gave the relation $E = mc^2$ between mass and energy; the General Theory of Relativity (1916) extended the theory to encompass gravitation. The General Theory has been borne out, for example, by its explanation

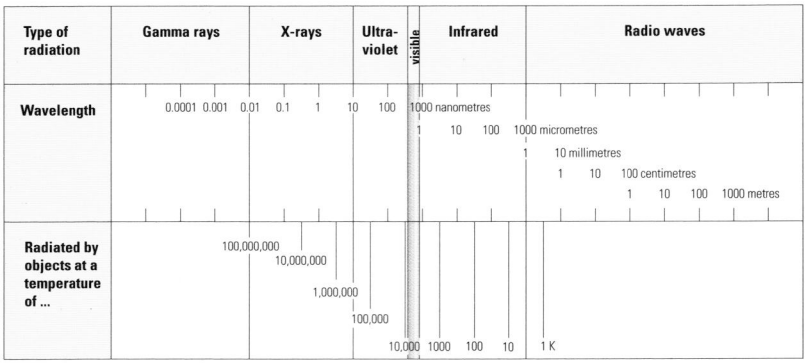

▲ **electromagnetic radiation** Spanning from short, gamma-ray wavelengths to long-wavelength radio emissions, the electromagnetic spectrum covers a wide range. Selection of appropriate wavelengths allows different processes and classes of objects to be observed. The visible region comprises only a tiny part of the electromagnetic spectrum.

of the advance of the perihelion of **Mercury**, the curvature of light by a gravitational field (SEE **Eddington**, Arthur), and the **gravitational redshift** of spectral lines. In the 1920s relativity made Einstein world-famous, but he never again matched the scientific work of his early years. He spent much time unsuccessfully seeking a unified field theory which would link relativity with electromagnetic forces. SEE ALSO **grand unified theory**.

Einstein–de Sitter universe SEE **flat universe**.

Einstein ring A circular image of a distant source produced by a **gravitational lens**. Albert Einstein realized that if the lensing object has its mass concentrated in a single point and lies directly in our line of sight to a distant source, then in theory the source's image should be a perfect circle. In practice this is rarely seen, but incomplete circles – arcs – are often observed.

ejecta SEE **rays**.

Elara A small **satellite** in Jupiter's intermediate group, discovered by Charles Perrine in 1905.

electromagnetic radiation (em radiation) The emission and propagation of energy in the form of periodic waves that can travel through a vacuum. It originates when charged atomic particles are accelerated, and consists of oscillating electric and magnetic fields. There is a continuum of electromagnetic radiation – from long-wavelength radio waves of low frequency and low energy, through visible light waves, to short-wavelength X-rays and gamma-rays of high frequency and high energy – known as the *electromagnetic spectrum*. The speed at which em radiation travels in a vacuum (c) is given by its wavelength (λ) times its frequency (ν):

$$c = \lambda\nu.$$

This speed is constant, and is commonly known as the speed of light (SEE **light, speed of**). In the *quantum theory*, an em wave (particularly if its wavelength is that of visible light or shorter) may also be regarded as a stream of particles called *photons*, each having an energy (E) related to the frequency (or wavelength) of the radiation:

$$E = h\nu = hc/\lambda.$$

With X-rays and gamma-rays, this energy, expressed in **electronvolts**, is a more important property than their wavelength. Photons also possess momentum, which means they can exert a pressure, called **radiation pressure**.

Almost all of our information about the Universe has been learned from studies of the electromagnetic radiation that reaches us from celestial objects. Much of the radiation that arrives at the Earth is absorbed by its atmosphere, but there are some bands of the electromagnetic spectrum that can reach the surface of the Earth and be studied by Earthbound instruments. The two main ones are known as the *optical window* and the *radio window*. Other wavelength bands of cosmic radiation are detected and analysed by instruments carried above the atmosphere by satellites or rockets.

electronvolt (symbol eV) A unit of energy used mainly in atomic physics for the energies of atomic particles. One electronvolt is defined as the kinetic energy acquired by a particle carrying a charge equal to that on one electron when it passes through a potential difference of 1 volt. 1 eV is equivalent to 1.6021×10^{-19} joules, and, from the equation $E = h\nu$ (SEE **electromagnetic radiation**), corresponds to a frequency of 2.42×10^{14} hertz, or a wavelength of 1240 nm. An electron with a kinetic energy of 1 eV has a velocity of about 580 km/s.

element, chemical One of the basic materials of which everything in the Universe is composed. Elements are chemically homogeneous substances – that is, they cannot be decomposed into any simpler substances by chemical means. The atoms of a particular element all have the same number of protons in their nucleus, but the number of neutrons can vary, giving rise to different *isotopes* of the same element. The number of known elements is over 100. Of these, 90 occur in nature, and the remainder have been prepared artificially, in laboratories. SEE ALSO **cosmic abundance**, **nucleosynthesis**.

element, orbital SEE **orbit**.

ellipse A closed curve, for which the sum of the distances of any point on it from two points within it – the *foci* (singular *focus*) of the ellipse – is a constant (SEE the diagram). In mathematics the ellipse is a type of conic section, so called because it is one of the intersections of a plane with a cone. The ellipse is important in astronomy because the closed **orbits** of planets, satellites, companion stars, comets, and so on are all close approximations to ellipses (**perturbations** make orbits depart from true ellipses) – SEE **Kepler's laws**. Important parameters of the ellipse, and therefore of orbits, are the **semimajor axis** (a) and the **eccentricity** (e). The shape of the ellipse is governed by its eccentricity, which is given by $e = c/a$, where c is the distance from the centre of the ellipse to one of its foci. For an ellipse, e is always less than 1.

elliptical galaxy A **galaxy** that is ellipsoidal in shape. Elliptical galaxies are composed of old stars and contain little gas or dust. They are often the central dominant galaxies in very rich clusters, and many are powerful radio sources. Ellipticals are

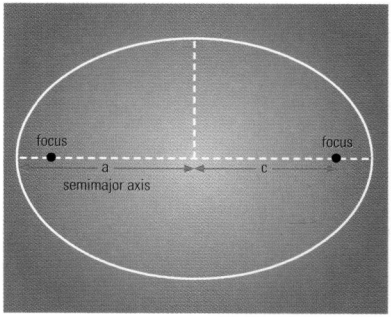

▲ **ellipse** The principal features of an ellipse are its two foci, its semimajor axis (a), its short (minor) axis, its eccentricity (e), and the distance of the foci from the centre of the ellipse (c). In the case of an ellipse used to define a planetary orbit, the Sun occupies one of the foci while the other is empty. The planets' orbits are reasonably close to circular, but for most comets the orbital ellipse is eccentric, with a very long major axis.

classified according to their degree of ellipticity (flattening), and given a designation from E0 to E7. They range widely in size, from 100 million stars for *dwarf ellipticals* to 10 million million stars for *supergiant ellipticals*.

ellipticity Another word for **oblateness**.

elongation The angular distance between the Sun and a planet or other Solar System body orbiting the Sun (or the angular distance between the Sun and the Moon). More accurately, it is the difference in the two bodies' celestial longitudes, measured in degrees. A planet with an elongation of 0° is at **conjunction**; at 90° or 270° it is at **quadrature**; and at 180° it is at **opposition**. The inferior planets, Mercury and Venus, when at their maximum angular distance from the Sun, are said to be at *greatest elongation east* when following the Sun (setting at their latest time after it) and at *greatest elongation west* when preceding the Sun (rising at their earliest time before it). SEE the diagram at **conjunction.**

emersion The reappearance of a celestial body from the shadow of another after it has been eclipsed or occulted.

emission nebula A cloud of gas in space that emits its own light. Ultraviolet radiation from nearby stars heats the gas of the nebula, causing its atoms to become ionized (lose their electrons). These electrons gain energy from the ultraviolet radiation. When they eventually recombine with the atoms, they re-emit the energy they gained, some of it at visible wavelengths. The free electrons in the gas also lose energy in the form of radio waves, so emission nebulae can also be detected by radio telescopes.

Interstellar gas is mostly hydrogen, which is easily ionized by ultraviolet light. The symbol for ionized hydrogen is H II, and such emission nebulae are therefore also called **H II regions**. Hydrogen gives off its strongest light in the red, so emission nebulae appear red in images. However, to the eye they often look green because the eye responds more readily to a prominent pair of spectral lines of this colour emitted by oxygen.

Hot stars that emit ultraviolet light are usually young, having recently formed from the gas and dust cloud around them. Thus the association between stars and gas clouds is not by chance. A famous example is the **Orion Nebula**, with the **Trapezium** stars inside it. Emission nebulae are very tenuous. Typically, every gram of material is spread through a volume of a million cubic km.

emission spectrum The **spectrum**, consisting of a series of bright lines, produced by a highly energetic source. Ener-

◄ emission nebula
The Lagoon Nebula, M8, in Sagittarius is a classic example of an emission nebula. It shows strong H II emission, stimulated by ultraviolet radiation from young stars forming in its midst.

▲ **Enceladus** While heavily cratered in some regions, the surface of Enceladus has also been smoothed over by episodes of cryovolcanism. Various types of terrain can be seen in this image from Cassini.

gy in the form of heat or electromagnetic radiation can be absorbed by an electron orbiting the atomic nucleus, causing it to jump to a higher orbit. When it returns to its old orbit, the electron emits radiation at a characteristic wavelength. The series of lines made up of emissions from different types of atom in the source is the emission spectrum. The hottest stars have many strong emission lines in their spectra. SEE ALSO **forbidden lines**.

Enceladus A **satellite** of Saturn, diameter 498 km (309 miles), discovered in 1789 by William Herschel. It is the most reflective body of any significant size in the Solar System, with an albedo approaching 1. Its surface consists of older, cratered terrain and more recently formed plains traversed by straight and curved grooves, and there are indications that **tectonic** and **volcanic** processes may still be operating. Tidal interaction with the larger satellite **Dione** probably explains why such a small body has been so geologically active. Material from Enceladus may be the source of Saturn's tenuous E Ring, in which the satellite is embedded; the satellite does possess a tenuous atmosphere.

Encke, Johann Franz (1791–1865) German mathematician and astronomer. In 1818 he computed the orbit of **Encke's Comet**, and predicted its return, and in 1837 he was the first to record the division in **Saturn**'s rings now named after him. Encke also calculated the solar parallax from measurements of the 1769 transit of Venus.

Encke's Comet A periodic **comet** discovered by Pierre Méchain in 1786 and independently by Jean Louis Pons in 1818. In 1819 **Johann Encke** computed its orbit and period, and proved that it was identical with a comet that had previously been observed on three returns. It was the second periodic comet to be discovered, and has the shortest-known period, at 3.3 years. It is the parent of the **Taurid** meteor shower.

ephemeris (plural **ephemerides**) A table giving the predicted positions of a celestial object, such as a planet, satellite or comet, at given intervals.

epicycle SEE **Ptolemy**.

Epimetheus A small inner **satellite** of Saturn discovered during the Voyager missions. It is **co-orbital** with **Janus**.

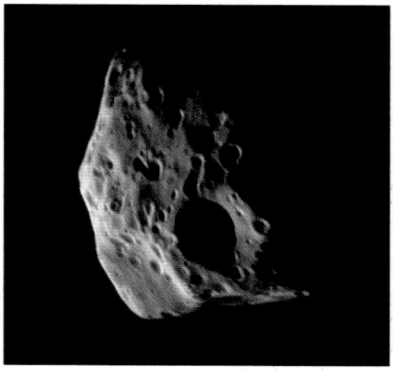

▲ **Epimetheus** The Cassini spacecraft obtained this image on 30 March 2005. The large crater just below centre is called Hilairea, and has a diameter of 33 km (21 miles). The moon itself is 116 km (72 miles) across.

epoch (1) A point in time used as a fixed reference for comparison of astronomical data in, for example, star catalogues. **Precession** and the **proper motions** of stars cause their positions on the celestial sphere to change gradually, and observations made over any considerable period of time have to be reduced to a common epoch for them to be comparable. SEE ALSO **standard epoch**.

epoch (2) One of the elements of an **orbit**, defined as the time of perihelion passage.

Epsilon Aurigae An **eclipsing binary** star with a very long period, 27 years, magnitude 3.0 at maximum and a minimum of 4.1 during eclipse. The primary component is a supergiant of type F0. The eclipses last 700 days, the total phase taking 400 days; the last one was in 1983–4. No light is detected from the secondary component, which is thought to be surrounded by an obscuring disk of gas and dust which causes the eclipses; the secondary itself may actually be a close binary consisting of two white dwarfs. At over 7000 light years away, Epsilon Aurigae is the most distant of the brighter naked-eye stars.

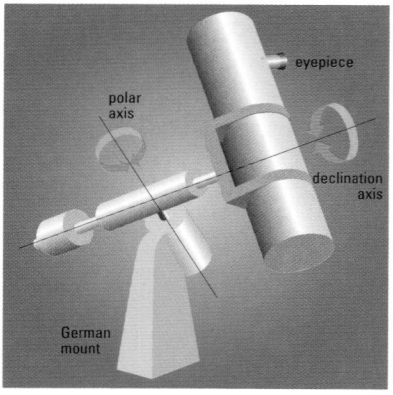

▲ **equatorial telescope**. The equatorial mounting allows a telescope to be driven so that it can follow the apparent motion of the stars due to Earth's rotation. The equatorial is based around two main axes. The polar axis, around which the driving motor turns the telescope, is aligned parallel to Earth's axis of rotation. North and south movement is made along the declination axis.

equant SEE **Ptolemy**.

equation of the centre An irregularity in the motion of a body in an elliptical orbit. It is the difference between the true anomaly and the mean anomaly (SEE **anomaly**).

equation of time The difference between mean solar time, as shown on a clock, and apparent solar time, as shown on a sundial, caused by the combined effect of the eccentricity of the Earth's orbit and the **obliquity of the ecliptic**. The difference is zero four times a year: on 15 April, 14 June, 1 September and 25 December. The maximum difference is 16 minutes and occurs in early November.

equatorial coordinates SEE **coordinates**.

equatorial telescope A telescope mounted such that it has one axis, the *polar axis*, parallel to the Earth's axis and therefore pointing to the celestial pole, and the other axis, the *declination axis*, at right angles to the polar axis and therefore parallel to the plane of the Earth's equator. Once an object is in the field of view the telescope needs only to be moved about the polar axis to follow the object. If the telescope is driven at the sidereal rate to counter the apparent rotation of the celestial sphere, the object will remain in the field of view.

There are various forms of equatorial mounting. The *German mounting*, so named because the first of the type was made by Joseph von **Fraunhofer**, is still used for small instruments. The telescope is attached at its centre of mass to one end of the declination axis, which pivots about the top end of the polar axis; balance is achieved by attaching counterweights to the other end of the declination axis. Large, professional reflecting telescopes before the age of computer-controlled **altaz-**

imuth mountings required massive equatorial mounts in which the telescope could be supported on both sides of the declination axis and sometimes at the top and bottom of the polar axis, yet still have enough freedom of movement to access most of the sky. Such designs included the *fork mounting* and the *horseshoe mounting*.

A departure from traditional equatorial mounts is the equatorial platform or *Poncet mounting*. This is a flat surface on whose underside are casters that sit on inclined planes. It allows amateur instruments like the traditionally altazimuth **Dobsonian telescope** to be given an equatorial capability.

equinoctial colure The **great circle** that passes through the north and south celestial poles and the equinoxes (which are also, in this context, called the equinoctial points).

equinox Either of the two points at which the Sun crosses the celestial equator, or the times at which these events occur. The *vernal equinox* (or *spring equinox*) is when the Sun crosses the equator from south to north on or near 21 March each year. The point at which this happens is called the *First Point of Aries*. The other equinox is the *autumnal equinox*, which is on or near 23 September, when the Sun crosses the equator from north to south. The point at which this happens is called the *First Point of Libra*. At the equinoxes, the Sun rises due east and sets due west.

Equuleus A very small and obscure constellation, representing a foal, next to Pegasus. Its brightest star, Alpha (Kitalpha), is only of magnitude 3.9. It has no features of particular interest.

Eratosthenes (*c.*275–195 BC) Greek geographer, mathematician and astronomer. He is best known for making the first scientific calculation of the Earth's circumference, for which he measured the altitude of the Sun at Alexandria on a day when he knew the Sun to be overhead at a place a considerable distance away. How accurately he knew this distance, and hence how accurate his result was, is uncertain, but the method was sound. Eratosthenes also compiled a star catalogue and measured the **obliquity of the ecliptic**.

▲ **Eros** The rotation of Eros can be seen in this series of images obtained from the NEAR–Shoemaker spacecraft as it made its close approach to the asteroid in February 2000.

Erfle eyepiece A telescope eyepiece with low power and wide field of view – over 60°. It consists of three elements, at least one of which is a doublet. (After German optician Heinrich Erfle, 1884–1923.)

Eridanus One of the largest constellations in the sky, representing a river; it extends from near Rigel in Orion to the south polar area. **Achernar** is its brightest star. Theta (Acamar) is a fine double, with components of magnitude 3.4 and 4.3. Epsilon, magnitude 3.7, is a nearby K2 main-sequence star, 10.5 light years away.

Eros **Asteroid** no. 433, discovered by Gustav Witt in 1898. It measures 33 × 13 km (21 × 8 miles) and rotates in 5.3 hours. Its orbit crosses that of Mars, and it is the largest of the **Amor asteroids**. The **NEAR Shoemaker** probe entered orbit around Eros in February 2000 and studied the asteroid for 12 months.

eruptive variable Any of a broad group of intrinsic variable stars that show abrupt changes in magnitude with no obvious period. The class includes many different types, such as **flare stars**, **R Coronae Borealis stars**, **shell stars**, **T Tauri stars**, certain young stars just settling onto the **main sequence**, as well as various other **irregular variables**, some associated with nebulosity.

ESA ABBREVIATION FOR **European Space Agency**.

escape velocity The minimum velocity that a body, such as a rocket or a space probe, must attain to overcome the gravitational attraction of a larger body and leave on a trajectory that does not bring it back again. It depends on the mass (m) and radius (r) of the larger body, and is given by $\sqrt{(2GM/r)}$, where G is the gravitational constant (see **gravity**). It is escape velocity rather than mass that determines the gases that a body is able to retain as an atmosphere.

Eskimo Nebula A bright, compact planetary nebula, designated NGC 2392, in the constellation Gemini. Lying at a distance of 3000 light years, it was discovered by William Herschel in 1787. The Eskimo Nebula takes its name from its telescopic appearance of a 'face' surrounded by a hazy periphery resembling a fur hood.

ESO ABBREVIATION FOR **European Southern Observatory**.

Eta Aquarids A **meteor shower** active from 24 April to 20 May, with a broad peak centred on May 4–5 reaching a **ZHR** of about 50. Its radiant is in Aquarius, near the star Eta Aquarii. Best seen from southerly latitudes, the shower is, like the Orionids, associated with Halley's Comet.

Eta Carinae A peculiar variable star lying within the nebula NGC 3372 – the Eta Carinae Nebula – in the constellation Carina, about 8000 light years away. It was originally catalogued at third magnitude, but from 1833 began to vary irregularly, becoming at its brightest second only to Sirius. It then faded to just below naked-eye visibility. Material thrown off in the outburst has formed the *Homunculus Nebula*, so called because to early observers its shape resembled a human figure, which obscures its light. Eta Carinae is the brightest infrared source in the sky, showing that it is still shining as brightly as in the 19th century. Studies of the light reflected from the dust cloud show that the central object varies in its emission lines with a period of 5.52 years. This suggests that the central object is in fact a binary system consisting of two stars, each of about 80 solar masses, orbiting each other at a distance of about 20 AU.

ESCAPE VELOCITIES OF BODIES IN THE SOLAR SYSTEM

Body	Escape velocity (km/s)
Sun	617.0
Mercury	4.4
Venus	10.4
Earth	11.2
Moon	2.4
Mars	5.0
Jupiter	59.5
Saturn	35.5
Uranus	23.5
Neptune	21.3
Pluto	1.1

Eudoxus of Cnidus (*c.*408–355 BC) Greek mathematician and astronomer. His theory of 27 'homocentric' (centred on the Earth) spheres was the first ever attempt to account for the movements of the Sun, Moon and planets on what we would now call a properly scientific basis. This picture of 'heavenly spheres' would survive for two thousand years. According to the Roman historian Pliny, Eudoxus fixed the length of the year at 365.25 days.

Euler, Leonhard (1707–83) Swiss mathematician. He made outstanding contributions to many areas of applied mathematics, especially theoretical astronomy. His analysis of the Moon's motion laid the foundations of lunar theory, and he made important advances in the study of the tides, planetary perturbations and the orbits of comets. Euler's work on optical systems strongly influenced the technical development of telescopes and microscopes.

Euro50 A design for a 50-m (164-ft) optical/infrared reflecting telescope by Scandinavian, British and Spanish astronomers to be constructed in the 2010s, probably at the **Roque de los Muchachos Observatory**. It was originally to be called the *Swedish Extremely Large Telescope*. With a segmented primary mirror, it would use **adaptive optics** to achieve a resolution of 2–3 milliarcseconds.

Europa The smallest of Jupiter's **Galilean satellites**, with a diameter of 3122 km (1940 miles), and the second-closest to the planet. Its density of 2.97 g/cm^3 indicates that it is a predominantly rocky body. Europa's smooth crust of water ice is crisscrossed by a network of light and dark linear markings. There are very few craters, so old ones must have been removed by some form of geological activity.

A kind of ice **tectonics**, driven by internal heating resulting from tidal forces, could be operating on Europa. Images from the **Galileo** probe support this idea: they suggest that the dark lines may be deposits from geysers that erupted along long fissures, and that a layer of liquid water or 'slush' lies below the surface. This layer may lie about 20–25 km (12–16 miles) down – deeper than previously believed – and a layer of ice this thick will present a formidable barrier to penetrate for possible future exploration. Some scientists have suggested that a sub-surface liquid environment on Europa might have favoured the development of primitive life, though this remains a matter of debate. Europa orbits at the outer edge of **Io**'s plasma torus, in which energetic particles break down water molecules to release oxygen, producing a thin atmosphere. SEE ALSO **satellite**.

European Southern Observatory (ESO) A consortium of ten European countries, including the UK. Its telescopes include a 3.6-m (142-inch) reflector, completed in 1975, and the **New Technology Telescope**, both at La Silla Observatory, and the **Very Large Telescope** at the Paranal Observatory.

European Space Agency (ESA) An organization of European nations to promote space research and technology for peaceful purposes, founded in 1975. ESA has 16 member states: Austria, Belgium, Denmark, Finland, France, Germany, Greece, Ireland, Italy, the Netherlands, Norway, Portugal, Spain, Sweden, Switzerland and the UK. Its headquarters are in Paris. ESA has designed and built the Ariane series of launch rockets, and operates a launch site at Kourou in French Guiana, on the Atlantic coast of South America.

EUVE ABBREVIATION FOR **Extreme Ultraviolet Explorer**.

evection An irregularity in the Moon's motion caused by perturbations by the Sun and planets. It amounts to a maximum of 1° 16′ in a period of 32 days.

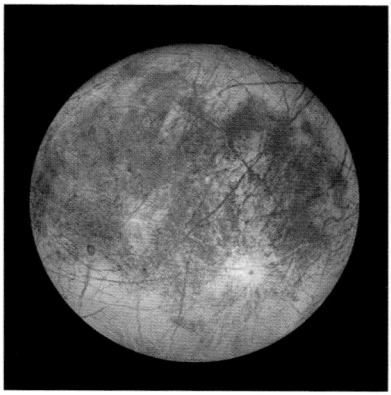

▲ **Europa** Rocky regions on Europa's surface are shown in brown. The pale areas towards the poles are icy plains. Long dark fractures can also be seen.

Evening Star Name sometimes given to the planet Venus when it is visible in the west after sunset.

event horizon The boundary surface of a **black hole** from within which nothing can escape. Observers outside the event horizon can therefore obtain no information about the black hole's interior. The radius of the event horizon is known as the **Schwarzschild radius**.

exit pupil For a telescope, the image of the primary mirror or lens as seen through the eyepiece. The point at which this image is formed is the optimum position for the observer's eye, so for comfort it should not be too close to the eyepiece itself (SEE **eye relief**).

exobiology Another name for **astrobiology**.

exoplanet Another name for **extrasolar planet**.

Exosat A European Space Agency satellite for X-ray astronomy, launched on 26 May 1983. It operated for three years, studying objects such as supernova remnants, X-ray binaries and galactic nuclei.

expanding Universe SEE **cosmology**, **redshift**.

extinction SEE **atmospheric extinction**, **interstellar matter**.

extrasolar planet (exoplanet) A planet orbiting a star other than the Sun. Stars with irregular **proper motions**, as often revealed by radial-velocity measurements, were long suspected of having planets orbiting them; **Barnard's Star** was a famous candidate. In 1992 the first confirmed extrasolar planets were detected, orbiting the pulsar PSR 1257+12. In 1995 the first planet orbiting another main-sequence star, 51 Pegasi, was identified. By late 2004 over a hundred planetary systems had been discovered around a wide diversity of stars up to 180 light years away. Many of these *substellar* objects are in very close orbits and are many times the mass of Jupiter. Masses and periods vary widely: around 0.1–17 Jupiter masses and 1.2–3090 days. The lower theoretical mass limit for a **brown dwarf** is about 0.013 solar masses (approximately 13 Jupiter masses), which by one convention is taken as the upper mass limit for a giant planet.

Extreme Ultraviolet Explorer (EUVE) A NASA satellite launched on 7 June 1992 to survey the sky and study individual sources at very short ultraviolet wavelengths, a region of the spectrum not covered by previous satellites. It operated until December 2000.

extrinsic variable SEE **variable star**.

eyepiece (ocular) The system of lenses in an optical instrument nearest the observer's eye. Its function in a telescope used for visual observation is to magnify the image formed at the focus by the primary lens or mirror. Telescope eyepieces usually have two lenses, the *field lens*, farthest from the eye, and the *eye lens*. The magnification yielded by an eyepiece is given by the ratio of the **focal length** of the telescope's primary lens or mirror to the focal length of the eyepiece. There are numerous designs of eyepiece, varying in focal length, field of view and magnification: SEE **Erfle eyepiece**, **Huygenian eyepiece**, **Kellner eyepiece**, **Nagler eyepiece**, **orthoscopic eyepiece**, **Plössl eyepiece**, **Ramsden eyepiece**. SEE ALSO **Barlow lens**.

eye relief In an optical instrument such as a telescope or binoculars, the distance between the eye lens of the **eyepiece** and the **exit pupil**. When the observer's eye is placed at the exit pupil, the whole of the eyepiece's field of view is visible. The eye relief must be sufficiently large for comfortable viewing.

F

Fabry–Pérot interferometer A type of **optical interferometer** for making accurate measurements of close spectral lines (SEE **spectrum**). The two parts of the incoming beam are recombined after passing through an arrangement of parallel plates called a *Fabry–Pérot etalon*. The beams undergo multiple reflections between the plates, which are highly reflective and accurately parallel. The instrument was developed by the French physicists Charles Fabry (1867–1945) and Alfred Pérot (1863–1925).

faculae (singular **facula**) Bright areas in the Sun's upper photosphere that herald the appearance of sunspots. They appear some days before the spots, in the same place, but can remain for months after the sunspots have gone. Faculae are at a higher temperature than their surroundings, and so appear brighter. The word is Latin for 'little torches'.

falling star A popular (US) name for a **meteor**.

False Cross A pattern of stars made up of Iota and Epsilon Carinae, and Kappa and Delta Velorum. It can be mistaken for the Southern Cross, but it is larger, less bright and more symmetrical.

Far Infrared and Submillimetre Space Telescope SEE **Herschel Space Observatory**

Far Ultraviolet Spectroscopic Explorer (FUSE) A NASA satellite launched on 24 June 1999 to study the sky at wavelengths of 90–120 nm, in the ultraviolet. One objective was to determine the amounts of deuterium in various cosmic sources in order to refine our knowledge of conditions in the very early Universe.

F corona SEE **corona**.

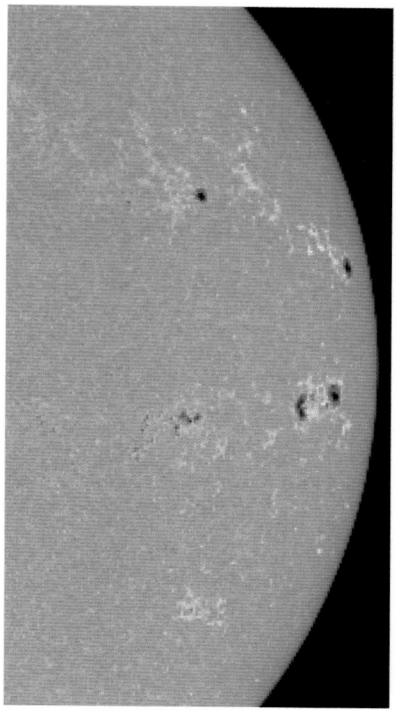

▲ **faculae** Bright regions marking the location of clouds of hot hydrogen above active regions on the Sun, faculae are best seen, as here, close to the limb. Faculae are commonest around sunspot maximum.

field of view The angular diameter of the area of sky visible through an optical instrument. For a telescope or binoculars it depends on the eyepiece that is used: a higher-power eyepiece will increase the magnification and thus decrease the field of view. Wide-angle eyepieces such as the **Nagler eyepiece** have fields of 80° or more.

filament A solar **prominence** seen in silhouette against the Sun's bright disk as a dark, thread-like marking in images taken

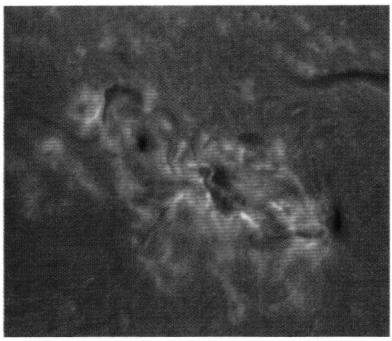

▲ **filament** The dark marking at top right in this image is a filament. It is seen in silhouette against the bright solar disk.

in the light of hydrogen alpha or the K line of calcium.

filar micrometer A device, used in conjunction with the eyepiece of a telescope, that incorporates cross-wires for measuring the angular size of an object or the separation between two objects. Superimposed on the field of view the observer sees two parallel wires and a third wire perpendicular to them. One of the parallel wires is movable, and the observer adjusts its position until the intersections of the two parallel wires with the third wire mark off the distance being measured. The angular distance may then be read from the micrometer scale.

filter A device, used in conjunction with an instrument for receiving electromagnetic radiation, that allows some wavelengths to pass but blocks others. Filters for use with optical instruments such as telescopes are at their simplest a thin sheet of material placed over the full aperture of the instrument. Simple colour filters made of gelatin bring out planetary features such as Jupiter's Great Red Spot. More sophisticated filters that transmit only a narrow band of wavelengths (said to have a narrow *passband*) are used for observing the Sun, or objects such as nebulae. Multilayer or interference filters are used to combat **light pollution**. In **CCD** imaging, colour pictures can be created by combining separate images obtained through blue, green and red filters. SEE ALSO **Sun**.

finder A small, low-power telescope mounted on a larger one and having a much larger field of view than the main telescope, enabling the observer to locate celestial objects more easily.

fireball An exceptionally bright **meteor**. Fireballs have been defined only loosely as meteors brighter than the planets; with the modern estimate of the maximum brightness of Venus, this can be taken to mean that all meteors brighter than magnitude −5 should be classified as fireballs. Some meteor showers, such as the Geminids, produce more fireballs than others.

first contact In a **solar eclipse**, the moment when the leading edge of the Moon's disk first touches the Sun's disk. In a **lunar eclipse**, it is the moment when the Moon begins to enter the umbra of the Earth's shadow. The term is also used for the corresponding stage in eclipses involving other bodies.

first light The occasion on which a newly constructed telescope is first used to observe a celestial object.

First Point of Aries An alternative name for the vernal **equinox**, so called because it originally lay in the constellation Aries. Because of **precession**, it has now moved into neighbouring Pisces.

first quarter One of the Moon's **phases**, when it is at quadrature on the way to being full and is half illuminated as seen from the Earth.

Flamsteed, John (1646–1719) English astronomer. He was appointed the first Astronomer Royal by King Charles II, with the task of obtaining accurate measurements of the Moon and stars for use in navigation. From Greenwich Observatory he used equipment fitted with sighting telescopes to measure with unprecedented precision the positions of thousands of stars. Flamsteed was a perfectionist and slow to release his results. Much to his anger, The Royal Society published his uncorrected observations. The catalogue of his corrected observations was not published until 1725, after his death; the numbers assigned to stars in this catalogue are still used (SEE **Flamsteed numbers**).

Flamsteed numbers A sequence of numbers allocated to the stars in each constellation in order of right ascension, for purposes of identification. The numbers were allocated by later astronomers to stars in John **Flamsteed**'s star catalogue.

flare A sudden and violent explosion in the Sun's inner atmosphere, usually above an active group of **sunspots**. In the *flash stage* a flare builds to a maximum in a few minutes, after which it gradually fades within an hour or so. Flares emit radiation across the whole electromagnetic spectrum, from gamma-rays to kilometre-wavelength radio waves. Particles are emitted, mostly electrons (some at half the speed of light) and protons, and smaller numbers of neutrons and atomic nuclei. A flare can cause material to be ejected in bulk, most spectacularly in the form of **coronal mass ejections**, at speeds that can exceed the Sun's escape velocity; **prominences** disturbed by the shock waves from flares may become detached and their material ejected into space. When energetic particles from flares reach the Earth they cause radio interference, magnetic storms and more intense **aurorae**. Although not well understood, the origin of flares is believed to be connected with local discontinuities in the Sun's magnetic field. It has even been suggested that flares are a result of coronal mass ejections rather than vice versa. SEE ALSO **flare star**.

flare star A variable star, usually a red dwarf (spectrum K–M), whose luminosity from time to time increases unpredictably by tenths of a magnitude to as much as 7 magnitudes in a very short time – only a few seconds or tens of seconds – and then decreases to its normal value, in a few minutes to tens of minutes. Flare stars are also known as *UV Ceti stars*, after the best-known example, UV Ceti, which is a pair of 13th-magnitude red dwarfs and lies just 8.7 light years away. Outbursts of flare stars are sometimes accompanied by an increase in radio emission. Large flares may give out ten to a hundred times as much light as the rest of the star. A form of flare star, the *BY Draconis stars*, show additional variations of a few tenths of a magnitude attributed to large 'starspots' passing across their disk as they rotate. They are therefore regarded as belonging to both **eruptive variables** and **rotating variables**. Flaring appears to be a consequence of instabilities in these young stars, allied with complex magnetic fields.

flash spectrum The **emission spectrum** of the solar chromosphere observed the few seconds just before and after totality during an eclipse of the Sun.

flatness problem A problem with the standard **big bang** model in that the **curvature of space** is observed to be

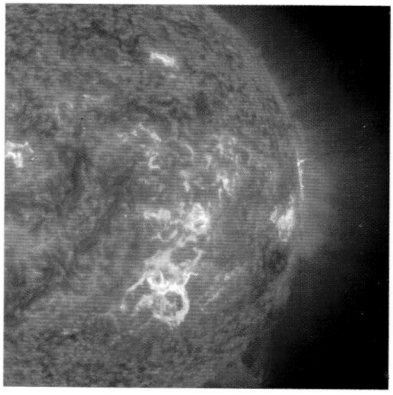

▲ **flare** Violent releases of magnetic energy in the inner solar atmosphere above active regions are seen in spectroheliograms as bright patches or ribbons, described as flares. Flares are most common at times of high sunspot activity.

improbably close to zero. Unless the Universe were flat to within 1 part in 10^{60} during the early stages of the big bang, space should now be strongly curved. **Inflation** solves the problem because the enormous expansion associated with inflation has flattened out any curvature on the 'small' scale of the visible Universe.

flat universe In cosmology, a mathematical model that describes the behaviour of a universe in which there is just enough matter to halt the expansion triggered in the **big bang**. It is often called the *Einstein–de Sitter model*. The universe will continue to expand, but at an ever-decreasing rate, theoretically coming to a halt an infinite time into the future. In a flat universe, the **curvature of space** is zero, the **deceleration parameter** equals 0.5 and the **density parameter** equals 1 (i.e. the mean density of the Universe equals the **critical density**). This model closely matches our observations of the real Universe. COMPARE **open universe**, **closed universe**.

flocculi (singular **flocculus**) Small features that give a mottled or granulated appearance to the Sun's chromosphere.

focal length The distance between the centre of a lens or curved mirror and its focus.

focal ratio The ratio of the **focal length** of a lens or curved mirror to its diameter. A focal ratio of, say, 8 is written as *f*/8.

focus (1) For a lens or curved mirror, the point at which the image is formed of a distant source lying on the axis of the lens or mirror.

focus (2) SEE **ellipse**.

following (abbreviation *f*) Describing the trailing edge, feature or member of an astronomical object or group of objects. For example, the *following limb* of the Moon is the edge facing away from its direction of motion; and the *following spot* (or *f*-spot) of a group of sunspots is the last of the group to be brought into view by the Sun's rotation.

Fomalhaut The star Alpha Piscis Austrini, magnitude 1.16, distance 25 light years, luminosity 17 times that of the Sun. It is a main-sequence star of type A3 surrounded by a **protoplanetary disk**.

forbidden lines Emission lines (SEE **emission spectrum**) in the spectra of some celestial objects which, when they were first observed, could not be identified with lines in spectra produced under laboratory conditions on Earth. It is now known that they are produced by atoms in what is called a metastable state. On Earth this state is very short-lived, but in highly rarefied nebulae, for example, it can last long enough for 'forbidden' lines to be emitted.

fork mounting A type of mounting used for a large **equatorial telescope**.

Fornax A southern constellation originated by **Lacaille**, representing a chemical furnace. The only star above fourth magnitude is Alpha, magnitude 3.9, a double. Fornax includes a cluster of galaxies.

fourth contact (last contact) In a **solar eclipse**, the moment when the trailing edge of the Moon's disk last touches the Sun's disk. In a **lunar eclipse** it is the moment when the Moon's trailing edge leaves the umbra of the Earth's shadow. The term is also used for the corresponding stage in eclipses involving other bodies.

Fraunhofer, Joseph von (1787–1826) German physicist and optician, the founder of astronomical spectroscopy. He studied the diffraction of light through narrow slits and developed the earliest form of diffraction grating. In 1814 he observed and began to map the dark lines in the Sun's spectrum that are now called **Fraunhofer lines**, and was the first to appreciate their significance. He later found similar but differently distributed lines in the spectra of other stars. Fraunhofer solved many of the scientific and technical problems of astronomical telescope-making. He manufactured achromatic lenses, was the first to build an **equatorial telescope** that realized the full potential of the design, and built the heliometer with which Friedrich **Bessel** measured the first stellar parallax.

Fraunhofer lines The dark **absorption lines** in the solar spectrum, caused by absorption at specific wavelengths in the upper, cooler layers of the Sun's atmosphere. The most prominent ones at visible wavelengths are caused by the presence of neutral hydrogen (the **hydrogen alpha** or Hα line), singly ionized calcium (the calcium H and K lines), sodium and magnesium. First observed in 1802 by William Hyde Wollaston, they were first carefully studied from 1814 by Joseph von **Fraunhofer**.

Fred L. Whipple Observatory An observatory located on Mount Hopkins, near Amado, Arizona, operated by the Smithsonian Astrophysical Observatory (SAO). It was opened in 1968, and later renamed to honour Fred **Whipple**, a former SAO Director. The main instruments are a 1.5-m (60-inch) and a 1.2-m (48-inch) reflecting telescope. The observatory runs the **MMT Observatory** jointly with the University of Arizona.

F star A star of spectral type F, with moderately strong hydrogen lines and strong ionized calcium and other metallic lines in its spectrum. Such stars have a surface temperature between 6100 and 7400 K on the main sequence; giants and supergiants of the same spectral type are about 300 K cooler. Main-sequence F stars range in mass from 1.2 to 1.7 solar masses, but most F-type giants and supergiants have evolved from considerably higher-mass stars. Bright examples are Canopus and Procyon.

▲ **Fraunhofer lines** Dark lines interrupt the continuum of the solar spectrum. These result from absorption of light at specific wavelengths by atomic species in the Sun's atmosphere.

full Moon The Moon's **phase** when it is at opposition, and its illuminated hemisphere is fully visible from the Earth.

fundamental stars Reference stars whose positions and **proper motions** have been determined with the greatest possible accuracy. Their coordinates are published in a *fundamental catalogue* and serve as the points to which positional measurements of other bodies can be related. The IAU adopted 1535 stars down to magnitude 7.5 as fundamental reference stars; relevant data are contained in the *Fifth Fundamental Catalogue*, published in Germany and known for short as the *FK5* from its German title. The *FK5 Extension* adds a further 3117 stars down to magnitude 9.5.

FUSE ABBREVIATION FOR **Far Ultraviolet Spectroscopic Explorer**.

G

galactic coordinates SEE **coordinates**.

galactic halo The spheroidal distribution of old stars and globular clusters that surrounds a galaxy. The halo around our Galaxy is about 50,000 light years in diameter, and there appears also to be a much more extensive halo of **dark matter** extending out to a radius of at least 150,000 light years.

Galatea One of the small inner **satellites** of Neptune discovered in 1989 during the Voyager 2 mission.

galaxy A huge assembly of stars, dust and gas, an example of which is our own Galaxy. There are three main types, as originally classified by Edwin **Hubble** in 1925. *Elliptical galaxies* are round or elliptical systems, showing a gradual decrease in brightness from the centre outward. They are given a designation from E0 to E7 in increasing degree of ellipticity. *Spiral galaxies* are flattened disk-shaped systems in which young stars, dust and gas are concentrated in spiral arms coiling out from a central bulge, the *nucleus*; they are designated S, with lower case letters added to show how tightly the arms are wound, from tight (a) to very loose (d). *Barred spiral galaxies* are distinguished by a bright central bar from which the spiral arms emerge; they are designated SB with letters from a to d appended, as for ordinary spirals. These three main classes are represented in the **tuning fork diagram**. In addition there are *lenticular galaxies*, systems intermediate between ellipticals and spirals, having a disk and nucleus, similar to spiral galaxies, but with no apparent spiral arms; they are classified S0. **Irregular galaxies** (designated Irr or Ir) are systems with no symmetry.

Current theories suggest that all galaxies were formed from immense clouds of gas at roughly the same time, soon after the big bang. In elliptical galaxies star formation took place rapidly over perhaps several hundred million years, using up all the interstellar gas. Spirals, however, are the result of a two-stage formation process. Gas was left over from the initial star formation which produced the bulge at their centres. The remaining gas rapidly settled into a disk, in which density waves formed spiral arms.

Galaxies can exist singly or in clusters that contain anywhere from just a few to thousands of members (SEE **cluster of**

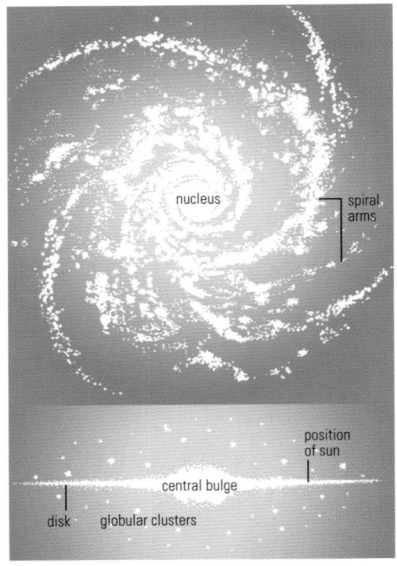

▲ **galaxy** The upper diagram shows the general outline for the structure of a spiral galaxy, of which our Galaxy is an example. Seen face-on (top), the galaxy has a fairly compact nucleus surrounded by an extensive disk containing star-rich spiral arms. Seen edge-on (below), the nucleus is revealed as a bulge, with the spiral arms confined to the flat plane of the disk. The galaxy as a whole is surrounded by a halo of globular clusters. As indicated, the Sun is located midway out in the disk of our Galaxy.

galaxies). Galaxies in clusters have a good chance of one near-collision in their lifetimes. Interactions between gas-rich galaxies such as spirals can produce bursts of star formation, and such interactions are a possible trigger mechanism for generating **active galactic nuclei**, as in **Seyfert galaxies** and **quasars**. These objects are generally believed to be galaxies with nuclei in extreme states of activity. The most massive elliptical galaxies are now believed to have been formed by repeated collisions and mergers of smaller, individual galaxies. SEE ALSO **disk galaxy**, **low-surface-brightness galaxy**.

Galaxy, the The star system that contains the Solar System; the capital 'G' distinguishes it from other **galaxies**. The Galaxy is spiral in shape and about 100,000 light years in diameter. Our Sun and Solar System are located at the edge of one of the spiral arms, 26,500 light years from the centre. The spiral arms form a disk-shaped system, with a bulging core (or nucleus) in the direction of Sagittarius. Old stars, of Population II, are found in the Galaxy's core; younger, hotter Population I stars, together with interstellar dust and gas, make up the spiral arms (SEE **stellar populations**). The stars of the spiral arms form the Milky Way, running around the sky. It is uncertain how many spiral arms the Galaxy has, but there is good evidence, from infrared studies for example, that there is a modest bar at the centre, and that it should be classified as a barred spiral galaxy. Surrounding it is the spheroidal **galactic halo**, which is about 100,000 light years in diameter. In it are found very old Population II stars, including those in **globular clusters**.

Between us and the Galaxy's central bulge passes a prominent spiral arm, seen as the bright areas of the Milky Way in Centaurus, Crux and Carina. This arm curls away from us in Carina where we see it virtually end-on, which accounts for the great accumulation of stars and nebulae in that constellation. Immediately outside this arm is a band of gas and dust, a region where stars have yet to form, which is visible as the dark rift in the Milky Way running from Cygnus to Crux. Beyond Carina is a fainter arm which curves round outside the Sun. It crosses the northern sky, gaining in brightness as it moves inward towards Cygnus.

THE GALAXY: DATA

Diameter of disk	100,000 l.y.
Thickness of nucleus	6000 l.y.
Average density	7×10^{-24} g/cm^3
Total mass	2×10^{45} g = 10^{12} solar masses
North pole of galactic plane (epoch 2000.0)	RA 12h 51.4m, dec. +27° 7′
Point of zero longitude (epoch 2000.0)	RA 17h 45.6m, dec. −28° 55′
Galactic longitude of north celestial pole	123°
Distance of Sun from centre	26,500 l.y.
Distance of Sun above galactic plane	14 l.y.
Rotational velocity of Sun	250 km/s
Period of Sun's revolution around centre	200 million years
Absolute magnitude	−20.5

The centre of the Galaxy is marked by the radio source **Sagittarius A**, which has been adopted as the zero point of galactic longitude. The *galactic equator* is the plane of the Galaxy, running along the Milky Way. The whole Galaxy is rotating, but the rotational rate varies with the distance from the centre. Our Sun circles the centre at about 250 km/s, taking 200 million years per orbit. The Galaxy's age is thought to be about 10,000 million years, during which time the Sun would have completed approximately 50 revolutions.

The Galaxy is a member of a cluster known as the **Local Group**, which includes the Andromeda Galaxy and the Magellanic Clouds.

Galaxy Evolution Explorer (GALEX) An astronomical observatory launched on 28 April 2003 on a planned 29-month mission to produce the first comprehensive ultraviolet map of galaxies and to identify celestial objects for study

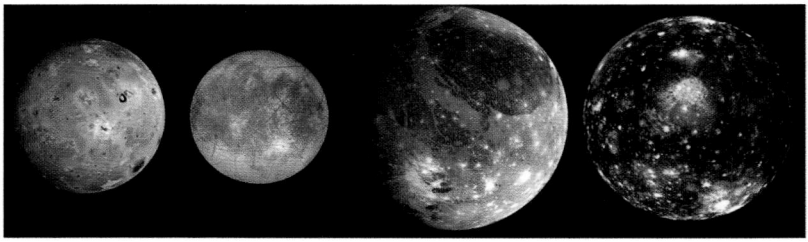

▲ **Galilean satellites** The four large satellites of Jupiter are arranged in order of distance from the planet: (left to right) Io, Europa, Ganymede and Callisto. The images were obtained by the Galileo orbiter.

by later spacecraft. GALEX should provide a better understanding of how galaxies evolve and change.

Galilean satellites The four chief satellites of Jupiter – **Io**, **Europa**, **Ganymede** and **Callisto** – named collectively after **Galileo**, who observed them in 1610. All are in low-inclination, near-circular, **synchronous orbits**, and their orbital periods are in simple ratios to one another as a result of **resonance**. Apart from the Earth's Moon, they are the brightest satellites in the Solar System, and all would be naked-eye objects were it not for the glare from Jupiter itself. They pass through cycles of **occultations** by Jupiter and eclipses in its shadow, and transits across its disk. SEE ALSO **Jupiter**, **satellite**.

Galileo Galilei (1564–1642) Italian astronomer, physicist and mathematician. He was one of the first to use the telescope for astronomical observations, and he improved its design. He discovered mountains and craters on the Moon, and found the four satellites of Jupiter now known as the **Galilean satellites**. He observed the phases of Venus and studied sunspots, from whose motion he deduced that the Sun rotates. He found that some celestial objects, such as the Milky Way and the Pleiades, are resolvable into many more stars than can be distinguished by the naked eye.

Galileo concluded that **Aristotle**'s picture of the world, then still widely believed, was wrong, and instead championed **Copernicus**'s heliocentric theory. This brought him into conflict with the Catholic Church, and led to his trial and house arrest for the last eight years of his life. His many advances in physics included the law of the pendulum and how to create a vacuum. His chief books were *Sidereus nuncius* (on discoveries with the telescope), *Dialogue on the Two Chief World Systems* (heliocentric Universe), and *Discourses and Mathematical Demonstrations on Two New Sciences* (mathematical physics).

Galileo A space probe to Jupiter, launched on 18 October 1989. En route, Galileo flew past and imaged the asteroids **Gaspra** and **Ida** in October 1991 and August 1993 respectively. Galileo entered orbit around Jupiter in December 1995, after dropping off a subprobe which entered the planet's atmosphere. This probe detected a hydrogen/helium ratio similar to the Sun's, and an intense radiation belt 50,000 km (30,000 miles) above the upper cloud deck. From 1995 to 2002 Galileo made over two dozen flybys of the Galilean satellites, returning highly detailed views. Its mission ended in 2003.

Galileo National Telescope (TNG) A 3.5-m (138-inch) reflecting telescope at **Roque de los Muchachos Observatory**, opened in 1997 as a national facility for Italian astronomers. (The initials are those of the Italian version of the name.)

Galle, Johann Gottfried (1812–1910) German astronomer. In 1846 he and

Heinrich D'Arrest were the first to locate the planet Neptune, close to positions calculated independently by John Couch **Adams** and Urbain **Le Verrier**. Galle discovered Saturn's faint C Ring, and determined the solar parallax (used to fix the scale of the Solar System) by observing the asteroid Flora.

gamma-ray astronomy The study of very high-energy **electromagnetic radiation** from space with wavelengths of about 0.01 nm and shorter. Gamma-rays are absorbed by the Earth's atmosphere, so all studies must be made by high-altitude balloons or spacecraft. Early surveys found a general concentration of emission in the galactic plane, with a peak towards the galactic centre, plus several strong individual sources such as the Crab Nebula, the Vela Pulsar and the quasar 3C 273. In 1991 NASA launched the **Compton Gamma Ray Observatory**, and in 2004 the **Swift Gamma Ray Burst Explorer**.

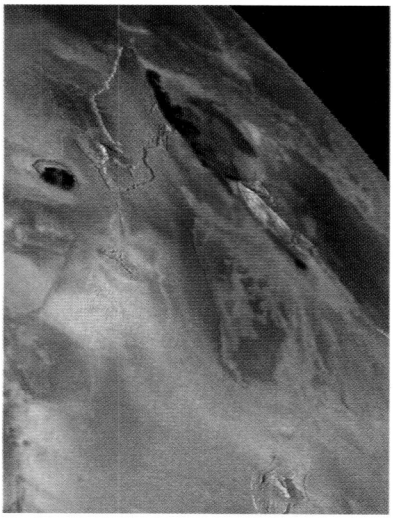

▲ **Galileo** Repeated passes close to the four large satellites of Jupiter by the Galileo orbiter since 1996 have revealed these in more detail than ever before. This image shows the volcano Zal Patera on Io.

Their very short wavelengths mean that gamma-rays have very high energies, and they are emitted from regions of very high temperature associated with some of the most violent events in the Universe. Most of the gamma radiation from the Galaxy is produced by cosmic rays interacting with gas in the interstellar medium. Certain gamma-ray lines at specific energies have been discovered, one attributed to the mutual annihilation of electrons and positrons in the galactic centre, and another to the isotope aluminium-26, which is produced by supernovae. Sporadic, low-energy (soft) gamma-ray flares, whose sources are known as *soft gamma repeaters* (SGRs), are now thought to be produced by magnetic-field events on **magnetars** – neutron stars with exceptionally strong magnetic fields.

Outside the Galaxy, the main sources of gamma-rays are those that produce **gamma-ray bursts**. There is also a gamma-ray background, a diffuse radiation from all over the sky, like the cosmic microwave background; its origin is unknown.

gamma-ray burst (GRB) An intense flash of gamma-rays, which may last from about a second to a few minutes. GRBs were first detected in 1967 by a satellite monitoring nuclear explosions on Earth. Because they occur randomly across the sky, at the rate of a few per day, they could not originate in our Galaxy (they would then be concentrated near the Milky Way). They had to be either relatively near or very distant. In 1997 it was found that GRBs are followed by an X-ray 'afterglow', which enabled their positions to be pinpointed with greater accuracy. Their optical counterparts were found to be young galaxies with high redshifts, which makes GRBs extremely energetic events.

Theorists have suggested exotic models such as colliding neutron stars or even **hypernovae** to account for the colossal energy output of GRBs. It has now been confirmed that some GRBs originate in distant supernovae, but it is suspected that there are also other sources. The **Swift Gamma Ray Burst Explorer** is designed

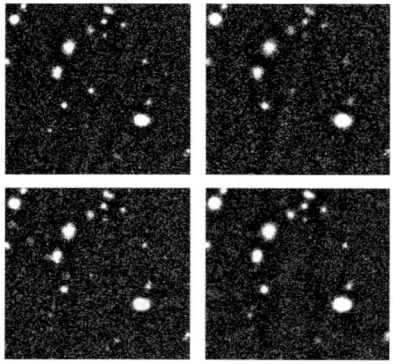

▲ **gamma-ray burst** GRB 00131 was detected by several spacecraft on 31 January 2000. This series of images obtained with the Very Large Telescope (VLT) shows the fading visible-wavelength afterglow, just left of centre.

to react rapidly, relaying an accurate position to a network of ground-based telescopes within seconds of detecting a burst.

Gamma Ray Large Area Space Telescope A satellite planned for launch in May 2007 to make the first detailed, high-definition observations of phenomena that generate high energies, including supermassive black holes, merging neutron stars and hot gases. Valuable data about the birth and early development of the universe should be generated.

Gamow, George (1904–68) Russian-American nuclear physicist and cosmologist. His most famous work, co-authored with Ralph Alpher and Hans **Bethe**, pictured a hot and dense early Universe in which the nuclei of elements were built up from protons, and was a cornerstone of **big bang** theory. Gamow's greatest prediction, borne out by the detection of the **cosmic microwave background** radiation, was that the Universe should by now have cooled to a few degrees above absolute zero.

Ganymede The largest of Jupiter's **Galilean satellites**, with a diameter of 5262 km (3270 miles), and the largest satellite in the Solar System – bigger, though less massive, than the planet Mercury. With a density of 1.94 g/cm^3, Ganymede contains a high proportion of water ice. Its surface is a mixture of dark, heavily cratered terrain and a brighter, more lightly cratered terrain covered with meandering, intersecting grooves that suggest recent geological activity. The most prominent feature is the dark Galileo Regio, some 4000 km (2500 miles) across. The **Galileo** probe detected a weak magnetic field, probably originating in a liquid layer deep below the surface. Ganymede's magnetosphere interacts with that of Jupiter. The satellite has an extremely thin oxygen atmosphere. SEE ALSO **satellite**.

gas giant SEE **giant planet**.

Gaspra Asteroid no. 951, discovered by Grigorii Neujmin in 1916. In 1991 it became the first asteroid to be imaged from close quarters when the probe **Galileo** passed it en route to Jupiter. It is irregular, with a maximum dimension of about 20 km (12 miles), and is pitted with small craters.

Gauss, Carl Friedrich (1777–1855) German mathematician. His many achievements include valuable contributions to astronomy, geodesy and physics. In 1801 he worked out how to calculate the orbit of the newly found asteroid Ceres from observations spanning just 9° of its orbit. This led him to the study of celestial mechanics, and on to a comprehensive study of the determination of cometary and planetary orbits, taking into account the effects of perturbations, and then to the theory of errors of observation, for which he developed the mathematical method of least squares. From 1807 Gauss was director of the observatory at Göttingen.

gegenschein A faint luminous patch on the ecliptic, visible usually only from the tropics on dark clear nights, directly opposite the Sun. It is caused by the scattering of sunlight back towards the Earth by par-

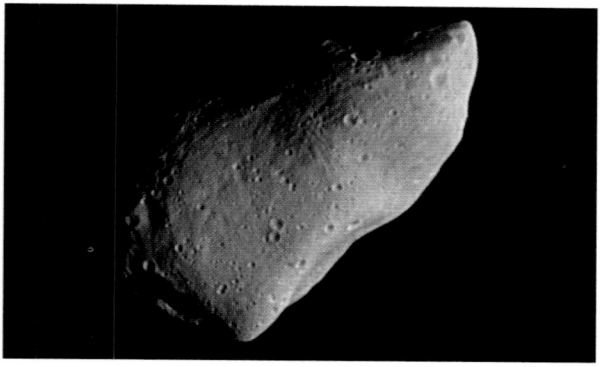

◀ **Gaspra** En route to Jupiter, the Galileo spacecraft made a flyby of asteroid (951) Gaspra in 1991, revealing it to be an irregular-shaped, heavily cratered body. Gaspra lies in the main belt.

ticles in the zodiacal dust cloud, and is part of the **zodiacal light**. The word is German for 'counterglow', by which name it was formerly also known.

Geminga One of the most powerful gamma-ray sources in the sky, in the constellation Gemini (hence the name). Geminga's energy is emitted as pulses of gamma-rays (and X-rays) with a period of 0.237 seconds, so it is most likely a spinning neutron star: a short-wavelength version of a **pulsar**.

Gemini A large, prominent constellation of the zodiac. Its brightest stars are **Castor** and **Pollux**, named after the twins of Greek mythology. Eta is a semiregular variable, magnitude 3.2 to 3.9, with a period averaging 233 days. Zeta is a Cepheid variable, range 3.6 to 4.2, period 10.15 days. The open cluster M35 contains about 200 stars.

Geminids A **meteor shower** which occurs between 7 and 15 December, peaking at **ZHR** in excess of 100 close to 13–14 December. Its radiant is in the constellation Gemini, near the star Castor. The shower, one of the most prolific, is associated not with a comet but with **Phaethon**, an asteroid.

Gemini Telescopes A pair of 8.1-m (320-inch) reflectors jointly funded by the USA, UK, Canada, Australia, Chile, Brazil and Argentina. Gemini North is sited at Mauna Kea, Hawaii, and was dedicated in 1999. Gemini South is at Cerro Pachón, Chile, and opened in 2002. Together they provide complete sky coverage at optical and infrared wavelengths.

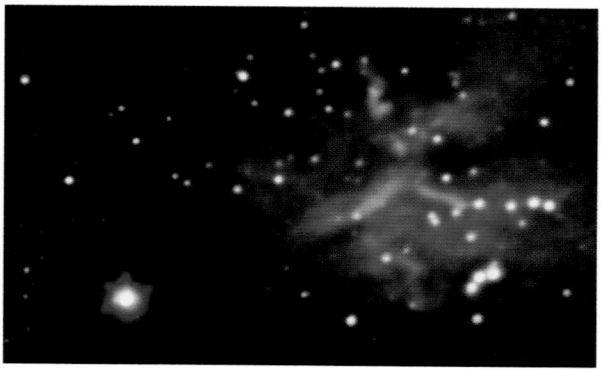

▶ **Gemini Telescopes** This infrared view of a star-forming region in the Milky Way was taken using the Gemini North telescope on Mauna Kea. Adaptive optics helped sharpen the image.

geocentric As viewed from or related to the Earth's centre, as in *geocentric parallax* or *geocentric system*. COMPARE **heliocentric**.

geocentric parallax SEE **parallax**.

geodesic SEE **curvature of space**.

German mounting A type of mount used for an **equatorial telescope**.

Ghost of Jupiter A bright planetary nebula, designated NGC 3242, in the constellation Hydra at a distance of 2000 light years. The object was named by William Herschel, who discovered it in 1785, from its visual telescopic appearance as a very dim version of the giant planet. Many observers report a pronounced blue-green colour.

Giacobinids A meteor shower associated with Comet **Giacobini–Zinner**. The shower is periodic, and produced significant displays on 9 October 1933, 9–10 October 1946 (a **meteor storm**) and 8 October 1985; a minor outburst occurred in 1998. Activity is seen only in those years when the parent comet is close to perihelion. The radiant lies close to the star Beta Draconis, and the shower is also known as the *Draconids*.

Giacobini–Zinner, Comet A short period **comet**, orbiting the Sun in 6.59 years. It was discovered in 1900 by Michel Giacobini (1873–1938) and recovered in 1913 by Ernst Zinner (1886–1970). In 1985 it became the first comet to be studied at close hand when the International Cometary Explorer space probe passed through the comet's ion tail. Debris from the comet produces the periodic **Giacobinid** meteor shower.

Giant Magellan Telescope (GMT) A design by a US consortium for an optical/infrared reflecting telescope with a primary made of seven mirrors with an equivalent aperture of 24.5 m (80.4 ft). The design is based on a **Gregorian tel-**

▲ **Giacobini–Zinner, Comet** This comet was imaged on 1 November 1998, by the Kitt Peak 0.9-m (36-inch) telescope. The teardrop-shaped coma shows strong bluish carbon monoxide emission.

▲ **Giotto** The ESA Giotto probe recorded this close-up view of the nucleus of Halley's Comet on 13 March 1986. Shortly after this image was obtained, collision with a dust particle from the comet disabled the camera.

escope and uses **active optics**, and **adaptive** optics for the secondary mirror.

giant molecular cloud (GMC) SEE **infrared astronomy**.

giant planet Any of the planets Jupiter, Saturn, Uranus and Neptune, also known as *gas giants*. They are so called because their masses are much greater than the Earth's. The terms are also used for large **extrasolar planets**

giant star A star that is placed well above the **main sequence** in the **Hertzsprung–Russell diagram**. Giant stars are larger than main-sequence (**dwarf**) stars of the same temperature, but giants of intermediate temperatures are smaller than the hottest dwarfs. Thus the term 'giant' really refers to the relative luminosity of a star rather than its dimensions. Giant stars evolve from dwarf stars that have run out of hydrogen in their central regions, and so represent later stages in the lives of stars. Because of their high luminosity, giants are quite common among naked-eye stars, examples being Arcturus, Aldebaran and Capella, but they are relatively rare overall. SEE ALSO **red giant**, **blue giant**, **supergiant**.

gibbous A **phase** of the Moon or a planet when more than half but less than all of its illuminated hemisphere is visible from the Earth.

Giotto A European Space Agency mission to investigate **Halley's Comet**. The probe was launched on 2 July 1985 and flew to within 600 km (375 miles) of the comet's nucleus on 14 March 1986. Giotto imaged the nucleus and analysed the gas and dust given off. After the encounter, the probe went on to rendezvous with Comet **Grigg–Skjellerup**, 10 July 1992.

glitch SEE **pulsar**.

globular cluster A near-spherical cluster of very old Population II stars (SEE **stellar populations**) in the halo of our Galaxy. Globular clusters contain a large number of stars (10^5 to 10^7), packed closely into a roughly spherical space typically 50 to 150 light years across. The stars are strongly concentrated toward the centre.

Most globulars are at least 10,000 million years old, as deduced from the highly evolved state of their stars. This extreme age, combined with their distribution in the galactic halo, indicates that globulars

▲ **globular cluster** Containing at least 300,000 ancient stars, the globular cluster G1 orbits the Andromeda Galaxy. This Hubble Space Telescope image shows G1 to have a rich, compact core and less densely packed periphery.

formed while our Galaxy was condensing from a huge cloud of gas. Globulars can also be seen around some other galaxies. They are more plentiful around elliptical galaxies than spirals – our Galaxy has about 140 known globular clusters, whereas the largest elliptical galaxies can have thousands.

Although they occupy the spheroidal galactic halo, most globulars lie no farther from the centre of the Galaxy than does the Sun. As a result they congregate in the general area of sky containing the centre of our Galaxy, in particular in the constellations Sagittarius, Ophiuchus and Scorpius. Globular clusters orbit the galactic centre, making periodic passages through the plane of our Galaxy. They suffer a slow loss of stars (known as *cluster evaporation*), because gravitational interactions accelerate some individual stars, mostly of low mass, into hyperbolic escape orbits. SEE ALSO **star cluster**.

globule SEE **Bok globule**, **cometary globule**.

GMC ABBREVIATION FOR giant molecular cloud (SEE **molecular cloud**).

GMT ABBREVIATION FOR **Giant Magellan Telescope** or **Greenwich Mean Time**.

Gold, Thomas (1920–) Austrian-American astronomer. With Hermann Bondi and Fred Hoyle he developed the **steady-state theory**, in which the continuous creation of matter drives the expansion of the Universe. In the 1960s he explained that the signals from pulsars were beams of synchrotron radiation from rapidly rotating neutron stars.

Goodricke, John (1764–86) English astronomer, born in Holland. With the encouragement of his neighbour and collaborator, Edward Pigott, the deaf-mute Goodricke began at the age of 17 to study the brightest variable stars. In 1782 he announced the cause of Algol's variability (that it was what is now called an eclipsing binary), and the length of its period. Two years later he discovered the variability and periods of Beta Lyrae and Delta Cephei, the prototype Cepheid variable. He died aged just 21 from pneumonia contracted while observing Delta Cephei.

Gould's Belt A region of bright stars and gas inclined at about 20° to the galactic plane. It contains the greatest concentration of naked-eye stars of spectral types O and B, and probably represents part of the local spiral arm of which the Sun is a member. It is named after the American astronomer Benjamin Gould (1824–96), who studied the region.

grand unified theory (GUT) A theory that attempts to 'unify' three of the four forces of nature, and thus to demonstrate that they are different manifestations of a single force. In descending order of strength, the four forces of nature are: the *strong nuclear force*, which binds atomic nuclei, the *electromagnetic force*, which holds atoms together, the *weak nuclear force*, which controls the radioactive decay of atomic nuclei, and **gravitation**. The weak and strong nuclear forces operate over tiny distances within atomic nuclei, while the other two forces are infinite in range.

In the 1970s and 1980s, the unification of the electromagnetic and weak nuclear forces was first predicted and then demonstrated experimentally. GUTs are attempts to unify this *electroweak force* with the strong nuclear force. Many physicists believe that a more elaborate theory, a so-called *theory of everything* (TOE), will eventually show how all four forces are unified.

Gran Telescopio Canarias (GTC) A Spanish 10.4-m (34-ft) reflecting telescope under construction at the **Roque de los Muchachos Observatory**, due to open in 2005. With its segmented mirror, the instrument closely resembles the 10-m telescopes of the **W.M. Keck Observatory**.

granulation The mottled appearance of the Sun's **photosphere** caused by gases rising from the Sun's interior. When seeing conditions are very good the granulations can be resolved into small, nearly circular patches, called *granules*, surrounded by darker areas. Individual granules are convection cells, typically 1000 km (600 miles) in diameter, and last only a few minutes.

graticule A system of reference marks or a measuring scale consisting of parallel vertical wires, or cross-wires, or a reticule (a grid of squares) placed in the focal plane of a telescope and covering the entire field of view. SEE ALSO **filar micrometer**.

grating SEE **diffraction grating**.

gravitation (gravity) The universal force of attraction between all particles of matter. *Newton's law of gravitation* states that the force of attraction (F) between two masses (m_1 and m_2) is inversely proportional to the square of their separation (r):

$$F = Gm_1m_2/r^2,$$

where G is the *gravitational constant*. This law is the basis of celestial mechanics.

Although at close range gravity is far, far weaker than the nuclear or electromagnetic forces, it has a greater range and is the most important force on the large, cosmological scale. The whole concept of gravitation was reinterpreted by Albert **Einstein** in his General Theory of Relativity in terms of the curvature of **spacetime**.

gravitational lens A massive body, such as a star, black hole or galaxy, that bends light or other radiation from a more

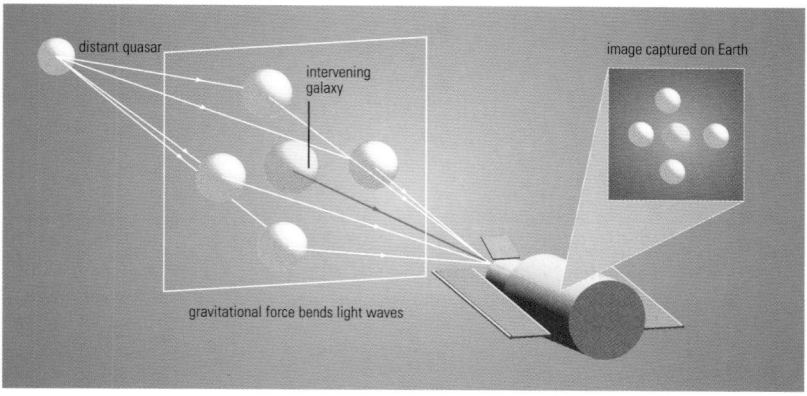

▲◀ **gravitational lens** (top) Light from a distant quasar may be bent in the gravitational field of a foreground galaxy to produce a distorted or multiple image. In this example, the quasar's light is split into four separate images, each arriving at the telescope along a slightly different path. (bottom) The massive galaxy cluster Abell 2218 in Draco acts as a gravitational lens, producing distorted, arc-shaped images of more distant objects in the same line of sight.

distant object in the same line of sight. The effect, which is predicted by Einstein's Theory of Relativity, was first detected during the total solar eclipse of 1919 by Arthur **Eddington**, who found that light from distant stars was bent as it passed the Sun. In 1979, the first example of gravitational lensing of light from a quasar was discovered. In such cases, the quasar's light may be split into two or four separate images by an intervening massive galaxy or cluster of galaxies. Other distant objects may have their light lensed into arcs, or even a complete **Einstein ring**. Gravitational lensing of quasars can be used to measure their distances. SEE ALSO **microlensing**.

gravitational redshift SEE **redshift**.

gravitational waves Waves emitted as the result of a violent disturbance in a gravitational field, such as a massive body being accelerated or distorted with a velocity that is an appreciable fraction of the velocity of light. Gravity waves are predicted by Einstein's General Theory of Relativity, but have proved elusive. Just as an oscillating charged particle emits electromagnetic waves, so a violently disturbed mass should emit gravity waves. Likely sources of gravity waves include close binary systems where at least one component is a neutron star or black hole, or the merger of two such dense bodies. Because gravity is by far the weakest of the forces of nature, gravity waves are extremely feeble. Detectors based on **interferometry** have been developed and are being commissioned, and should be sensitive enough to pick up gravitational waves.

Gravity Probe/Einstein A NASA spacecraft, launched on 20 April 2004, to test Einstein's General Theory of Relativity by measuring the precession of gyroscopes in Earth polar orbit.

Great Attractor A hypothetical distant concentration of matter with a mass about 100,000 times that of the Milky Way, 150 million light years away in the direction of the constellations Hydra and Centaurus. It was suggested to account for the observed motions of galaxies, including those of the **Local Group**. Although there is a cluster of galaxies in the place in question, it now seems more likely that the gravitational pulls of several different concentrations of **dark matter** are responsible, and there is no single Great Attractor.

◀ **Great Attractor** The central region of ACO 3627, a huge galaxy cluster thought to form part of the Great Attractor, is seen here in an image obtained by the European Southern Observatory. The Great Attractor's gravitational pull is dragging all the galaxies in our local cluster towards it.

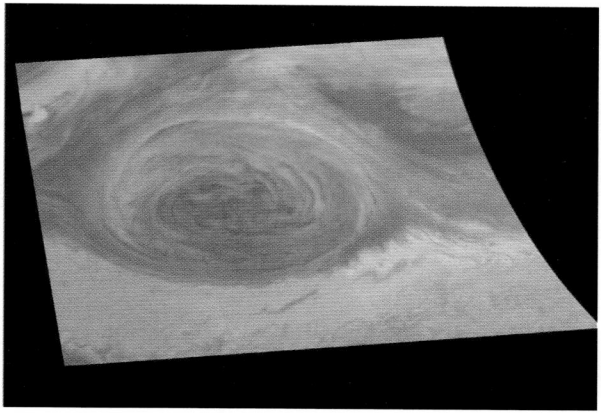

► **Great Red Spot**
A 1996 Galileo orbiter image shows this huge storm, which has raged in Jupiter's atmosphere for hundreds of years.

great circle A circle on a sphere whose plane passes through the centre of the sphere. On the **celestial sphere** the celestial equator is a great circle, as are meridians (circles passing through both celestial poles).

greatest elongation SEE **elongation**.

Great Observatories Umbrella term for four major NASA spaceborne telescopes: the **Hubble Space Telescope**, the **Compton Gamma Ray Observatory**, the **Chandra X-ray Observatory** and the **Spitzer Space Telescope**.

Great Red Spot (GRS) A large oval area in the southern hemisphere of Jupiter. Since it was first observed, by Heinrich Schwabe in 1831 (and possibly by Robert Hooke in 1664, though this may have been a different but similar spot), it has varied in size and hue – sometimes deep red, at other times pale grey. Even when the GRS is faint, its position is marked by a *hollow* on the southern edge of Jupiter's dark South Equatorial Belt. The GRS's red colour may be produced by phosphorus.

Green Bank The location in West Virginia, USA, of the **National Radio Astronomy Observatory** and its Green Bank Telescope.

green flash An atmospheric phenomenon, observable under ideal conditions during the last seconds before sunset or the first seconds after sunrise, in which the uppermost segment of the Sun separates and flashes with a green colour. At these moments most of the sunlight has been absorbed in its long passage through the atmosphere, and only two narrow bands of wavelengths in the red and green are left. The atmospheric refraction of light is greater for short wavelengths (green rather than red), so the green flash appears.

greenhouse effect The warming of a planet's surface by heat trapped by its atmosphere. Incoming sunlight is absorbed at the surface and re-radiated at longer wavelengths, in the infrared – that is, as heat. If the atmosphere contains *greenhouse gases*, such as carbon dioxide, which are not transparent to the infrared, the heat is reflected back to the ground. The effect is the same as is produced by the glass enclosing a greenhouse. The result is to raise the planet's average surface temperature by an amount that depends on the atmospheric composition and density. On **Venus** the dense carbon dioxide atmosphere has raised the surface temperature to around 750 K. On Earth a smaller greenhouse effect has led to a surface temperature that suits many life forms, but some

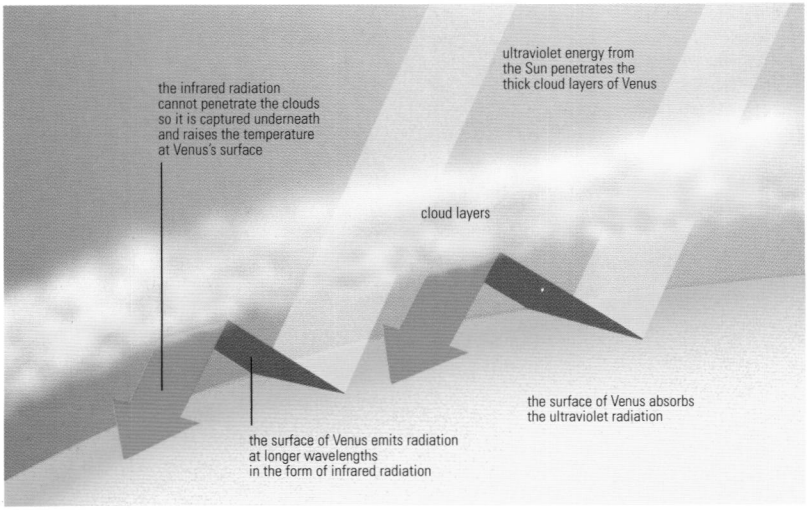

▲ **greenhouse effect** Short-wave radiation from the Sun readily penetrates the atmosphere to reach the ground. Longer wavelength radiation re-emitted from the ground is trapped by atmospheric constituents such as carbon dioxide, raising the planetary temperature. This greenhouse effect is significant in keeping Earth habitable, but has led to inhospitably high temperatures on Venus.

people believe that industrial emissions of carbon dioxide will increase the greenhouse effect, producing *global warming*, and should be halted.

Greenwich Mean Time (GMT) Mean solar time for the meridian of Greenwich. In 1928 the International Astronomical Union recommended that GMT should be known as **Universal Time** for scientific purposes. Today, that usage has extended to legal and other purposes, although the term GMT is still used in many contexts. Greenwich was chosen as the world's prime meridian for timekeeping and navigation at the International Meridian Conference in 1884.

Gregorian calendar SEE **calendar**.

Gregorian telescope A reflecting telescope described by Scottish mathematician James Gregory (1638–75) in 1663, five years before Isaac Newton constructed his first reflector. Light received by the parabolic primary mirror is reflected to a concave spheroidal secondary mirror, and then reflected back through a hole in the centre of the primary to an eyepiece. It differs from the Cassegrain telescope in having a concave instead of a convex mirror for the secondary. The configuration, long disused, has been revived in designs for giant instruments such as **Euro50**.

Grigg–Skjellerup, Comet The periodic **comet** with the second-shortest period known, 5.09 years. It was discovered in 1902 by New Zealander John Grigg (1838–1920) and recovered in 1922 by Australian John 'Frank' Skjellerup (1875–1952). In 1992 it became the third comet to be encountered by a spacecraft when the **Giotto** probe passed within 200 km (125 miles) of it.

GRO ABBREVIATION FOR **Compton Gamma Ray Observatory**.

GRS ABBREVIATION FOR **Great Red Spot**.

Grubb, Howard (1844–1931) Irish telescope-maker. In 1868 he took over the telescope-manufacturing business established by his father, **Thomas Grubb** (1800–1878), and went on to make seven of the so-called astrograph refractors used for the *Carte du Ciel* photographic sky survey, and many other medium-sized and large telescopes for leading observatories.

Grus A southern constellation, representing a crane. Its brightest star is Alpha (Alnair), magnitude 1.7. Mu and Delta are both optical pairs, wide enough to be separated with the naked eye.

G star A star of spectral type G, characterized by weak hydrogen lines, but strongly ionized calcium lines, and (at lower temperatures) strong neutral calcium and sodium lines in its spectrum. A G star has a surface temperature of between about 5000 and 6000 K if it is on the **main sequence**, and between 4600 and 5600 K if it is among the giants. The coolest G-type supergiants are about 4200 K. G-type dwarfs range in mass from 0.9 to 1.1 solar masses. The giants and supergiants have generally evolved from more massive stars further up the main sequence. Almost all G stars are dwarfs with compositions similar to the Sun. A few are subdwarfs with lower metal abundances. A small fraction of the G-type giants have peculiar chemical compositions that give them the name *barium stars*. Examples of G stars are the Sun, Alpha Centauri and Capella.

Gum Nebula An immense emission nebula spanning 30 to 40° of sky in the constellations Puppis and Vela, containing the Vela Pulsar. The nearest part is 450 light years away, and its farthest part 1500 light years. It is named after the Australian astronomer Colin Gum (1924–60).

GUT ABBREVIATION FOR **grand unified theory**.

H

H I region A cloud of neutral (i.e. not ionized) atomic hydrogen gas in interstellar space. H I (pronounced 'h one') regions are typically of 10 to 100 solar masses and 10–20 light years across, with a temperature of around 100 K. They are about ten times as dense as interstellar matter, which is too low for molecular hydrogen to form in significant quantities.

H I regions do not emit visible radiation, but may be detected from their radio emission at a wavelength of 21 cm (SEE **twenty-one centimetre line**). Radio telescopes can thus detect H I regions even in other galaxies. Different kinds of galaxy contain different proportions of H I. Our Galaxy and other spiral or irregular galaxies contain an appreciable fraction of their total mass in H I. Elliptical galaxies contain no (or extremely little) detectable H I.

In our Galaxy, H I regions are heavily concentrated in a flattened disk in which a spiral pattern occurs. Spiral arms are made conspicuous, even in very distant galaxies, by hot, massive blue stars lying along the arms. These represent recent star-forming activity from the H I regions in the arms. Not all H I regions are capable of forming stars, only the densest ones that also contain molecular cores (SEE ALSO **giant molecular cloud**). Vast H I clouds, apparently devoid of luminous matter, have also been detected in the space between galaxies (SEE **Magellanic Clouds**).

H II region A cloud of ionized hydrogen in interstellar space. H II regions (pronounced 'h two') are often embedded within **H I regions**, and have densities 10 to 100,000 times greater. Because of the high temperatures (8000–10,000 K) and densities of H II regions, the pressure inside them is far higher than in the surrounding H I region or interstellar gas, so all H II regions are expanding, at around 10 km/s. Visible radiation from the H II region results from the *recombination* of protons and electrons to form hydrogen atoms; hot dust emits at infrared wavelengths; and *free–free radiation* at radio wavelengths is emitted as electrons curve around protons. On images, H II regions often appear red because of the **hydrogen alpha line** at 656 nm, but they can look greenish when seen visually because of a prominent pair of **forbidden lines** of oxygen at 496 and 501 nm.

Regions of ionized hydrogen can be produced by ultraviolet emission from old hot stars, from nova or supernova explosions, or from high-velocity **jets** or stellar winds impacting on interstellar gas. Conventionally, however, these are not classified as H II regions but as, respectively, **planetary nebulae**, **supernova remnants** and **Herbig–Haro objects**.

H II regions are stellar nurseries, the stars within them having recently formed through gravitational collapse within a **giant molecular cloud**. The life of H II regions is only a few million years, and their presence along with the hot stars marks out the arms in spiral galaxies. Complex H II regions are found in **starburst galaxies**. SEE ALSO **emission nebula**.

HA ABBREVIATION FOR **hour angle**.

Hα SEE **hydrogen-alpha line**.

Hadar Another name for **Beta Centauri**.

Hale, George Ellery (1868–1938) American astronomer. He planned and secured finance for the 1-m (40-inch) refracting telescope at **Yerkes Observatory**, completed in 1897, the 2.5-m (100-inch) Hooker reflector at **Mount Wilson Observatory**, completed in 1917, and the 5-m (200-inch) reflector at the **Palomar Observatory**, which entered regular service in 1949. Each was in turn the largest telescope of its day. The 5-m (200-inch) was named the Hale telescope in his honour. Hale was an accomplished solar observer, discovering among other things the magnetic fields of sunspots via the Zeeman effect in their spectra. He also invented the spectroheliograph.

Hale–Bopp, Comet A long-period comet discovered in 1995 by American amateur astronomers Alan Hale (1958–) and Thomas Joel Bopp (1949–). It was exceptional for its size – the nucleus was about 40 km (25 miles) across – so that even though it approached the Earth no closer than 1.3 AU, it was still a brilliant object (magnitude −0.7 at maximum) in April 1997. Hale–Bopp emitted dust at a prodigious rate. As the nucleus rotated, with a period of 11.5 hours, the dust created a series of concentric shells, visible in the comet's inner coma. The gas and dust tails extended to 20° and 25° respectively, corresponding to actual distances of over 1 AU.

Hale Telescope The 5-m (200-inch) reflecting telescope at the **Palomar Observatory**.

Hall, Asaph (1829–1907) American astronomer. His planetary studies covered the orbital elements of satellites, the rotation of Saturn and the mass of Mars. On 11 and 17 August 1877 he discovered the two tiny satellites of Mars, which he named Deimos and Phobos. The largest craters on Phobos have been named Hall and Stickney (the maiden name of Hall's wife).

Halley, Edmond (1656–1742) English scientist. His most famous achievement in astronomy was to realize that comets could be periodic, following his observation of **Halley's Comet** in 1682. He catalogued the stars of the southern hemisphere, discovered the proper motions of stars, suggested that the nebulae were clouds of gas and speculated on the infinity of the Universe. He made the first complete observation of a transit of Mercury and showed how the results could be used to measure the Sun's distance. As the second Astronomer Royal, he embarked on a long series of lunar studies. It was Halley who persuaded Isaac **Newton** to write the *Principia*, and he also financed its publication. He founded modern geophysics, charting variations in the Earth's magnetic field, and establishing the magnetic origin of the aurora borealis. In meteorology he showed that atmospheric pressure decreases with altitude, and studied monsoons and trade winds.

Halley's Comet The brightest and best-known periodic **comet**. It takes 76 years to complete an orbit that takes it from within Venus's orbit to outside Neptune's. The orbit's inclination is 162°, making it **retrograde** (1). It was observed by Edmond **Halley** in 1682; later he deduced that it was the same comet as had been seen in 1531 and 1607, and predicted it would return in 1758. There are historical records of every return since 240 BC. Several space probes were sent to Halley on its 1986 return. The closest approach was by the probe **Giotto**, whose cameras showed the nucleus to be an irregular object measuring 15 × 8 km (9 × 5 miles). The main constituent was found to be water ice, but the surface was covered with a dark

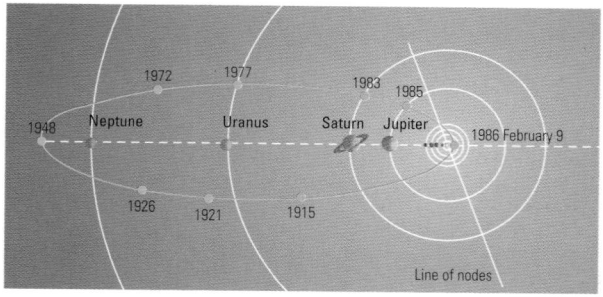

► **Halley's Comet** The elongated elliptical orbit of Halley's Comet takes it from aphelion beyond the orbit of Neptune to perihelion between the orbits of Venus and Mercury. The comet's orbit is retrograde, and it was last at perihelion on 9 February 1986.

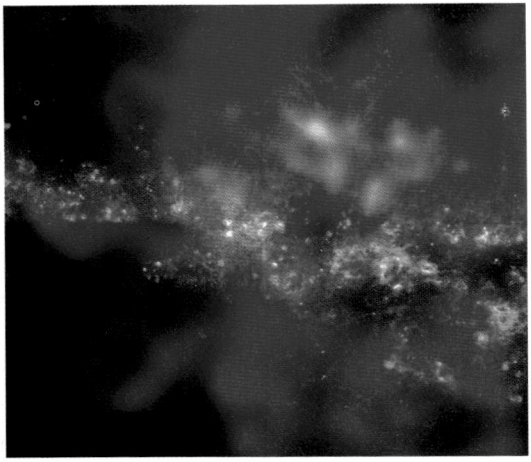

◀ **halo** The most prominent components of spiral galaxies like the Milky Way are the stars, gas and dust populating their nuclear bulge and disk. Many galaxies are also surrounded by tenuous halos of hot plasma. This X-ray image of a Milky Way-type galaxy reveals how vast such a halo can be.

deposit, giving an albedo of only about 0.03. Some craters were imaged.

halo (1) A bright whitish ring, complete or partial, seen around a celestial body, usually the Sun or the Moon. It is caused by the refraction of light by ice crystals suspended in the atmosphere.

halo (2) The **galactic halo**, a spheroidal distribution of stars and globular clusters around our Galaxy.

halo population stars SEE **stellar populations**.

Harvard classification The system of spectral classification of stars developed in the late 19th century by E. C. **Pickering** at Harvard College Observatory. Stars were originally placed in a sequence from A to P, indicating decreasing strength of hydrogen absorption lines in their spectra. The rearrangement and pruning of these classes gave the sequence O, B, A, F, G, K, M, based on decreasing surface temperature, used in the ***Henry Draper Catalogue***. It was later refined as the **Morgan–Keenan classification**.

harvest Moon The full Moon nearest the autumnal **equinox** (on 23 September). The Moon's path is then almost parallel to the horizon, and so it rises only about 15 minutes later each evening instead of the usual half-hour or more, providing a succession of moonlit evenings formerly of great help to harvesters. In the southern hemisphere the harvest Moon is the full Moon nearest the vernal equinox (on 21 March). SEE ALSO **hunter's Moon**.

Hawking, Stephen William (1942–) English theoretical physicist. He has used the General Theory of Relativity and quantum mechanics to study the **big bang** and **black holes**. He found that small black holes should lose energy – by emitting *Hawking radiation* – and eventually 'evaporate'.

Hayashi track The evolutionary path on the **Hertzsprung–Russell diagram** of a star beginning to form (SEE **protostar**) after it has collapsed from a gas cloud and before it reaches the main sequence. The Hayashi track lies above and to the right of the main sequence, and is nearly vertical. The star descends on its Hayashi track, decreasing in luminosity but with little change in effective temperature. (After Japanese astrophysicist Chushiro Hayashi, 1920– .)

heavy elements SEE **metals**.

Hektor Asteroid no. 624, discovered in 1907 by August Kopff. It measures roughly 300 × 150 km (200 × 100 miles) and is the largest and brightest of the **Trojan asteroids** sharing Jupiter's orbit. Its elongated profile may indicate that it is a binary asteroid.

Helene A small **satellite** of Saturn, discovered in 1980 by Pierre Lacques and Jean Lecacheux. It is an irregular body, and is **co-orbital** with Dione, orbiting near the L_4 **Lagrangian point** of Dione's orbit around Saturn.

heliacal rising In modern usage, the rising of a celestial body simultaneously with the rising of the Sun. Formerly it meant the moment when a bright star, such as Sirius, could just be seen rising before the Sun. Observation of this event was once of great importance for agricultural purposes, the heliacal rising of Sirius heralding to the ancient Egyptians the annual flooding of the Nile.

heliocentric As viewed from or in relation to the Sun's centre, as in *heliocentric coordinates* and the *heliocentric system*. COMPARE **geocentric**.

heliocentric parallax SEE **annual parallax**.

heliometer A refractor with the objective lens split across its diameter to give two images, formerly widely used for measuring very small angular distances.

heliosphere The region of space in which the Sun's magnetic field and the solar wind dominate the interstellar medium. It is similar in form to a planet's **magnetosphere**, with an outer *bow shock* and a boundary called the *heliopause*, where the energy of the solar wind particles has fallen to the level of galactic cosmic ray particles. Measurements continuing to be returned by the Voyager probes suggest that the heliosphere extends for 100 to 150 AU.

heliostat A mirror on an equatorial mount driven so that it follows the Sun's apparent motion across the sky and directs its light into a fixed instrument such as a solar telescope. A more sophisticated version is the **coelostat**.

helium flash The sudden commencement of helium fusion in the core of a red giant after the star's supply of hydrogen fuel has been exhausted. The helium flash is restricted to stars of less than about 2 solar masses; in more massive stars, helium fusion commences more gradually.

Helix Nebula A planetary nebula in Aquarius, NGC 7293. It has one of the largest angular diameters of any planetary nebula, a quarter of a degree, and at 300 light years away it is one of the nearest to the Sun.

Henry Draper Catalogue A catalogue of stellar spectra, compiled by Annie **Cannon** at Harvard College Observatory. It was named after Henry Draper (1837–82), an American pioneer of astrophotography, whose widow supported the work financially. The catalogue, completed in 1924, introduced the Harvard

▲ **Helix Nebula** One of the closest planetary nebulae, the Helix Nebula (NGC 7293) is visually faint because its light is spread over a fairly large area of sky.

system of classifying stellar spectral types: O, B, A, F, G, K, M, in order of decreasing surface temperature. Stars are still widely known by their HD numbers from this catalogue. SEE ALSO **spectral classification**.

Herbig–Haro object (HH object) One of a class of small, faint nebulae independently discovered by George Herbig (1920–) and Guillamero Haro (1913–88). They are irregular in outline, contain bright knots, and are found in regions rich in interstellar material. HH objects are thought to be produced by very young stars hidden behind clouds of gas and dust. Powerful stellar winds from the young stars excite the gas, producing emission lines, while the light from the star is reflected by dust in other parts of the cloud.

Hercules A constellation named after the great hero of Greek mythology, but not very conspicuous. Its brightest star is Beta (Kornephoros), magnitude 2.8. Alpha (Rasalgethi) is a semiregular red giant variable, range 2.7 to 4.0, with a period that varies from 50 to 130 days; there are also longer-term variations with a period of about 6 years. It lies 382 light years away. It is also a double star, with a magnitude 5.4 companion, period about 3600 years. Zeta is a fine close binary, with a period of 34 years, magnitudes 2.9 and 5.5. The most interesting deep-sky objects are the two globular clusters M13 and M92.

Hermes An asteroid, about 1 km (0.6 mile) in diameter, discovered by Karl Reinmuth in 1937. On 28 October of that year it came within 800,000 km (500,000 miles) of the Earth, up to then the closest known approach of any asteroid. Hermes was then 'lost' until October 2003, when Brian Skiff rediscovered it with the Lowell Observatory Near-Earth Object Search telescope. Hermes has now been assigned a number (69230). Radar observations suggest it may be a binary asteroid with components 300–450 m in diameter.

Herschel, Caroline Lucretia (1750–1848) German-born English astronomer, sister of William **Herschel**. In 1772 she joined her brother at Bath in England as his housekeeper and astronomical assistant. In the 1780s she became an observer in her own right, discovering many nebulae (as they were then called, including the galaxy NGC 253 in Sculptor) and eight comets. When her brother died in 1822, Caroline returned to her

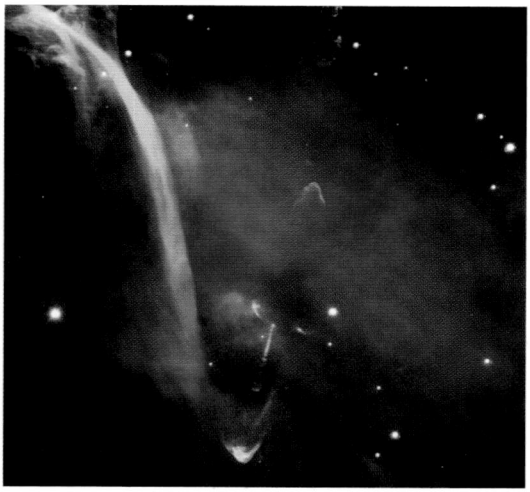

◀ **Herbig–Haro object** High-speed jets from the young object HH 34, near right centre, impact on the surrounding interstellar material to produce V-shaped shockwaves. HH 34 lies in Orion, in a region of ongoing star formation.

native Hanover, where she received many honours. She was the first woman to achieve real distinction in astronomy.

Herschel, John Frederick William (1792–1871) English scientist and astronomer, son of William **Herschel**. He extended his father's work on double stars and nebulae, discovering over 500 more nebulae and clusters. In 1834 he took one of William's telescopes to the Cape of Good Hope and undertook a systematic survey of the southern sky, discovering over 1200 doubles and 1700 nebulae and clusters. He combined these and his father's observations into a *General Catalogue of Nebulae and Clusters*, which formed the basis for the later **NGC**. He made the first good direct measurement of solar radiation, and inferred the connection between solar and auroral activity. He was also a pioneer of photography and its application in astronomy.

Herschel, (Frederick) William (originally Friedrich Wilhelm Herschel) (1738–1822) German-born English astronomer. In 1757 he came to England as a musician and became a naturalized Englishman. He turned to astronomy in the 1770s, making telescopes and mirrors for his own use and later for sale. Herschel became famous in 1781 by discovering the planet Uranus, which he wanted to name after King George III; the next year he was appointed the king's private astronomer. He discovered two satellites of Uranus (1787) and two of Saturn (1789). He observed and catalogued many double stars, hoping to measure their distances by detecting the parallactic movement of the nearer against the farther if they were optical doubles. He observed over 2000 nebulae and clusters, and published catalogues of them which later would be incorporated in the **NGC**. Herschel realized that the Milky Way is the plane of a disk-shaped stellar universe, whose form he calculated by counting the numbers of stars visible in different directions, and also noted the motion of the Sun towards a point in the constellation Hercules.

Herschel Space Observatory A European Space Agency satellite, formerly known as the Far Infrared and Submillimetre Space Telescope, planned for launch in 2007. It will carry a cooled 3.5-m (138-inch) Cassegrain telescope for photometry and spectroscopy in the 57–670 μm wavelength range and will orbit the Earth's L_2 Lagrangian point.

Hertzsprung, Ejnar (1873–1967) Danish astronomer. He showed that the colours of the stars were related to their brightnesses, plotting one against the other on what is now called the **Hertzsprung–Russell diagram** and discovering the existence of giant and dwarf stars. All studies of stellar evolution stem from this diagram, but Hertzsprung's work remained largely overlooked until Henry Norris **Russell** independently developed the diagram in a slightly different form.

Hertzsprung gap The region on the **Hertzsprung–Russell diagram** between the main sequence and the giant branch where there are few stars. The gap exists because stars evolve rapidly through this region.

Hertzsprung–Russell diagram (HR diagram) A plot of the **absolute magnitude** of stars against their **spectral type**; this is equivalent to plotting their **luminosity** against their surface temperature or their **colour index**. Brightness increases from bottom to top, and temperature increases from right to left. The diagram was devised by Henry Norris **Russell** in 1910, independently of Ejnar **Hertzsprung**, who had had the same idea some years earlier.

On the HR diagram most stars, including our Sun, lie on a diagonal band known as the **main sequence**. A region populated by **giant stars**, mainly K and M spectral types, lies above the main sequence. There are further groupings above the main sequence called **supergiants** and **subgiants**. Below the main sequence are **subdwarfs** and **white dwarfs**.

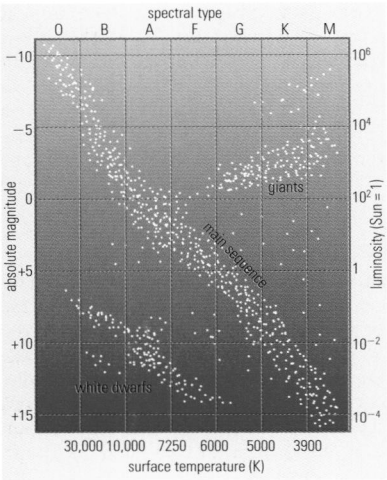

▲ **Hertzsprung–Russell diagram** A plot of stars' luminosity against temperature. Most stars lie on the main sequence running from top left to bottom right. This diagram shows the principal regions of the Hertzsprung–Russell diagram, with giant stars above the main sequence to the top right, and dwarfs to the lower left.

The HR diagram is an important aid for interpreting astrophysical data. For example, the total range of stellar brightness is 27 magnitudes, corresponding to a factor of 10^{11} in luminosity, and stars' surface temperatures range from 2200 to 50,000 K. Luminosity and surface temperature are both related to the size of a star. Most main-sequence stars are about the size of the Sun. White dwarfs are about the size of the Earth, while red giants and supergiants can be as big as the Earth's orbit, and supergiants even exceed that of Saturn. About 90% of the stars in our neighbourhood are main-sequence stars, about 10% are white dwarfs, and only about 1% are red giants or supergiants.

The existence of the three main groupings on the HR diagram indicates that a typical star passes through three very different stages in its life, and the progress of an evolving star can be followed on an HR diagram, from **protostar** to main sequence, then to giant star, and ending as a white dwarf (SEE **stellar evolution**). The diagram can also provide information about the ages of stars, and such dating is especially useful for establishing the ages of star clusters. Because the main sequence is a progression in stellar mass, the stars at the upper (most massive) end of the main sequence are the first to leave. As the cluster ages, its main sequence 'burns down' like a candle as stars of progressively decreasing mass and luminosity become red giants. The stage which this process has reached reveals the cluster's age. SEE ALSO **Hayashi track**, **Hertzsprung gap**, **instability strip**.

HET ABBREVIATION FOR **Hobby–Eberly Telescope**.

Hevelius, Johann (1611–87) German astronomer. He was the last great astronomer to make observations with naked-eye sighting instruments for positional measurements, from which he produced a catalogue of 1500 stars. He also made a map of the Moon, and discovered the Moon's **libration** in longitude.

Hewish, Antony (1924–) English radio astronomer. He and Martin Ryle developed the technique of **aperture synthesis** in 1960. In 1967 his student Jocelyn **Bell Burnell** obtained the first signal from a pulsar, now known as CP 1919. For subsequent work on pulsars, Hewish shared (with Ryle) the 1974 Nobel Prize for Physics.

Hidalgo Asteroid no. 944, discovered by Walter Baade in 1920. Its diameter is about 50 km (30 miles), and its eccentric comet-like orbit takes it from just outside Mars's orbit out as far as Saturn. Although it has an asteroidal designation, it could well be a large extinct cometary nucleus.

high-velocity star A star in the galactic halo that does not share the rotation of the majority of stars in the Galaxy, including the Sun, around the galactic centre. Consequently, such stars seem to have a large proper motion and a high velocity relative to the Sun. The orbits of high-velocity stars are usually eccentric, with a large inclination to the galactic plane. Most of them are Population II stars (SEE **stellar populations**).

Himalia The largest of Jupiter's intermediate group of **satellites**. It was discovered by Charles Perrine in 1904.

Hipparchus of Nicea (*c.*190–125 BC) Greek astronomer, geographer and mathematician. He made many accurate astronomical observations and compiled a star catalogue (now lost but incorporated in Ptolemy's ***Almagest***) giving coordinates and magnitudes. Hipparchus discovered the **precession** of the equinoxes. He found irregularities in the motions of the Sun and Moon, and from observations of eclipses made commendable estimates of their distances and sizes.

Hipparcos A satellite launched on 8 August 1989 by the European Space Agency, designed for **astrometry**. The name is a contraction of High Precision Parallax Collecting Satellite, and is also an allusion to **Hipparchus of Nicea**, who recorded accurate star positions more than two millennia previously. Hipparcos operated until August 1993, measuring parallaxes and proper motions of 120,000 stars to a precision of 1 to 2 milliarcseconds, and of a million further stars (though not to such a high accuracy). These two sets of observations were published in 1997 as the *Hipparcos Catalogue* and the *Tycho Catalogue*.

Hirayama family A group of **asteroids** having closely similar orbital elements (SEE **orbit**), in particular the **semimajor axis**. About a hundred such families are known, named usually after the largest asteroid in them. The *Flora*, *Eos* and *Themis* families are three of the main ones. The fact that in many families the members have similar spectra suggests that each family originates from one larger body which has been broken up by impact. These groupings were discovered by Japanese astronomer Kiyotsugu Hirayama (1874–1943) in 1928.

Hobby–Eberly Telescope (HET) A reflecting telescope opened in 1997 at McDonald Observatory, Texas, with an effective aperture of 9.2 m (360 inches). The main mirror consists of 91 hexagonal segments 1 m (40 inches) across, the images from which are directed to a common focus by a complex corrector system. The mounting is considerably simplified by having the telescope permanently aimed at an altitude of 55°, and moving only in azimuth. Even so, 70% of the sky is visible to it over the course of the year. It is named after William P. Hobby and Robert E. Eberly, American supporters of public education.

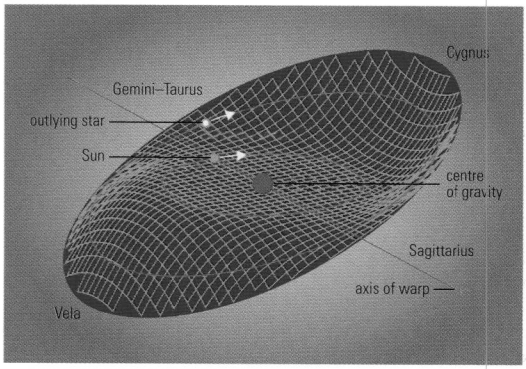

► **Hipparcos** Astrometric data from the Hipparcos satellite revealed warping of the disk of the Galaxy. The scale of the warping is exaggerated in this artist's rendition.

homogeneity SEE **cosmology**.

Homunculus Nebula SEE **Eta Carinae**.

horizon The **great circle** on the **celestial sphere** 90° from the observer's overhead point, the zenith.

horizontal coordinates SEE **coordinates**.

horizon problem A problem with the standard **big bang** model in that the **cosmic microwave background** radiation is too uniform. Areas of the sky now separated by as little as 2° would not have been causally connected to each other at the time of the origin of the radiation, so there would be no reason for them to have the very similar temperatures they are indeed observed to have. **Inflation** solves the problem because the enormous initial rapid expansion means that all parts of the visible Universe were causally connected when the radiation originated.

Horologium An obscure constellation of the southern sky, representing a pendulum clock, introduced by **Lacaille**. Alpha, which is of magnitude 3.9, is the constellation's only star above magnitude 4.

Horsehead Nebula A dark nebula with a very distinctive shape, like the head and mane of a horse. It is silhouetted against the bright nebula IC 434 south of Zeta Orionis. IC 434 is colliding with an extensive dusty cloud on the east side of Orion, and the Horsehead is the most distinctive intrusion of that dark cloud. The Horsehead is 1600 light years away and measures some 3 light years across.

horseshoe mounting A type of mounting used for a large **equatorial telescope**.

Horseshoe Nebula SEE **Omega Nebula**.

hot dark matter SEE **dark matter**.

◀ Horsehead Nebula
Silhouetted against the emission nebula IC 434, the Horsehead Nebula in Orion is a very difficult object visually. Photographs such as this show the striking profile of the intruding dark nebula in the foreground.

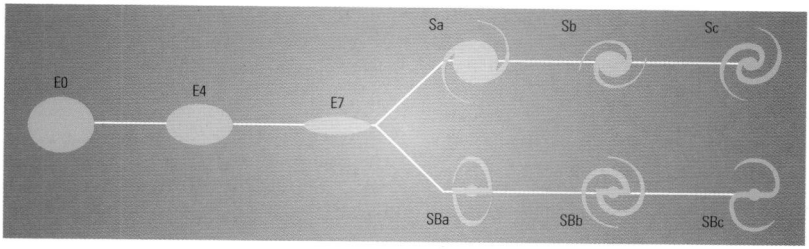

▲ **Hubble classification** A useful way of describing galaxy shapes, the Hubble classification splits these into three main classes. Elliptical galaxies (E) are subdivided on the basis of how far they are from spherical. The classification branches (and is hence sometimes known as the 'tuning-fork diagram') to afford description of spiral (S) and barred spiral (SB) galaxies.

hour angle (HA) The angle that the **hour circle** of an object makes with the **celestial meridian**. It is measured westward from the meridian, along the celestial equator to the point where the equator and hour angle first intersect.

hour circle A **great circle** on the **celestial sphere** passing through both celestial poles. **Declination** is measured along hour circles. On each half of an hour circle running from pole to pole, all points have the same **right ascension**.

Hoyle, Fred (1915–2001) English astrophysicist and cosmologist. With Hermann Bondi and Thomas Gold he developed the **steady-state theory**, in which the continuous creation of matter drives the expansion of the Universe. Although it was subsequently displaced by the big bang theory (which owes its name to a disparaging remark by Hoyle), it sparked research by Hoyle and others into **nucleosynthesis** in stars. Hoyle attracted controversy with unorthodox ideas, such as his suggestion that life was brought to Earth by comets.

HR diagram ABBREVIATION FOR **Hertzsprung–Russell diagram**.

HST ABBREVIATION FOR **Hubble Space Telescope**.

Hubble, Edwin Powell (1889–1953) American astronomer. At Mount Wilson Observatory he used the 2.5-m (100-inch) telescope to examine nebulae. He found that although some nebulae were clouds of gas lying within our Galaxy, many others – the so-called spiral nebulae – were resolvable as independent star systems. His **Hubble classification** of galaxies is still in use. In 1924, he measured the distance to the Andromeda Galaxy by identifying **Cepheid variable** stars in it. He attributed the **redshift** in the spectra of galaxies to indicate that they were receding, which was interpreted by others as evidence that the Universe was expanding. Hubble established the velocity–distance relationship of recession (SEE **Hubble constant**, **Hubble law**).

Hubble classification A classification of **galaxies** according to their shape or structure, introduced by Edwin Hubble in 1925. The three major categories are elliptical (E), spiral (S) and barred spiral (SB); each category has several subdivisions depending on the observable shape or structure of the galaxy. Irregular galaxies (Ir) were not included in Hubble's original classification. SEE ALSO **tuning fork diagram**.

Hubble constant (symbol H_0) The rate at which the velocity of recession of galaxies increases with distance from us, according to the **Hubble law**. The zero subscript specifies that it is the expansion rate at the present time that is meant, for, according to the theory of the **big bang**, the rate changes with time (SEE **deceleration parameter**). Depending on what assumptions are made, recent observations suggest that H_0 probably lies within $\pm 5\%$ of 71 km/s/megaparsec although some determinations as low as 50 km/s/megaparsec are still being made.

The inverse of the Hubble constant is called the *Hubble time*, which gives a maximum age for the Universe on the assumption that there has been no slowing of the expansion. The Hubble time for $H_0 = 71$ km/s/megaparsec is 14 billion years, assuming that all the galaxies started to recede from a common point in the past, at the time of the big bang (SEE **age of the Universe**).

Hubble diagram A graph in which the apparent magnitude of galaxies is plotted against the redshift of their spectral lines. It is a straight line, demonstrating the linear relation between redshift and distance, as embodied in the **Hubble law**.

Hubble law The law proposed by Edwin **Hubble** in 1929 claiming a linear relation between the distance (r) of galaxies from us and their velocity of recession (v), deduced from the redshift in their spectra. The figure linking velocity with distance is the **Hubble constant**. Hence $v = H_0 r$.

Hubble Space Telescope (HST) The largest astronomical telescope so far put into space, of Cassegrain design with a main mirror 2.4 m (94 inches) in diameter. It was launched by the space shuttle *Discovery* in April 1990 into an orbit about 600 km (375 miles) high. It was found to suffer from **spherical aberration**, attributed to a manufacturing error in its main mirror. Astronauts fitted corrective optics to the telescope during a servicing mission in December 1993.

The HST is designed to be repaired and serviced in orbit by space shuttle crews, and to allow its main and auxiliary instruments to be replaced with new ones; servicing missions took place in 1997, 1999 and 2002. Its present instrumentation consists of the Advanced Camera for Surveys (ACS), a battery of three cameras capable of imaging from the mid-ultraviolet to the near-infrared regions of the spectrum; the Wide Field and Planetary Camera 3 (WF/PC-3, replacing two earlier versions); the Cosmic Origins Spectrograph (COS); the Space Telescope Imaging Spectrograph (STIS); and the Near-Infrared Camera and Multi-Object Spectrometer (NICMOS).

A final servicing mission scheduled for 2004 was cancelled, leaving the HST's future capability in some doubt; its successor, the **James Webb Space Telescope**, is not planned for launch until 2011.

Hubble's Variable Nebula An emission nebula in the constellation Monoceros, associated with the variable star R Monocerotis. Variations in the light output of this young star, and motions of dark obscuring material in the star's vicinity, cause the nebula (designated NGC 2261) to vary in both brightness and outline on timescales of weeks. Discovered by William Herschel in 1783, NGC 2261 lies 2600 light years away, and was the subject of detailed study by Edwin Hubble in the 1910s.

Huggins, William (1824–1910) English amateur astronomer and spectroscopist. He built an observatory and constructed the first stellar spectroscope, which he used in a series of investigations aided by neighbour and chemist William Miller (1817–70). In 1863 he showed that the stars contain the same chemical **elements** that are present on Earth, and the next year he confirmed the gaseous nature of a diffuse nebula. He investigated the motion of Sirius, and made the first discovery of a **moving cluster** (the Ursa Major cluster). In 1876 he took one of the first photographs of a star's (Vega's) spectrum. Huggins used spectroscopy to investigate many other objects, including comets, meteors and the 1892 nova in Auriga.

Hulst, Hendrik Christoffel van de (1918–2000) Dutch astronomer. He car-

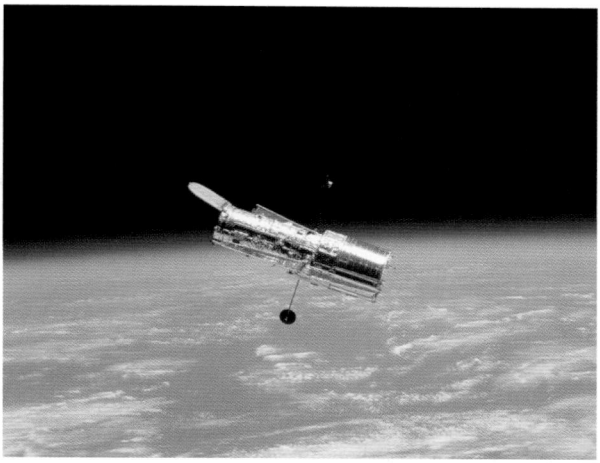

► **Hubble Space Telescope** Launched in 1990, the Hubble Space Telescope has provided astronomers with many results of fundamental importance. Here it is seen soon after release following the second servicing mission, carried out by Space Shuttle astronauts in 1997.

ried out research on interstellar matter and the solar corona. In 1944 he predicted that hydrogen should emit radio waves with a wavelength of 21 cm (SEE **twenty-one centimetre line**), and in 1951 he and Jan **Oort** used Doppler shifts at this wavelength to map the Galaxy.

hunter's Moon The first full Moon after the **harvest Moon**, providing a succession of moonlit evenings in early October. In the southern hemisphere the hunter's Moon is in April or May.

Huygenian eyepiece A basic telescope lens commonly found on small refractors. It consists of two simple elements and is relatively free of **chromatic aberration**. (After Christiaan Huygens.)

Huygens, Christiaan (1629–95) Dutch mathematician, physicist, and astronomer. In 1655 he discovered Saturn's largest satellite, Titan, and the next year explained that the planet's telescopic appearance was due to a broad ring surrounding it. He helped to develop the telescope and introduced the convergent eyepiece. Huygens' many contributions to physics include the idea that light is a wave motion, and the theory of the pendulum; he built the first pendulum clock.

Huygens SEE **Cassini-Huygens**.

Hyades A large open cluster several degrees across, 151 light years distant in the constellation Taurus. The brightest members are visible to the naked eye; in all it contains about 200 stars, most within a radius of 20 light years. Its estimated age is 650 million years. It is a **moving cluster**, receding from us at 43 km/s. The star Aldebaran lies in the same field as the Hyades but is less than half as distant.

Hyakutake, Comet An exceptional comet, discovered by Japanese amateur astronomer Yuuji Hyakutake (1950–2002) in January 1996, which passed within 0.1 AU of the Earth. At its best, around 25 March, it was of zero magnitude. The gas tail, which some observers reported to be over 90° long, underwent several **disconnection events**, and activity on the Sunward side of the nucleus was apparent in the form of jets and fans.

Hydra The largest constellation in the sky, representing the water snake killed by Hercules; it extends from Canis Minor to the south of Virgo. Its brightest star is Alpha (Alphard), magnitude 2.0. The Mira variable R Hydrae can reach magnitude 3.5 at its brightest, and fades to 10.9;

its period is 389 days. There is one bright open cluster, M48.

hydrogen-alpha line (Hα line) The absorption or emission line of hydrogen in the red portion of the spectrum at a wavelength of 656.3 nm. The first line in the **Balmer series**, it is the third most conspicuous absorption line in the Sun's spectrum, and is responsible for the pink colour of the solar chromosphere when viewed during a total eclipse. Light of this wavelength is often used for monochromatic study of the Sun with the spectrograph, and Hα filters are used by amateur observers for solar imaging.

hydrogen spectrum The **spectrum** of atomic hydrogen. The hydrogen atom has the simplest structure of any atom (one proton as nucleus with a single orbiting electron), and the hydrogen spectrum is the simplest to interpret. As over 90% of the atoms in the Universe are hydrogen atoms, hydrogen lines feature prominently in the spectra of many celestial objects. The spectrum contains five principal series of lines associated with different levels of energy in the hydrogen atom, among them the **Balmer series** and the **Lyman series**. Each line corresponds to a quantum-mechanical process called a transition in which the electron 'jumps' between orbits (SEE **emission spectrum**).

Hydrus A far southern constellation representing a small water snake. There are few interesting telescopic objects. Beta Hydri, magnitude 2.8, is the brightest star in the constellation and the nearest fairly bright star to the south celestial pole.

Hygeia **Asteroid** no. 10, discovered by Annibale de Gasparis in 1849. It orbits in the main belt, and is the fourth-largest asteroid, with a diameter of 408 km (254 miles).

hyperbola An open curve. In mathematics the hyperbola is a type of conic section, so called because it is one of the intersections of a plane with a cone, and is formed when a plane cuts a cone at an angle to the cone's axis that is less than the cone's slope. In astronomy a hyperbolic **orbit** is the path followed by a celestial body which passes another body but is not captured into an orbit around it. For a hyperbolic orbit the **eccentricity** is greater than 1.

hyperboloidal Having the form of the surface obtained by rotating a **hyperbola** about its axis. The hyperboloid is the shape used for the convex secondary mirror in a **Cassegrain telescope**.

Hyperion An outer **satellite** of Saturn, discovered by William Cranch Bond and his son George Philips Bond, and independently by William **Lassell** in 1848. It is irregular in shape, measuring 370 × 225 km (230 × 140 miles), and displays chaotic rotation – tumbling about no fixed axis as it follows an eccentric orbit which is in a 4:3 resonance with that of **Titan**. Although Hyperion is a low-density, icy body, its surface is dark. There are several large craters and a scarp, named Bond–Lassell, which extends for 300 km (200 miles). It seems most likely that Hyperion is a fragment of a larger body.

hypernova An exceptionally violent **supernova** event that appears to be an order of magnitude greater than a normal supernova explosion. Such events have been observed in starburst regions of irregular galaxies.

I

Iapetus A large outer **satellite** of Saturn discovered by Giovanni Domenico Cassini in 1671. Its diameter is 1440 km (895 miles), and its density of 1.21 g/cm^3 indicates that it is predominantly icy. Like most of Saturn's satellites, Iapetus has **synchronous rotation**. Its trailing hemisphere is bright, but its leading hemisphere (the one facing the direction of orbital travel) is extremely dark, being completely coated by a deep-red material. It seems more likely that the dark material was deposited from outside rather than expelled from within. It may be dust ejected from **Phoebe**, the next satellite out; or material from the satellite itself which has fallen back to the surface following ejection by major impacts. The **Cassini** spacecraft revealed at least three large impact structures on Iapetus, along with a narrow, steep mountain range reaching a height of 20 km (12 miles) and extending for 1300 km (800 miles) along the satellite's equator in the dark hemisphere. As a result of the difference in **albedo** between the hemispheres, Iapetus appears markedly brighter when west of Saturn than when east from our terrestrial viewpoint.

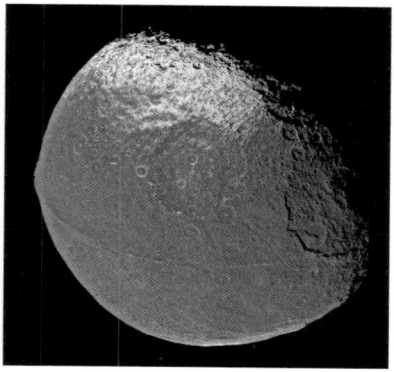

▲ **Iapetus** Cassini flew past Iapetus on 31 December 2004. The resulting images show the heavily cratered region named Cassini Regio; the topographic ridge extending along the moon's equator is conspicuous.

▲ **Ida** Asteroid (243) Ida was imaged by the Galileo spacecraft in August 1993. Its tiny satellite Dactyl is visible at right.

IAU ABBREVIATION FOR **International Astronomical Union**.

IC ABBREVIATION FOR the *Index Catalogue*, consisting of two supplements to the *New General Catalogue* (SEE **NGC**) produced in 1895 and 1908 by J. L. E. **Dreyer**. Objects in the *Index Catalogues* are still referred to by their IC numbers.

Icarus Asteroid no. 1566, discovered by Walter **Baade** in 1949. It is an **Apollo asteroid** with a diameter of about 1.5 km (1 mile), and has a perihelion distance of 30 million km (19 million miles), within the orbit of Mercury.

Ida Asteroid no. 243, discovered by Johann Palisa in 1884. In 1993 it became the second asteroid to be examined from close quarters, by the probe Galileo en route to Jupiter. It is an elongated, cratered body measuring 56 × 24 × 21 km (35 × 15 × 13 miles), and has a 1.3-km (0.8-mile) satellite named Dactyl.

immersion The entry of a celestial body into the shadow of another at the beginning of an eclipse or occultation.

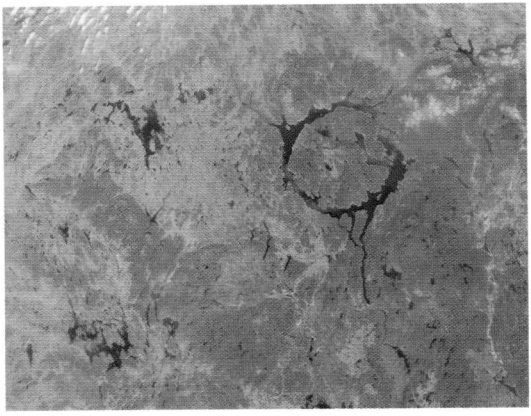

◀ **impact feature** The circular Manicouagan impact structure in Quebec, Canada, is revealed in this satellite image. The 70-km (44-miles) diameter feature was formed by an asteroid impact about 212 million years ago.

impact feature Any surface feature of a planet, satellite or asteroid, or other body with a solid surface, that results from the past impact of another body. The most obvious are impact **craters**. *Basins* are large shallow depressions marking the impact of a large body, which have often been filled in by upwelling fluid from below; examples are the lunar maria (SEE **mare**). *Multiringed basins*, such as **Valhalla** on Callisto, are surrounded by concentric ringed structures representing 'ripples' from the impact. Other impact features include **rays** radiating from craters, and *palimpsests* – the 'ghosts' of craters flattened out or filled in since the impact that caused them. The term was also applied to the visible atmospheric disturbances that were caused by the collision of Comet **Shoemaker–Levy 9** with Jupiter.

inclination (1) (symbol *i*) The angle between the orbital plane of a body orbiting the Sun and the plane of the ecliptic.

inclination (2) (symbol *i*) The angle between the orbital plane of a satellite and the equatorial plane of the planet it orbits.

inclination (3) (symbol *i*) The angle between the orbital plane of a double star and the plane at right angles to the line of sight. The angle is zero if the orbit is seen in plan, and 90° if seen in profile.

inclination (4) (symbol *i*) The angle between the equatorial plane of a planet or satellite and its orbital plane, equivalent to the angle between the axis of rotation and the perpendicular to the orbital plane.

Indus A small southern constellation representing an American Indian. The only star above magnitude 3.5 is Alpha (3.1). Epsilon is the least luminous star visible to the naked eye: magnitude 4.7, distance 11.8 light years, type K4, absolute magnitude 6.9.

inequality The departure from uniform orbital motion of a body, due to **perturbations** by other bodies and the eccentricity of its orbit.

inferior conjunction SEE **conjunction**.

inferior planet Either of the planets Mercury and Venus, both of whose orbits lie inside the Earth's.

inflation A proposed early stage of the Universe, immediately after the **big bang**, in which for a brief fraction of a second the Universe expanded faster than the speed of light. Relativity is not contravened because it does not involve the movement of objects through space faster than light, but rather the expansion of space itself.

Inflation was proposed in 1981 by Alan Guth to explain the **flatness**, **horizon** and **monopole problems** that arise with the standard big bang model. Inflation requires the universe to expand exponentially by at least a factor of 10^{50} and perhaps by as much as $10^{1,000,000,000,000}$ over the interval from about 10^{-34} to 10^{-32} seconds after the start of the big bang. After inflation the Universe could be as large as a football or many light years in diameter, in either case continuing to expand 'normally'. The driving mechanism for inflation is uncertain, but one suggestion is that immense quantities of energy were released when the strong nuclear force came into existence as a separate entity (SEE **grand unified theory**), and this not only drove the expansion but also created all the matter and radiation in the Universe.

Although inflation may seem an unlikely process, observational support for it came with the observation by the **Wilkinson microwave anisotropy probe** of polarization in the **cosmic microwave background** radiation. Inflation predicts that polarization with exactly the observed properties should be produced during the last scatterings of photons at the end of the **decoupling era**. SEE ALSO **cosmology**.

Infrared Astronomical Satellite (IRAS) A satellite, a collaboration between the USA, Netherlands, and UK, that surveyed the sky at infrared wavelengths. It was launched on 26 January 1983 and operated for nine months, cataloguing nearly 250,000 sources. Its discoveries included several comets and asteroids, possible planetary systems forming around nearby stars, and **starburst galaxies**.

infrared astronomy The study of **electromagnetic radiation** falling between the red end of the visible spectrum and radio wavelengths, between about 700 nm (or 0.7 μm) and 1 mm. The Earth's atmosphere is opaque to much of this wavelength range, although there are a number of atmospheric *windows* through which infrared (IR) observations can be made from high mountains. To avoid the Earth's atmosphere, observations are made from high-flying aircraft (SEE **Stratospheric Observatory for Infrared Astronomy**), from uncrewed balloons and from satellites such as the **Infrared Space Observatory**.

Astronomical sources of IR radiation include the planets and other Solar System bodies, **extrasolar planets**, the stars and dusty regions of the Milky Way, and other galaxies (including **starburst galaxies**). Most of the interstellar dust and gas in our Galaxy is contained in *giant molecular clouds* (GMCs, SEE **molecular cloud**), such as the one in Orion, part of which is visible as the Orion Nebula. Maps of GMCs at IR wavelengths show the warm or hot regions where star formation is in progress or is about to begin. On a much smaller scale, highly compact and opaque **Bok globules** with just a few tens of solar masses are also potential star-forming regions. Infrared observations show the temperature of the globules to be very low, 10–20 K, indicating that they are at an early stage of gravitational collapse. Faint, wispy structures in interstellar space termed *infrared cirrus*,

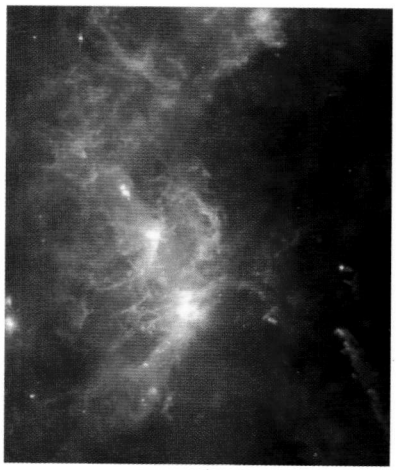

▲ **infrared astronomy** Produced using data from IRAS, this image shows the constellation Orion. The bright patch at lower centre is the Orion Nebula. Gas and dust glow brightly in the infrared region.

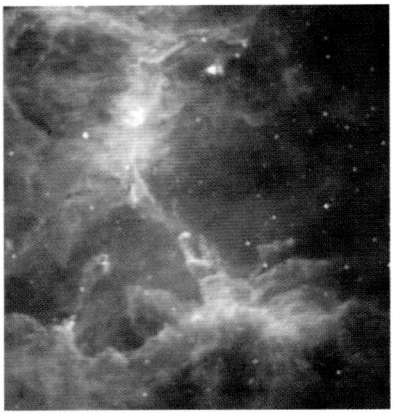

▲ **Infrared Space Observatory** (ISO) The Eagle Nebula is shown here in a composite ISO image. Shorter infrared wavelengths are rendered in blue, longer wavelengths in red.

discovered by IRAS, are probably caused by infrared emissions from dust grains. These grains are composed of carbon in the form of graphite at a temperature of 30–40 K, ejected by the stellar winds of cool stars or supernova explosions.

Infrared wavelengths can penetrate the otherwise opaque clouds of dust in the plane of our Galaxy, so we can see sources towards the galactic centre that are completely obscured at optical wavelengths. The observations show a high density of late-type stars and a large ring of molecular gas and dust, as well as many regions of active star formation. Beyond the Milky Way, IR observations are a key to understanding the energy output of active galaxies and quasars and the physical processes operating within them.

In the Solar System, IR observations allow us to work out the basic chemistry of the planets, their satellites and the asteroids. Molecules in planetary atmospheres exhibit detailed IR absorption spectra, which permit the composition and temperature of the atmospheres to be established. Spectral reflectance observations of asteroids, planetary satellites and small planets without substantial atmospheres have been used to discover the mineralogical compositions of their surfaces. IR techniques have established the presence of water and carbon dioxide ices on Mars, water ice on many of the satellites of Jupiter, Saturn and Uranus, and methane ice on Triton and on Pluto. Measurement of the black-body emission of small Solar System bodies has given us information on their sizes and surface albedos where no other technique would succeed.

Infrared Space Observatory (ISO) A European Space Agency satellite with a 0.6-m (24-inch) telescope for infrared astronomy, launched on 17 November 1995. It operated until 1998.

inner planet Any of the planets Mercury, Venus, Earth and Mars, which orbit inside the main asteroid belt.

instability strip The narrow region of the **Hertzsprung–Russell diagram** where pulsating stars are located. It extends from the brightest Cepheid variables down to pulsating white dwarf stars, and also includes RR Lyrae stars, Delta Scuti stars and dwarf Cepheids.

Integral A joint European, US and Russian satellite for gamma-ray astronomy launched on 17 October 2002; the name is short for International Gamma-Ray Astrophysics Laboratory.

interferometry The study of a celestial object by analysing its interference pattern. When electromagnetic radiation emitted by a source is split into two and recombined, a pattern of interference fringes is produced that can be used to reveal more detailed information about the source than is possible from a single beam.

An *interferometer* is an instrument consisting of two or more detectors (telescopes or antennas) which collect signals from a single source. The signals are then combined to form the interference pattern. The longer the wavelength of the radiation being detected, the farther apart the detectors need to be.

Various types of **optical interferometer** are used to measure the diameters of stars and resolve fine detail in their spectra. Interferometry is particularly important in radio astronomy, where the long wavelengths of the radiation mean that a single detector gives only a coarse resolution (SEE **radio interferometer**).

intergalactic matter Matter in the space between galaxies. There are no significant quantities of intergalactic dust, but there are clear signs of intergalactic gas. X-ray emission from clusters of galaxies comes from a hot, tenuous gas with a temperature of 10 to 100 million K between the galaxies in the cluster. Radio-emitting clouds of electrons ejected by radio galaxies are drawn out into a tadpole shape as if they were ploughing through a resisting medium. How much intergalactic gas is primordial hydrogen and helium from the big bang that is falling into the clusters, and how much is matter expelled from the galaxies, is uncertain. Tidal interactions (SEE **tides**) between galaxies are common, and radio observations reveal streams of neutral hydrogen around interacting galaxies, as, for example, with M81 and M82. A similar gas stream, the Magellanic Stream, links the **Magellanic Clouds** with our Galaxy. It is possible that **dark matter** exists in the space between galaxies and galaxy clusters.

International Astronomical Union (IAU) The controlling body of world astronomy, founded in 1919, with its headquarters in Paris. It is the authority for naming celestial bodies and their surface features.

International Atomic Time (TAI) The international reference scale of atomic time, formed by comparing data from around 200 atomic clocks around the world. (The abbreviation TAI is derived from the name in French.) The fundamental unit of time is the SI second, which is defined in terms of the resonance of the caesium atom. SEE ALSO **Universal Time**.

International Space Station (ISS) A space station being built as a collaborative venture between NASA, the Russian

▲ **International Space Station** Photographed by a crew member on the Space Shuttle *Endeavour* in December 2001 is the International Space Station, with the Earth providing a dramatic background. The robot manipulator arm is clearly visible, as are several modules and solar arrays.

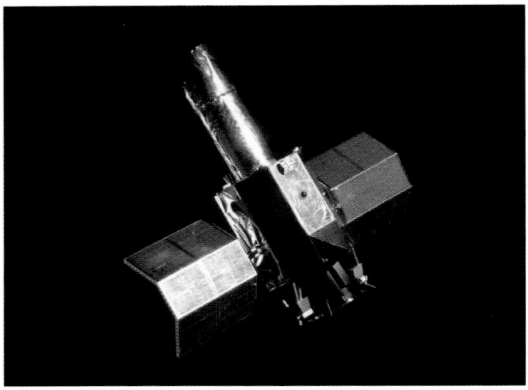

◀ **International Ultraviolet Explorer** The IUE, seen here in an artist's impression, studied comets, asteroids and the atmospheres of the giant planets. It also provided insights into stellar winds and hot, massive stars.

Space Agency, ESA, Japan and Canada. The first section was launched in 1998. More than 90 missions will be required to launch all the ISS's modules, which will total nearly 420 tonnes. It is unlikely that the assembly of the ISS will be completed before 2010.

International Ultraviolet Explorer (IUE) A joint NASA, European Space Agency and UK satellite, carrying a 0.45-m (18-inch) telescope. It was launched on 26 January 1978 and operated until 1996, taking ultraviolet spectra of stars and galaxies.

interplanetary matter The material in the space between the planets. It is made up of atomic particles (mainly protons and electrons) escaping from the Sun via the solar wind, and dust particles (mainly from **comets**, but some possibly of cosmic origin) and neutral gas in the plane of the ecliptic. The particles in the inner Solar System constitute the *zodiacal dust cloud*, scattered sunlight from which is responsible for the **zodiacal light**. An extremely tenuous ring of dust, over 50 million km (30 million miles) wide, orbits the Sun just outside the Earth's orbit. There is also the dust in **meteor streams**.

interstellar matter The matter contained in the regions between objects in the Galaxy. It includes clouds of *interstellar gas*, mainly hydrogen with some helium, and a small percentage (about 1% by mass) of interstellar dust grains. The hydrogen is mainly in a relatively cool neutral form at a temperature of 10–100 K (SEE **H I region**). Some very much hotter and more diffuse regions of ionized gas (**H II**) also exist, together with dense clouds of molecular hydrogen and other molecules. The dust grains are elongated in shape, about 100 nm long, and aligned by the galactic magnetic field. Most grains are thought to consist of silicates or forms of carbon, and some have icy coatings.

The clouds of gas and dust are largely confined to the plane of the Galaxy and tend to be concentrated in the spiral arms. Our Galaxy contains about 10^{10} solar masses of material distributed between the stars, making up about 10% of its visible mass. The density of these clouds is less than that of a laboratory vacuum on Earth, but they nevertheless have an obscuring effect on the light of stars behind them. The clouds scatter and absorb starlight passing through them, a phenomenon known as *interstellar extinction*. On average, starlight is dimmed by one magnitude for every kiloparsec it travels in the plane of the Galaxy. Red light is dimmed less than blue light, which causes a reddening of starlight.

The spectra of remote stars show **absorption lines** superimposed on the stellar spectra, produced by interstellar dust and gas. Sodium, calcium, potassium, iron, titanium and other elements have been detected from these absorption lines. These elements are ejected from stars. Heavier elements, including lead, gallium and krypton, are manufactured in the extreme conditions of a **supernova** explosion. SEE ALSO **nebula**.

interstellar molecules Molecules that exist in interstellar **molecular clouds**. Most have been identified by their emission lines at millimetre and centimetre wavelengths, although the first, including the cyanogen radical (CN) and the methylidyne radical and ion (CH, CH^+), were detected in the 1940s at optical frequencies. The hydroxyl radical (OH) was the first (in 1963) to be identified by radio astronomy. The list of interstellar molecules has since been extended to more than 130, and includes hydrogen cyanide (HCN), ammonia (NH_3), methane (CH_4) and methyl alcohol (CH_3OH), as well as more complex organic molecules such as the amino acid glycine.

intrinsic variable SEE **variable star**.

inverse-square law A reduction in the intensity of a physical quantity in proportion to the square of the distance from its source. For example, a planet three times the Earth's distance from the Sun would experience one-ninth the gravitational attraction of the Sun and receive one-ninth as much sunlight.

Io The innermost of Jupiter's four **Galilean satellites**, and the second-smallest, with a diameter of 3643 km (2264 miles). It has the highest density, at 3.53 g/cm^3, indicating a lack of water ice in its make-up. Io's tidal interaction with Jupiter, **Europa** and **Ganymede** raises a bulge up to 100 m (330 ft) high in its surface, creating sufficient flexing and heating to make it by far the most volcanically active world in the Solar System. The surface, multicoloured by the presence of different forms of sulphur, is being

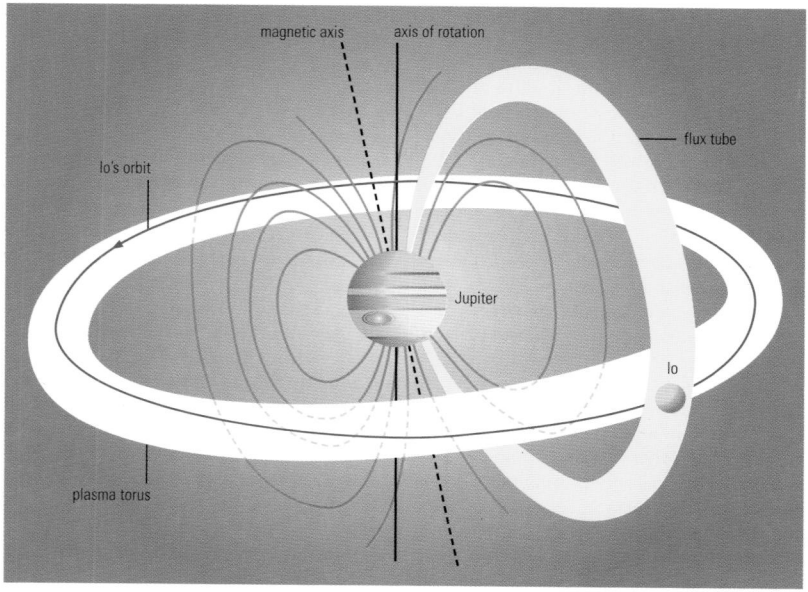

▲ **Io** Material ejected by Io's volcanoes forms a torus around Jupiter. A magnetic flux tube links Io and Jupiter: particles ejected during the satellite's volcanic eruptions can lead to enhancements of Jupiter's aurorae.

◀ **Io** A composite of images from the Galileo orbiter shows the volcanically active surface of Jupiter's moon Io. Intense volcanism driven by tidal stresses leads to rapid resurfacing of Io.

continually renewed. The active regions contain volcanic vents and calderas, lava flows and volcanic plumes which send material to heights of up to 300 km (200 miles). Plumes prominent when the Voyager probes imaged Io included those named Pele, Loki and Prometheus. In the active regions there may be lakes of molten sulphur or even molten rock; elsewhere there are plains formed by volcanic deposits, and mountains. A thin atmosphere of sulphur dioxide surrounds Io, and a doughnut-shaped *plasma torus* of ionized (electrically charged) sulphur and oxygen surrounds its whole orbit and interacts with Jupiter's **magnetosphere** to create a *flux tube* carrying an electric current of about 5 million amperes (SEE the diagram on page 117). In 1996 the **Galileo** probe detected an iron core and magnetic field, and found the surface features to have changed significantly since the satellite was first imaged during the 1979 Voyager missions. SEE ALSO **satellite**.

ionosphere A region in the atmosphere of a planetary body in which there are free electrons and ions produced by ultraviolet radiation and X-rays from the Sun. The degree of ionization is greatly affected by solar activity.

The Earth's ionosphere extends from about 60 to 300 km above the surface, and reflects radio waves of long wavelengths, although shorter wavelengths can pass through undisturbed. The reflecting power of these layers makes long-range broadcasting and telecommunications possible up to frequencies of about 30 MHz. The radio waves and microwaves from around 30 MHz to about 100 GHz that pass through the ionosphere are used in radio astronomy and satellite communications; they make up the *radio window*. The free electrons formed as a result of ionization do at times have serious adverse effects on the reception of radio signals.

IRAS ABBREVIATION FOR **Infrared Astronomical Satellite**.

Iris Asteroid no. 7, discovered by John Russell Hind in 1847. Although it ranks only about 25th in size, with a diameter of around 200 km (125 miles), it is the fourth-brightest.

iron meteorite A meteorite consisting mainly of iron (90–95%) and some nickel, with traces of other metals. Iron meteorites are classified into a number of groups according to the proportions of these other metals that they contain, and each group is thought to correspond to a different parent asteroid.

irradiation A physiological phenomenon in which a bright object viewed against a dark background appears larger than it really is. The effect can introduce errors into measurements made by an observer using a telescope.

irregular galaxy A galaxy that shows little or no symmetry. Some irregular galaxies are smaller than spirals and ellipticals, contain much gas and are undergoing star formation. Some are classified as irregular for the only reason that they do not fit into other categories of galaxy.

irregular variable A pulsating variable star whose variation in brightness does not follow any regular or predictable pattern. Irregular variables are divided into two broad groups: **eruptive variables** such as **shell stars**, **T Tauri stars** and young stars associated with nebulosity; and **pulsating stars** that are red giants or supergiants of late spectral class, and exhibit slow variations.

Isaac Newton Group of Telescopes (ING) Group of instruments consisting of the 4.2-m (165-inch) **William Herschel Telescope**, the 2.54-m (100-inch) Isaac Newton Telescope and the 1.0-m (39-inch) Jacobus Kapteyn Telescope, at the **Roque de los Muchachos** observatory on the island of La Palma. The ING is funded by the UK, the Netherlands and Spain.

ISO ABBREVIATION FOR **Infrared Space Observatory**.

isotropy SEE **cosmology**.

ISS ABBREVIATION FOR **International Space Station**.

IUE ABBREVIATION FOR **International Ultraviolet Explorer**.

J

James Webb Space Telescope An infrared telescope to succeed the Hubble Space Telescope equipped with three main instruments, to study the earliest moments after the big bang. It is scheduled to be launched in August 2011 into a special orbit 1.5 million km (900,000 miles) from the Earth. The telescope's 6.5-m (20-ft) diameter mirror will be shaded by a sunshield the size of a tennis court.

Jansky, Karl Guthe (1905–50) American radio engineer. In 1931, while looking for sources of 'noise' that interfered with radio communications, he discovered one located in the Milky Way, in Sagittarius. Jansky himself did not follow this up, but it led directly to the development of radio astronomy. The unit of flux density used in radio astronomy is named the *jansky* in his honour.

Janssen, Pierre César Jules (1824–1907) French astronomer. He specialized in solar spectroscopy and developed special instruments, including a spectrohelioscope for observing prominences without waiting for an eclipse. He discovered the helium line in the solar spectrum, and produced a photographic atlas of the Sun.

▲ **Jodrell Bank** The Lovell Radio Telescope at Jodrell Bank was built in 1957 and originally known as the Mark I Telescope. It is still one of the world's largest fully steerable radio telescopes.

Janus A small inner **satellite** of Saturn, **co-orbital** with Epimetheus. The two were discovered in 1978 when John Fountain and Stephen Larson re-examined photographs taken some years before. They are both irregular in shape, and are probably fragments of the same larger body. Either satellite may be the object reported in 1966 by Audouin Dollfus and given the name Janus, but not subsequently confirmed.

Jeans, James Hopwood (1877–1946) English mathematician and physicist. In astronomy he worked on stellar dynamics, the formation and evolution of stars, binary systems, and spiral galaxies. In 1917 he proposed that the Solar System was formed from matter pulled out of the Sun by the gravitational attraction of a passing star, a theory later developed by Harold Jeffreys (1891–1989), but since abandoned. Jeans wrote many popular books on astronomy.

jet A stream of very high-energy particles emitted by objects such as **active galactic nuclei** and protostars. They are detectable by their radio emissions and may also be visible at optical wavelengths. In galaxies there are often two jets, emerging in opposite directions from the nucleus perpendicular to the galactic plane. SEE ALSO **bipolar flow**.

Jewel Box The open cluster NGC 4755 in the constellation Crux, around the star Kappa Crucis. The name derives from the varied colours of the stars in the cluster. It lies 7600 light years away.

Jodrell Bank The site at Macclesfield, Cheshire, UK, of the University of Manchester's Nuffield Radio Astronomy Laboratories, founded after World War II by Bernard **Lovell**. Its main instrument is the 76-m (250-ft) fully steerable dish opened in 1957 and named the Lovell Telescope in 1987; a major upgrade was completed in 2003. A second dish, elliptical in shape, 38 × 25 m (125 × 83 ft), was opened in 1964. These telescopes can be used individually or as part of the **MERLIN** array.

Joint Astronomy Centre (JAC) Organization based in Hilo, Hawaii, that operates the James Clerk Maxwell Telescope and the UK Infrared Telescope on behalf of the UK, Canada and the Netherlands.

Jones, Harold Spencer (1890–1960) English astronomer. He was concerned with refining values for the distances and periods of Solar System bodies, and in 1941 obtained what was then the most accurate value of the Sun's distance. As the tenth Astronomer Royal, he planned the Royal Greenwich Observatory's move to Herstmonceux in Sussex.

Jovian The adjective for Jupiter. *Jovian planet* is another term for **giant planet**.

Julian calendar SEE **calendar**.

Julian Day A system of numbering days continuously, without division into years and months. It is used in astronomy when calculating events that recur over long periods, such as the solar cycle of sunspot variations. The Julian Day Number is the number of days that have elapsed since noon GMT on 1 January 4713 BC, which is defined as day zero. The idea was conceived in 1582 by the French chronologist Joseph Justus Scaliger. The name 'Julian' was given by Scaliger in memory of his father, Julius Caesar Scaliger (1484–1558), and is not to be confused with the Julian calendar devised by the Roman Emperor Julius Caesar.

The Julian Date is given by the Julian Day Number for the preceding noon at Greenwich followed by the fraction of the mean solar day that has elapsed. Thus, the Julian Date of an observation made at 6 p.m. on 1 January 1995, is 2,449,719.25, the Julian Day Number being 2,449,719.

In historical usage the year 1 BC is followed by the year AD 1, whereas in astronomical usage AD 1 is preceded by year 0, which in turn is preceded by year −1 (corresponding to the historical year 2 BC). Thus to convert years BC to astronomical years, subtract 1 and prefix a minus sign.

Juliet One of the small inner **satellites** of Uranus discovered in 1986 during the Voyager 2 mission.

Juno Asteroid no. 3, discovered by Karl Harding in 1804. It is the eighth-largest, with a diameter of 268 km (167 miles).

Jupiter The fifth major planet from the Sun, and the largest of the giant planets. It is one of the brightest objects in the sky, with a magnitude at opposition of between −2.3 and −2.9. Its maximum apparent diameter is 50 arc seconds. Following centuries of telescopic observation, our knowledge of

JUPITER: DATA

Globe	
Diameter (equatorial)	142,800 km
Diameter (polar)	133,500 km
Density	1.33 g/cm^3
Mass (Earth = 1)	317.8
Volume (Earth = 1)	1320
Sidereal period of axial rotation (equatorial)	9h 50m 30s
Escape velocity	59.5 km/s
Albedo	0.73
Inclination of equator to orbit	3° 07′
Temperature at cloud-tops	125 K
Surface gravity (Earth = 1)	2.69
Orbit	
Semimajor axis	5.203 AU = 778.3 × 10^6 km
Eccentricity	0.048
Inclination to ecliptic	1° 18′
Sidereal period of revolution	11.86y
Mean orbital velocity	13.06 km/s
Satellites	63

▲ **Jupiter** This view of Jupiter was assembled from four images obtained by the Cassini spacecraft on 7 December 2000. The long-lived Great Red Spot is seen at lower right. The shadow of Europa, the smallest of the Galilean satellites, is the dark spot at lower left.

the planet and its satellites owes most to visits by space probes, including Pioneers 10 and 11, Voyagers 1 and 2, and most notably **Galileo**. The main data for Jupiter are given in the table on page 121.

Telescopically, Jupiter shows an oblate disk, somewhat flattened by its rapid rotation, crossed by a pattern of dark *belts* and light *zones*. Most conspicuous are the North Equatorial Belt and South Equatorial Belt, between whose confines faster rotation (9 hours 51 minutes, known as *System I*) is observed than in the rest of the atmosphere (*System II*, 9 hours 55 minutes): in other words, Jupiter's atmosphere shows **differential rotation**. Other atmospheric features include short-lived dark and light spots and streaks. More persistent features have included white ovals in the South Temperate Belt, and also in this region the long-lived **Great Red Spot**. The ovals are cyclonic and anticyclonic storms.

Turbulence can cause major disturbances, particularly in the South Tropical Belt and Zone, and in the South Equatorial Belt. Jupiter's dynamic atmosphere is influenced by the planet's internal heat (Jupiter emits about 1.7 times as much heat as it receives from the Sun), and shows wind speeds of up to 400 km/h (250 mile/h). The bright zones are high-altitude clouds supported by upwelling gases, while the dark belts are regions of descent.

Jupiter's predominant atmospheric constituents are hydrogen (almost 90%) and helium. Chemical species detected include ammonia, methane, water and hydrogen cyanide. Chemical reactions may be driven by ultraviolet radiation from the Sun, lightning and auroral discharges. At 1000 km (600 miles) below the cloud-tops the atmosphere gives way to an ocean of liquid molecular hydrogen. At a depth of 20,000 to 25,000 km (12,500 to 15,000 miles) the immense pressure causes the hydrogen to behave as a metal. At the centre of the planet there is probably an iron–silicate core perhaps 15 to 20 times the Earth's mass.

The deep metallic hydrogen 'mantle' gives Jupiter a powerful magnetic field (nearly 20,000 times as strong as the Earth's) and a very extensive **magnetosphere**. High-energy plasma is funnelled into extremely intense radiation belts. Most of Jupiter's trapped plasma comes from volcanoes on **Io** and occupies a *plasma torus* and a *flux tube* linking satellite and planet. The magnetosphere is a source of the planet's powerful radio emissions.

Jupiter has 63 known **satellites**, as of early 2005. The four major ones – Io, Europa, Ganymede and Callisto – are often referred to collectively as the **Galilean satellites**. Most of the others are minor objects, perhaps captured asteroids or comets, orbiting at distances of up to 28 million km (18 million miles) from the planet. Jupiter became the focus of public attention in 1994 when the fragmented Comet **Shoemaker–Levy 9** plunged into its atmosphere, producing a succession of impact scars. Jupiter also possesses a modest ring system (SEE **ring, planetary**), the brightest component of which is about 50,000 km (30,000 miles) above the cloud-tops.

K

Kapteyn's Star A red subdwarf, 12.7 light years away in the southern constellation Pictor, which has the second-largest proper motion known: 8.7 arc seconds per year. It is of magnitude 8.9, type M1, and is over 250 times less luminous than the Sun. It was discovered by the Dutch astronomer Jacobus Kapteyn (1851–1922).

K corona SEE **corona**.

Keck Telescopes A pair of identical telescopes at the W. M. Keck Observatory on Mauna Kea, Hawaii. Both have mirrors 10 m (396 inches) in diameter, consisting of 36 hexagonal segments. Keck I was completed in 1992, and its twin, Keck II, in 1996. They are run by the University of California and Caltech.

Kellner eyepiece An eyepiece with large **eye relief**, and so widely used for binoculars as well as telescopes. It is essentially a **Ramsden eyepiece** with one element replaced by an achromatic doublet (SEE **achromatic lens**). (Invented in 1849 by German optician Carl Kellner, 1826–55.)

Kepler, Johann (1571–1630) German mathematician and astronomer. He supported the Sun-centred Solar System put forward by **Copernicus**. In 1600 he went to Prague, becoming assistant to Tycho **Brahe**, on whose death in 1601 he succeeded as Imperial Mathematician to the Holy Roman Emperor, Rudolf II, with the task of completing tables of planetary motion begun by Tycho. From Tycho's accurate observations he concluded in *Astronomia nova* (1609) that Mars moved in an elliptical orbit, and went on to establish the first of his laws of planetary motion (SEE **Kepler's laws**). Erroneous reasoning led Kepler nevertheless to the second of

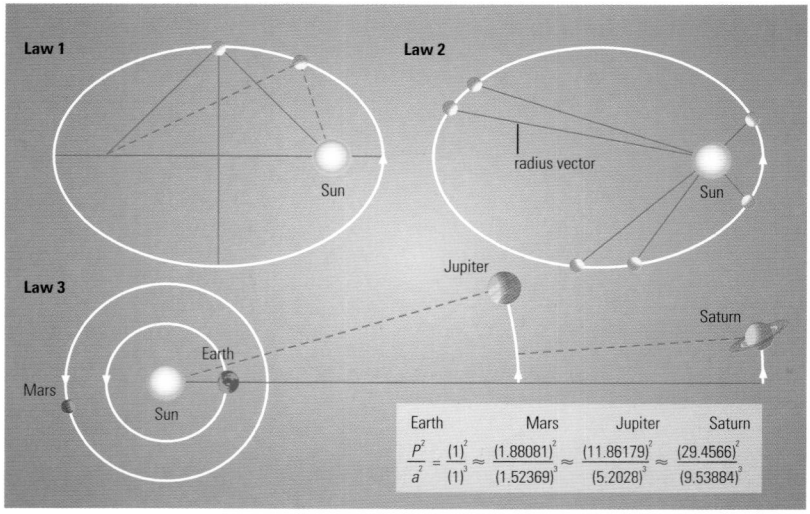

▲ **Kepler's laws** Planetary motions around the Sun are described by Kepler's three laws. The laws apply to all bodies in closed orbits around the Sun, and to satellites orbiting planets.

► **Kitt Peak National Observatory** This aerial view of Kitt Peak shows the many telescopes sited there. From bottom left to top right are: the 3.5-m (138-inch) WIYN Telescope; the 0.9-m (35-inch) WIYN Telescope; and the 2.1-m (83-inch) KPNO Telescope. The triangular building at top right is the McMath–Pierce Solar Facility.

these laws, and his desire to match celestial and musical harmony led to the third law, published in *Harmonices mundi* (1619). The *Rudolphine Tables*, based on Tycho's observations and Kepler's laws, appeared in 1627 and remained the most accurate until the 18th century. Other works of his include *De stella nova*, on the supernova of 1604 (SEE **Kepler's Star**), and *Dioptrice* (1611), on optics and the theory of the telescope.

Kepler's laws The three fundamental laws governing the motions of the planets around the Sun (and other celestial bodies in closed orbits), first worked out by Johann **Kepler** between 1609 and 1619, and based on observations made by Tycho Brahe (SEE the diagram).

Law 1 deals with the shape of a planetary orbit: the orbit of each planet is an **ellipse** with the Sun at one of the foci.

Law 2 explains the varying speed of planetary motion such that the planet moves faster the nearer it is to the Sun: the line joining the planet to the Sun (the *radius vector*) sweeps out equal areas in equal times. This law is known as the *law of areas*.

Law 3 relates the size of the orbit to the period of revolution: the square of the period (P) is directly proportional to the cube of the mean distance of the planet from the Sun (a). In practice this means that P^2/a^3 is approximately constant for all bodies in orbit around the Sun.

Kepler's Star A **supernova** that appeared in October 1604 in the constellation Ophiuchus, reaching magnitude −3 and remaining visible to the naked eye until March 1606. It was studied by Johann **Kepler**. The smooth light curve is like that of a Type I supernova. The remnant is a radio source, and some faint nebulosity is visible optically. It was the last supernova seen in our Galaxy.

Kirkwood gaps Regions in the main **asteroid** belt where, because of **resonances** with Jupiter, no stable orbit is possible, and from which asteroids are therefore absent (or nearly absent). The gaps are at distances at which the revolution period of an orbiting body would be **commensurable** with Jupiter's, and Jupiter would perturb the body. There are pronounced gaps at $\frac{1}{3}$, $\frac{2}{5}$ and $\frac{1}{2}$ of Jupiter's period, for example. The gaps in the ring system of **Saturn** are kept clear by a similar perturbing effect of its satellites, in particular Mimas, whose period is twice that of a body at the distance of the Cassini Division. The American astronomer Daniel Kirkwood (1814–95) explained how the gaps came about.

Kitt Peak National Observatory An observatory near Tucson, Arizona, part of the US National Optical Astronomy Observatories, containing the world's largest collection of telescopes. Its largest instrument is the 4-m (158-inch) Mayall Telescope, opened in 1973. It is also the site of the McMath–Pierce Solar Telescope, the world's largest solar telescope; the 3.5-m (138-inch) WIYN Telescope, opened in 1994; and the 2.3-m (90-inch) Steward Reflector of the University of Arizona.

Kohoutek, Comet A **comet** discovered in 1973 by the Czech astronomer Luboš Kohoutek (1935–). When first detected it was nearly as far from the Sun as Jupiter, and was predicted to become as bright as the Moon at perihelion in December 1973. Although it failed to live up to these expectations, it was studied extensively and advances in comet science were made as a result.

K star A star of spectral type K, characterized by numerous absorption lines, especially of iron, titanium and calcium, and the presence of molecular bands of CH and CN in its spectrum. Main-sequence K stars have surface temperatures ranging from 3550 to 4900 K. Giants are about 400 K cooler than this, and supergiants 300 K cooler still. At their lowest point on the main sequence, K stars are about half a solar mass, rising to 0.8 of a solar mass at their highest. Because K-type giants and supergiants may be either evolved old stars of about one solar mass or evolved stars of higher masses, a wide range is represented. However, most of the K-type giants are stars of a few solar masses. Only a small fraction of K stars are variable: the coolest Cepheids have K-type spectra, as do the hottest of the **Mira** stars. A few per cent of K-type giants show spectral peculiarities with an overabundance of carbon and certain **heavy elements**. Bright examples of K stars are Pollux, Arcturus, one component of Alpha Centauri and Epsilon Eridani.

Kuiper, Gerard Peter (1905–73) Dutch-American astronomer. After World War II he embarked on an observational programme which revitalized lunar and planetary astronomy. He discovered the satellites Miranda (of Uranus) in 1948 and Nereid (of Neptune) in 1949, and found methane in the atmospheres of Uranus, Neptune and Titan, and carbon dioxide in the atmosphere of Mars. He suggested the existence of what is now called the **Kuiper Belt**. Kuiper was involved with many US missions to the Moon and planets, in particular with the Ranger and Mariner probes. He advocated carrying infrared telescopes on high-flying aircraft; the Kuiper Airborne Observatory was named in his honour.

Kuiper Belt (Edgeworth–Kuiper Belt) A flattish zone stretching from the orbit of Neptune at 30 AU to about 1000 AU from the Sun, populated by **planetesimals** currently classed as asteroids. Its existence was proposed by Kenneth Edgeworth (1880–1972) in 1949 and by Gerard Kuiper in 1951 as a repository for comets or planetesimals left over from the formation of the Solar System. The concept was resurrected in the 1980s as a possible source of short-period comets, to account for the generally low orbital inclinations of such comets, which the spherical **Oort Cloud** could not.

The first member of the Kuiper Belt was discovered in 1992, and by late 2004 almost 850 had been found. Most are between 100 and 500 km (60–300 miles) in size, although larger ones exist: for example, **Varuna** may be as large as 900 km (560 miles). There are almost certainly many smaller members. Members of the belt are often called *trans-Neptunian objects* (TNOs). Many, like Pluto, are in a 3 : 2 orbital resonance with Neptune, and are called *plutinos*.

Physically, Kuiper Belt objects are probably similar to comets. They show no comet-like activity because they are so far from the Sun. If they move inward into planet-crossing orbits they may become **Centaurs**, and then outgassing can start, as has been observed with **Chiron**.

L

Lacaille, Nicolas Louis de (1713–62) French astronomer. He spent the years 1751 to 1753 surveying the southern-hemisphere skies from the Cape of Good Hope, and introduced 14 new southern constellations. In 1761 he made an accurate measurement of the Moon's distance that took into account the Earth's oblateness; this made possible a more accurate method of determining terrestrial longitude.

Lacerta A small, obscure northern constellation between Andromeda and Cygnus, representing a lizard, introduced by **Hevelius**. Alpha, its brightest star, is magnitude 3.8.

Lagoon Nebula A bright nebula in Sagittarius, designated M8 or NGC 6523, surrounding the open star cluster NGC 6530. The nebula is cut by a dark cloud which gives it its name. The nebula and cluster lie at a distance of 5200 light years.

Lagrange, Joseph Louis de (1736–1813) French mathematician, born in Italy. He made many contributions to celestial mechanics, studying the Moon's **libration**, the motion of Jupiter's satellites, and perturbations and stability in the Solar System. From his investigation in 1772 of the **three-body problem** he found special solutions of the equations which suggested the existence of the equilibrium positions now known as **Lagrangian points**.

Lagrangian point One of the points at which a celestial body can remain in a position of equilibrium with respect to two much more massive bodies orbiting each other. At these points the forces acting on the smaller body cancel out. There are five Lagrangian points, in the orbital plane of the two large bodies (SEE the diagram). Bodies at points L_1, L_2 and L_3, on the line joining M_1 and M_2, are less stable than bodies at points L_4 and L_5, which form equilateral triangles with M_1 and M_2. The **Trojan asteroids** lie in stable orbits near the L_4 and L_5 Lagrangian points of Jupiter's orbit around the Sun. (After Joseph Louis de **Lagrange**.)

Lalande, Joseph Jérôme (Le Français) de (1732–1807) French astronomer. His main achievement was a catalogue of over 47,000 stars. He did much to popularize astronomy in France. Lalande observed Neptune in 1795, 51 years before it was located by Johann Galle, without realizing that it was a planet.

La Palma Location of the **Roque de los Muchachos Observatory**.

Laplace, Pierre Simon de (1749–1827) French mathematician and astronomer. He used Isaac Newton's law of **gravitation** to interpret the perturbations of planets and satellites, the shape and rotation of Saturn's rings, and the stability of the Solar System. In 1796 he put forward his *nebular hypothesis* of the origin

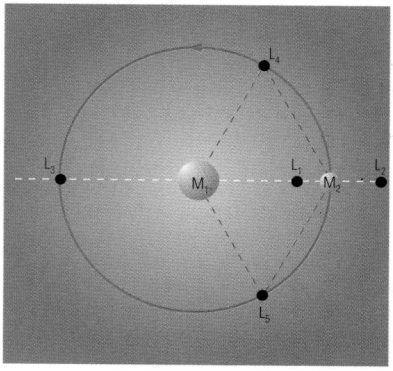

▲ **Lagrangian point** Many spacecraft are stationed around the L_1 point between Earth and the Sun. The L_4 and L_5 (60° ahead of and behind the planet) Lagrangian points of Jupiter's orbit are occupied by Trojan asteroids.

of the Solar System, which, although wrong in detail, is broadly in agreement with the presently accepted view. His five-volume *Traité de mécanique céleste* published over the period 1799–1825 summarized 18th-century advances in celestial mechanics.

Large Binocular Telescope A twin reflecting telescope, sited at **Mount Graham International Observatory**, scheduled for completion in 2005. Each component has an 8.4-m (330-inch) primary mirror, giving a combined effective aperture of 11.8 m (465 inches). The centres of the mirrors are 14.4 m (567 inches) apart.

Large Magellanic Cloud (LMC) SEE **Magellanic Clouds**.

Larissa One of the small inner **satellites** of Neptune, first detected in 1981 by Harold Reitsema and colleagues when it occulted a star, and confirmed by the 1989 visit of Voyager 2. Larissa appears to be the inner **shepherd moon** to the planet's Adams Ring.

Las Campanas Observatory An observatory near La Serena, Chile, run by the Carnegie Institution of Washington. Its main telescopes are the twin **Magellan Telescopes** and the 2.5-m (100-inch) Du Pont Telescope, in operation since 1977.

Lassell, William (1799–1880) English brewer and amateur astronomer. From the observatory he built, he discovered several planetary satellites: Neptune's **Triton** – only 17 days after the discovery of Neptune itself, Saturn's **Hyperion**, and Uranus's **Ariel** and **Umbriel**. He also discovered 600 nebulae. Lassell was the first to use the equatorial system of mounting for reflecting telescopes.

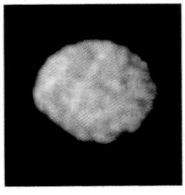

◀ **Larissa** Neptune's inner satellite Larissa was imaged during Voyager 2's flyby in August 1989.

last contact SEE **fourth contact**.

last quarter One of the Moon's **phases** when it has almost reached western quadrature and is half illuminated as seen from the Earth.

late-type star A cool star of spectral type K, M, C or S. The name was given when it was believed that the spectral types represented an evolutionary sequence, hot stars being the youngest and cool stars the oldest. Although that is now known not to be the case, the term has remained in use.

latitude A coordinate on the celestial sphere, or on the surface of a celestial body. *Celestial* (or *ecliptic*) *latitude* is the angular distance of a celestial body from the ecliptic; it is measured at right angles to the ecliptic, from 0° at the ecliptic to 90° at the ecliptic poles, positive to the north and negative to the south. *Galactic latitude* is the angular distance of a body from the galactic equator, measured from 0° to 90° at right angles to the galactic equator. *Heliocentric latitude* is the celestial latitude of a body as seen by a hypothetical observer at the centre of the Sun, rather than the centre of the Earth. *Heliographic latitude* is the latitude of a point on the Sun's surface, north or south of the Sun's equator.

leap second The adjustment of one second in the radio time signal at midnight on 30 June or 31 December. By international agreement the rate of radio time signals has been maintained since January 1972 in agreement with the International Atomic Time scale. However, the Earth's rotational period is slowing by about 0.003 of a second per day, so the radio time signals gradually diverge from Coordinated **Universal Time** (UTC), which is based on the Earth's daily rotation. To correct for this divergence, the radio time is retarded by the addition of one leap second as necessary.

Leavitt, Henrietta Swan (1868–1921) American astronomer. She specialized in variable stars, noticing that the brighter of the **Cepheid variables** tended to have the longer periods. From studies of Cepheids in the Small Magellanic Cloud she formulated the **period–luminosity law**, which was important in establishing the distance scale for other galaxies.

Leda A small **satellite** in Jupiter's intermediate group, discovered by Charles Kowal in 1974.

Lemaître, Georges Edouard (1894–1966) Belgian priest and astrophysicist. In 1927 he argued that an expanding Universe would have originated in the radioactive explosion of a 'primeval atom' or 'cosmic egg' into which all mass and energy was originally concentrated. This was the first statement of what would become known as the **big bang** theory.

lens An optical component of an instrument, made of a transparent material (such as glass) which, by the material's property of refraction, modifies an incoming beam of light to help form an image of the light source. Single lenses have either two curved surfaces, or one curved and one plane surface. Lenses are used in various combinations in astronomical instruments. SEE **aberration** (2), **eyepiece**, **objective**, **telescope**.

lenticular galaxy A type of **galaxy** intermediate in form between spiral and elliptical galaxies. Lenticular galaxies are designated S0, which indicates that they have the flattened form of spirals but lack spiral arms. They appear lens-shaped, hence their name.

Leo A large and easily identified constellation of the zodiac, in mythology representing the lion killed by Hercules. Its brightest star is first-magnitude **Regulus**, part of the curved figure of stars, known as the Sickle, that forms the lion's head and chest. Gamma (Algieba) is a fine, wide binary, magnitudes 2.2 and 3.5, one K1-type giant and one G7-type giant, period 620 years. The Mira variable R Leonis can reach magnitude 4.4 (minimum 11.3, period 310 days). There are five galaxies in Leo on Messier's list: M65, M66, M95, M96 and M105.

Leo Minor A small constellation between Leo and Ursa Major. It has no star above magnitude 3.8.

Leonids A **meteor shower** that occurs between 15 and 20 November. Its radiant is in the constellation Leo, in the 'Sickle'

► **Lenticular galaxy** M85 (also designated NGC 4382) is a lenticular galaxy of type S0. It is the most northerly member of the Virgo Cluster, and is situated in the constellation Coma Berenices.

asterism. The shower is periodic, with modest displays (**ZHR** = 15–20) in most years, but intervals of elevated activity, including **meteor storms**, are seen for about a third of the 33-year orbital period of the parent comet, Comet Tempel–Tuttle, around the time of its perihelion. Notable storms occurred in 1799, 1833, 1866 and 1966. Most recently, intense Leonid activity was observed in 1999, 2001 and 2002; in addition to these storms, a spectacular display of Leonid fireballs occurred in 1998. SEE ALSO **meteor stream**.

Lepus A small constellation representing a hare hiding under the feet of Orion. Its brightest star is Alpha (Arneb), magnitude 2.6. It contains the very red Mira variable R Leporis (range 5.5 to 11.7, period 427 days) and the globular cluster M79.

Le Verrier, Urbain Jean Joseph (1811–77) French astronomer. In 1846, independently of John Couch **Adams**, he predicted that an unknown planet (Neptune) was responsible for discrepancies between the calculated and observed orbital motions of Uranus. In 1859 he suggested that a planet which he prematurely named Vulcan lay within the orbit of Mercury, but despite some claimed sightings its existence was never confirmed.

Libra An obscure constellation of the zodiac, representing the scales of justice, but formerly pictured as the claws of the scorpion, Scorpius, which it adjoins. The brightest star, Beta, is magnitude 2.6. Delta is an eclipsing binary of the Algol type, range 4.9 to 5.9, period 2.33 days.

libration A small oscillation of a celestial body about its mean position. The term is used most frequently in connection with the Moon. As a result of libration it is possible to see, at different times, 59% of the Moon's surface. However, because the areas that pass into and out of view are close

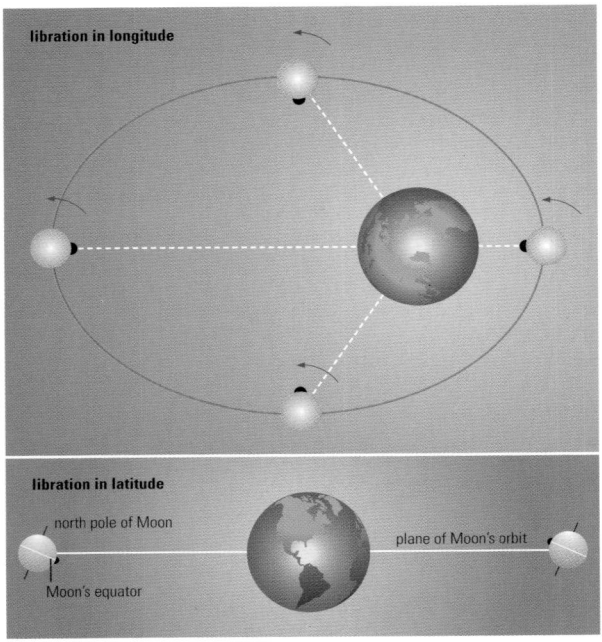

◀ **libration** Although the Moon is in synchronous rotation with respect to the Earth, ground-based observers can still, over the course of several lunations, see significantly more than just half of its surface. Libration in longitude (top) allows the observer to see slightly around the mean east or west limb, while the tilt of the lunar axis relative to the Moon's orbital plane allows regions beyond either pole to be seen on occasion.

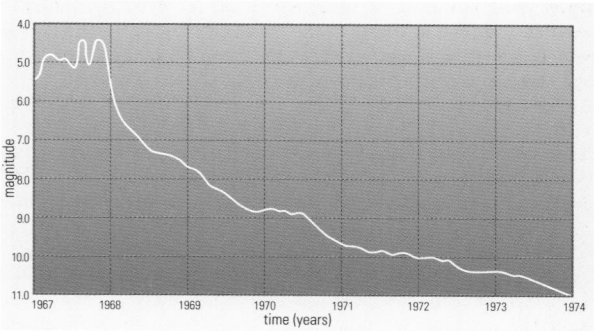

► **light curve** An example of a light curve is this plot of the magnitude (brightness) of Nova Delphini 1967 against time. The nova was slow to arrive at peak brightness, showing several maxima over a twelve-month period, before its light curve entered a long, gradual decline.

to the limb and therefore extremely foreshortened, in practice libration has less of an effect on the features that can be clearly made out on the Moon's disk than this figure might suggest.

Physical libration results from slight irregularities in the Moon's motion produced by irregularities in its shape. Much more obvious is *geometrical libration*, which results from the Earth-based observer seeing the Moon from different directions at different times. There are three types of geometrical libration. *Libration in longitude* arises from a combination of the Moon's synchronous rotation and elliptical orbit (SEE the diagram). As a result, at times a little more of the lunar surface is visible at the eastern or western limb than when the Moon is at its mean position. *Libration in latitude* arises because the Moon's equator is tilted slightly from its orbital plane, so that the two poles tilt alternately towards and away from the Earth. A smaller effect is *diurnal libration*, by which the Earth's rotation allows us to see more of the Moon's surface at its western or eastern limb when it is rising or setting.

Lick Observatory The observatory of the University of California, sited on Mount Hamilton near San Jose, California. Its oldest telescope is a 0.91-m (36-inch) refractor, in operation since the observatory opened in 1888. The largest telescope is the 3-m (120-inch) C. Donald Shane Telescope, completed in 1959.

light, speed of (symbol *c*) The constant speed at which light and other **electromagnetic radiation** travels in a vacuum. Its value is 299,792,458.0 m/s. The first reasonably good estimate of *c* was made by Ole **Römer** in 1675. Albert **Einstein** showed with his Special Theory of Relativity that the speed of light in a vacuum is always the same, whatever the relative speed of the observer and the source of the light.

light curve A graph of the change in magnitude with time of a variable star or other body showing brightness variations.

light pollution The detrimental effect of artificial lighting on astronomical observation and imaging. The problem has become much worse in recent years with the proliferation of such lighting and the development of new types of streetlamp. For amateurs, light pollution reduces the **limiting magnitude** (1) and makes deep-sky objects and comets, for example, difficult or impossible to see, but the development of special **filters** and of **CCDs** and related technology is providing solutions. For professional astronomers it can impede or distort scientific observation, particularly where the area surrounding a once-isolated observatory has become built up.

light-time The time required for light to travel from any celestial object to the Earth. For example, light takes 8.3 minutes to reach the Earth from the Sun.

light year (symbol l.y.) A unit of distance used in astronomy, equal to the distance travelled by light in a vacuum in one tropical year. 1 l.y. = 9.4607×10^{12} km = 0.3066 parsecs = 63,240 AU.

limb The edge of a celestial body which is visible as a disk, such as the Sun, the Moon or a planet. SEE ALSO **following, preceding**.

limb darkening The decrease in the brightness at visible wavelengths of a celestial body towards its limb, giving it a vignetted appearance. The term is used most often in connection with the Sun, where, towards the limb, the observer is looking at the deeper, brighter levels of the Sun through a greater thickness of its atmosphere. The giant planets also show limb darkening.

limiting magnitude (1) The faintest magnitude of objects visible or recordable by photography or electronic means on a given occasion. The limiting magnitude depends on sky conditions, instrument aperture and sensitivity of the eye or recording apparatus.

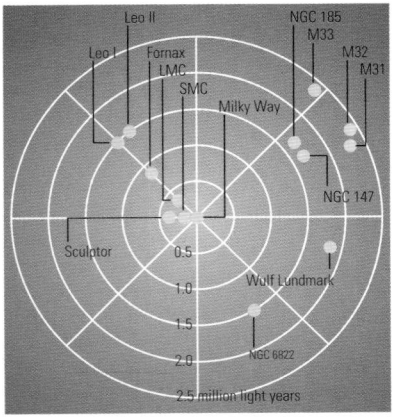

▲ **Local Group** Members of the Local Group are gravitationally bound to one another. As shown in this plot, which is centred on our Galaxy, the closest neighbours are the Magellanic Clouds and Sculptor Dwarf.

limiting magnitude (2) The faintest magnitude of objects recorded in a star atlas.

line of apsides SEE **apsides**.

line spectrum SEE **spectrum**.

Little Dumbbell A planetary nebula in the constellation Perseus, lying 3400 light years away. Discovered by Pierre Méchain in 1780, the object is designated M76, and takes its popular name from its twin-lobed appearance, resembling a smaller, fainter version of the **Dumbbell Nebula**.

LMC ABBREVIATION FOR **Large Magellanic Cloud**.

local arm SEE **Orion Arm**.

Local Group The small group of about 40 galaxies that includes our Galaxy, the two Magellanic Clouds, the Andromeda Galaxy and the Triangulum spiral; most are dwarf galaxies. They are distributed over a roughly ellipsoidal space about 5 million light years across. All the members are gravitationally bound so that, unlike more distant galaxies, they are not receding from us or from one another. The ten brightest (in absolute magnitude) are given in the table.

local time Time as measured at a given place on Earth, based on the mean solar day. Local time changes with longitude. A difference of 15° in longitude corresponds to a difference of one hour in local time. To avoid the inconvenience of a multiplicity of local times, the Earth has been divided into 24 **time zones**. In addition, *daylight saving time* (summer time) one hour in advance of standard time is kept in some countries for part of the year.

longitude A coordinate on the celestial sphere or on the surface of a celestial body. *Celestial* (or *ecliptic*) *longitude* is measured eastwards from the vernal equinox (the First Point of Aries) to the **great circle**

THE LOCAL GROUP: THE MAIN MEMBERS

Galaxy	Apparent magnitude	Absolute magnitude	Distance (thousand l.y.)
Andromeda Galaxy (M31)	3.4	−21.2	2510
Milky Way	—	−20.9	—
Triangulum Galaxy (M33)	5.9	−18.9	2770
Large Magellanic Cloud	0.2	−18.5	160
Small Magellanic Cloud	2.0	−17.1	200
M32	8.1	−16.5	2510
IC 10	10.3	−16.4	2150
M110	8.4	−16.4	2710
Barnard's Galaxy (NGC 6822)	8.3	−16.0	1630
NGC 185	9.0	−15.6	2020

passing through the pole of the ecliptic and the object to be measured. It is reckoned in degrees from 0 to 360°. *Galactic longitude* is the angle measured along the galactic equator from the direction of the galactic centre to the point at which the galactic equator intersects the galactic circle of longitude of the object to be measured. *Heliocentric longitude* is the celestial longitude of a body as seen by a hypothetical observer from the centre of the Sun. *Heliographic longitude* is the longitude of a point on the surface of the Sun. It is measured from the solar meridian that passed through the ascending node of the Sun's equator on the ecliptic on 1 January 1854, at Greenwich mean noon, from 0 to 360° in the direction of the Sun's rotation.

long-period comet SEE **comet**.

long-period variables Pulsating red giants or supergiants with periods ranging from about 80 to 1000 days, usually of spectral type M or C with characteristic emission lines. They are also known as **Mira** stars after the first of the type to be discovered. Each star has an average period which can vary by about 15%, so they are to some extent irregular. Their amplitudes range from 2.5 to 11 magnitudes, but the amplitude can also change appreciably from one cycle to the next. (Stars with lesser amplitudes are generally designated **semiregular variables**.) Long-period variables fall into three main divisions. Those that have a rise steeper than the fall tend to have wide minima and short, sharp maxima; as the asymmetry becomes greater, the period lengthens. Stars with symmetrical curves have the shortest periods. A third group show humps on their curves, or have double maxima and have long and short periods.

Lovell, (Alfred Charles) Bernard (1913–) English radio astronomer. From 1951 to 1981 he was director of the **Jodrell Bank** experimental station for radio astronomy, and oversaw the construction there of the world's first large steerable radio telescope. He used the telescope to pioneer the study of radio emission from space.

Lowell, Percival (1855–1916) American astronomer. In 1894 he built an observatory, now called the Lowell Observatory, at Flagstaff, Arizona. From there he observed Mars, producing intricate charts of the so-called **canals**, which he ascribed to the activities of intelligent beings. He calculated that the orbital irregularities of Uranus and Neptune were caused by an undiscovered **Planet X**, and his predictions of its position led to the discovery of the planet Pluto in 1930 by Clyde **Tombaugh**. It was at Lowell's urging that Vesto Slipher carried out the observations that led to the discovery of galactic **redshifts** in 1916.

lower culmination SEE **culmination**.

low-surface-brightness galaxy (LSB galaxy) A galaxy that is very faint at optical wavelengths either because it contains few stars or because its stars are spread out. LSB galaxies of many types are known, from dwarf irregulars to giant spirals. It is possible that, though difficult to detect, they are very common, and thus might account for some of the **missing mass** of the Universe.

L star A star of spectral type L, characterized by the presence of lithium lines in its spectrum. Because lithium is quickly destroyed in nuclear reactions, its presence indicates that no thermonuclear processes are taking place within these stars. The spectra also show absorption lines of metal hydrides, such as FeH and CrH, and alkali metals such as sodium and potassium. With surface temperatures typically of 1500–2000 K, L stars are cooler than M stars. The L type was introduced in 1998 for **brown dwarfs** and faint stars detectable at infrared wavelengths. Even cooler objects (below 1500 K) are classified separately, as **T stars**.

luminosity (symbol *L*) The absolute brightness of a star, given by the amount of energy radiated from its entire surface per second. Luminosity is expressed in watts (joules per second), or in terms of the Sun's luminosity. *Bolometric luminosity* is a measure of the star's total energy output, at all wavelengths. **Absolute magnitude** is an indication of luminosity at visual wavelengths.

luminosity class SEE **Morgan–Keenan classification**.

luminosity function A measure of the distribution of stars in space according to their **absolute magnitude**. It is usually expressed as the number of stars per cubic parsec for a given range of absolute magnitude. The value of the luminosity function in the solar neighbourhood rises to a maximum at absolute magnitudes of 14–15, then falls away for fainter magnitudes, implying that faint M-type dwarfs of about one ten-thousandth of the Sun's luminosity are the most abundant type of star locally. Luminosity functions can be plotted for other categories, such as particular types of star, star clusters, galaxies or clusters of galaxies.

Luna A series of lunar probes launched by the former Soviet Union. The first three in the series were originally named *Lunik*. Luna 2 became the first man-made object to hit the Moon, on 13 September 1959. Luna 3, launched on 4 October 1959, obtained the first photographs of the farside of the Moon. Luna 9 achieved the first soft landing on the Moon on 3 February 1966. Luna 10, in March 1966, was the first craft to enter lunar orbit. In September 1970 Luna 16 made the first automatic return of a soil sample to Earth. Luna 17 carried the first uncrewed lunar roving vehicle, named Lunokhod 1.

lunar eclipse An **eclipse** of the Moon by the Earth's shadow. Since the Moon's orbital plane is inclined to the plane of the ecliptic, a lunar eclipse can happen only when the Moon is at opposition (i.e. at full Moon) and at the same time is at or near one of its nodes (SEE **ecliptic limits**). The Sun, Earth and Moon are then very nearly in a straight line. Lunar eclipses are visible from anywhere on Earth where the Moon is above the local horizon. The Moon does not become completely invisible during an eclipse because it is partially lit by sunlight that has been refracted by the Earth's atmosphere, and it appears reddish because the blue component is removed from the refracted sunlight by **scattering**.

A *total lunar eclipse* occurs when the Moon is entirely within the umbra, the dark central part of the Earth's shadow (SEE the diagram). The eclipse is *partial* when the Moon is partly within the umbra, and *penumbral* when the Moon passes through the penumbra but completely misses the umbra. The overall duration of a total lunar eclipse, from **first contact** to **fourth con-**

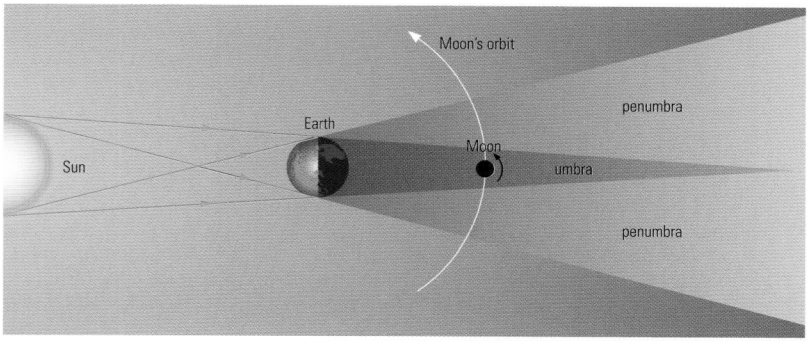

▲ **lunar eclipse** Earth casts a shadow to the right in this diagram. The shadow has two components – a dark central umbra, and a lighter area of partial shadow, the penumbra. If the full Moon – opposite the Sun in Earth's sky – passes through the central cone of the umbra, a total lunar eclipse will result.

tact, can be $3\frac{1}{2}$ hours; totality, the interval from **second contact** to **third contact**, can last for $1\frac{3}{4}$ hours.

Lunar Orbiter A series of US probes put into orbit around the Moon to photograph its surface in detail in preparation for the Apollo landings. There were five probes in the series, all successful, from Lunar Orbiter 1 in August 1966 to Lunar Orbiter 5 in August 1967.

Lunar Prospector A NASA probe to the Moon, launched on 7 January 1998. From polar orbit, its neutron spectrometer found strong evidence for the presence of water ice in the lunar regolith in both polar regions. The probe also found that the Moon has a small iron core.

lunar transient phenomenon (LTP) A short-lived, localized change in the appearance of a lunar surface feature, such as a glow or an obscuration. LTPs were reported by lunar observers for many years but treated with scepticism, but they are now generally regarded as real events. In 1958 a spectrum of activity in the crater Alphonsus was obtained, and instruments set up by the Apollo astronauts established a connection between LTPs and moonquakes. Since they are commonest when the Moon is at perigee, LTPs are most likely to be releases of gas and dust induced by tidal forces (SEE **tides**).

lunation The time interval between two identical phases of the Moon, for example from one new Moon to the next new Moon. This corresponds to the Moon's **synodic period** or **synodic month** of 29.53059 days.

lunisolar precession The main component of **precession**, resulting from the gravitational effects of the Moon and Sun on the Earth. It amounts to about 50.4 arc seconds per year.

Lupus A southern-hemisphere constellation, adjoining Centaurus, representing a wolf. Its brightest star is Alpha, magnitude 2.3, but it has no objects of particular interest.

Lyman series A series of lines in the ultraviolet region of the **hydrogen spectrum**, analogous to the **Balmer series** in the visible part. The first Lyman line, $L\alpha$, is at 121.6 nm, and was first detected in the solar spectrum by photography from a rocket. The second, $L\beta$, is at 102.6 nm; $L\gamma$ is at 97.2 nm. In the spectra of **quasars**, Lyman lines are redshifted into the visible region. (After American physicist Theodore Lyman, 1874–1954.)

Lynx A dim northern constellation between Auriga and Ursa Major, introduced by **Hevelius**. Its leading star is Alpha, magnitude 3.1, but the only other star above fourth magnitude is 38 Lyncis, at magnitude 3.9.

Lyot, Bernard Ferdinand (1897–1952) French astrophysicist. He invented the **coronagraph** for observing the Sun's corona, and the **Lyot filter**. He also pioneered astronomical polarimetry, the measurement of the degree of polarization of light from celestial bodies, and invented a number of devices for the purpose.

Lyot filter A **filter** for observing the Sun at a particular wavelength, developed by Bernard **Lyot** in the 1930s. Sunlight is directed through a succession of alternating layers of quartz plates and polaroid sheets, and each separate passage reinforces some wavelengths and cancels others. The result is a number of very narrow (0.1 nm) and widely spaced wavebands which are easy to isolate.

Lyra A small but prominent constellation. In mythology, it represents the harp given to Orpheus by Apollo. Its brightest star is **Vega**. Lyra contains the famous 'Double Double', the quadruple star Epsilon; the two main components, magnitudes 4.6 and 4.7, can be separated with the naked eye and each is again double. Beta (Sheliak) is an eclipsing variable, range 3.3 to 4.4, period 12.9 days; it is the prototype of the **Beta Lyrae stars**. Between Beta and Gamma (Sulafat), magnitude 3.2, lies the **Ring Nebula**, M57, a planetary nebula.

Lyrids A **meteor shower** active from 19 to 25 April, with its radiant on the border of the constellations Lyra and Hercules, south-west of the star Vega. Peak activity, around 22 April, is usually modest, with a **ZHR** of about 10. The Lyrids are associated with Comet Thatcher (1861 G1).

Lysithea A small **satellite** in Jupiter's intermediate group, discovered by Seth Nicholson in 1938.

M

McDonald Observatory The astronomical observatory of the University of Texas, on Mount Locke, Texas. It is home to the **Hobby–Eberly Telescope** and also houses a 2.7-m (107-inch) reflector, opened in 1968.

MACHO Abbreviation for 'massive compact halo object', an as-yet undetermined type of invisible body assumed to pervade in the Galaxy's halo (outer regions) to account for the **dark matter** that it is supposed to contain. MACHOs could be brown dwarfs, extinct white dwarfs or Jupiter-sized planets. SEE ALSO **missing mass**.

Mach's principle The philosophical proposition that the properties of a body, particularly its inertia, depend on the existence of all the other objects in the universe. Proposed by Ernst Mach (1938–1916), it is often illustrated by a 'thought experiment' involving two cylindrical spacecraft partly filled with water. One of the spacecraft spins; the other does not. The water in the spinning spacecraft will accumulate against the spacecraft's walls, while in the non-spinning spacecraft the water will float around randomly. The only way the water in the spinning spacecraft can 'know' that it is spinning is by reference to the rest of the Universe.

McNeil's Nebula A variable nebula in the constellation Orion, close to the bright M78, illuminated by a young star. It was first brought to wider attention by the American amateur astronomer Jay McNeil in early 2004, following his observations of it in a bright phase. Subsequent examination of archival images revealed the nebula's long-term variability, over timescales of decades.

Maffei Galaxies Two nearby galaxies so close to the plane of the Milky Way that at optical wavelengths they are almost completely obscured by galactic dust. Maffei I is an S0 (lenticular) galaxy over 3 million light years away, on the edge of the Local Group. Maffei II is an Sb spiral but, at a distance of 15 million light years, it is unrelated to either Maffei I or the Local Group. They were discovered as infrared sources in 1968 by the Italian astronomer Paolo Maffei (1926–).

Magellan A NASA space probe to Venus, launched on 4 May 1989. It went into orbit around Venus in August 1990 and made a detailed radar map of 98% of the planet, revealing impact craters, volcanic mountains, lava flows and other features. It operated until 1994.

Magellanic Clouds The two nearest galaxies to our own, visible to the naked eye as isolated patches of the Milky Way in the southern sky and named after the Portuguese explorer Ferdinand Magellan (1480–1521), who described them. The *Large Magellanic Cloud* (LMC) lies in the constellations Dorado and Mensa and is about 6° across. The *Small Magellanic*

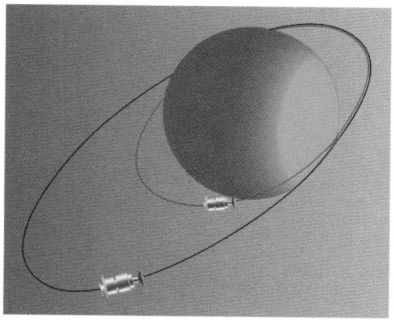

▲ **Magellan** During the first phases of NASA's Magellan mission, its orbit around Venus was highly elliptical (blue), and it used radar to map almost the whole planet. After three years, the orbiter used aerobraking techniques to reach a lower, circular orbit (red); it then mapped Venus's gravity field.

Cloud (SMC) is in Tucana and is about 3° across. Their distances are about 160,000 and 200,000 light years. Although usually classified as irregular galaxies, both Clouds are somewhat bar-shaped with faint outer structure that can be interpreted as rudimentary spiral arms. The mass of the LMC is about 10^{10} solar masses, and that of the SMC about 2×10^9 solar masses.

At the north-east end of the LMC lies the **Tarantula Nebula**, one of the largest and brightest groupings of hot stars and bright nebulosity known in any galaxy. Near here erupted **Supernova 1987A**, the first naked-eye supernova since 1604. The irregular distribution of star clusters and diffuse nebulae in the LMC contrasts with the more uniform stellar distribution in the SMC. Young stars in the LMC lie in a thin disk, which is seen nearly face-on (about 27° to the plane of the sky). There is no evidence for a spherical halo of the kind that surrounds our Galaxy.

The SMC is greatly extended in our line of sight, with a depth of about 60,000 light years, over five times the dimension it presents to us. It appears to have been distorted by tidal forces, possibly as a result of a close passage of the SMC and LMC some 200 million years ago.

The *Magellanic Stream* is a vast cloud of neutral hydrogen (SEE **H I region**) stretching between the Clouds and extending towards the Galaxy. It was probably torn out from the Clouds by gravitational interaction with the Galaxy.

Magellan Telescopes A pair of reflecting telescopes, both of aperture 6.5 m (255 inches), at **Las Campanas Observatory** in Chile. The telescopes were completed in 2000 and 2002, and can be used independently or together.

magnetar A neutron star with an exceptionally strong magnetic field (10^{14}–10^{15} gauss). Magnetic-field events involving these exceptionally strong fields account for the previously unexplained gamma-ray outbursts observed in the *soft gamma repeaters* and the similar X-ray events observed in *anomalous X-ray pulsars*. SEE ALSO **gamma-ray astronomy**, **pulsar**.

magnetohydrodynamics (MHD) The study of the behaviour of plasma (a

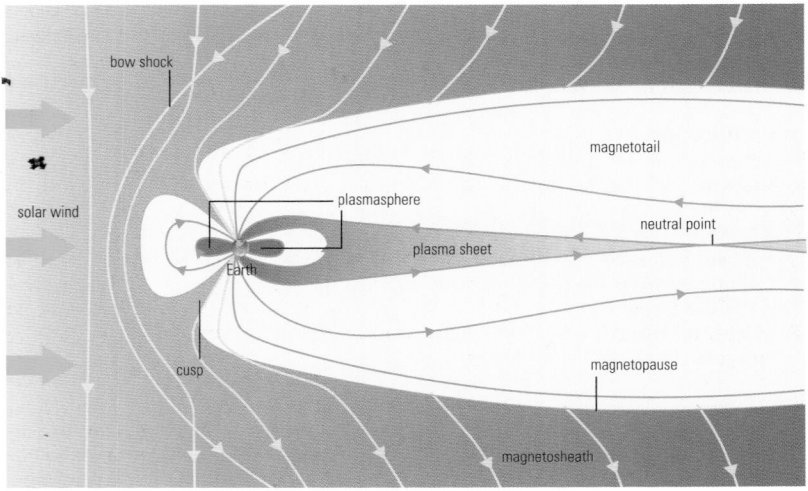

▲ **magnetosphere** The Earth's moving iron core gives it an active magnetic field that extends well beyond the planet. The arrows indicate the direction of the magnetic field.

mixture of ions and electrons) in a magnetic field. Magnetic-field lines are 'trapped' by a plasma, so the motion of a plasma distorts a magnetic field and, conversely, a magnetic field exerts a force on a plasma, modifying its motion. Plasmas occur in many astronomical contexts, so MHD is significant in the study of solar phenomena such as sunspots, the interaction of the solar wind with planetary **magnetospheres**, neutron stars, star formation and supernovae.

magnetosphere The region of space surrounding a planet in which the planet's magnetic field predominates over the **solar wind** and controls the behaviour of plasma (charged particles) trapped within it. The boundary of the magnetosphere is called the *magnetopause*, outside which is a turbulent magnetic region called the *magnetosheath*. Where the steady outward flow of the solar wind is first interrupted there is a **bow shock**. Downwind from the planet, the solar wind draws the magnetosphere out into a long, tapering *magnetotail*.

Mercury, the Earth and the giant planets possess magnetospheres. The Earth's contains the **Van Allen Belts** of charged particles. Jupiter's magnetosphere is many times larger than the Sun, and contains a vast disk of plasma and **Io**'s plasma torus.

magnification The factor by which an image produced by an optical system increases the angular size of an object. It is normally denoted by a multiplication sign, as in ×50 or 50× (SEE **binoculars**). The magnification given by a telescope is found by dividing its **focal length** by the focal length of the eyepiece being used with it. The higher the magnification, the dimmer the image and the more prone it is to the effects of bad seeing and defects of the instrument.

magnitude A measure of the brightness of celestial objects. In the 2nd century BC the Greek astronomer **Hipparchus of Nicea** classified the stars into six magnitudes, first magnitude being the brightest and sixth magnitude the faintest visible to the naked eye. In the 19th century it was found that a star of first magnitude was about 100 times as bright as one of sixth magnitude. It was decided to adopt a scale of magnitudes in which this ratio is exactly 100:1. The ratio of brightness between one magnitude and the next is therefore the fifth root of 100, which is 2.512. Thus, differences of 2.5, 5 and 10 magnitudes correspond to brightness ratios of 10, 100, and 10,000 respectively. Very bright bodies, e.g. Sirius, Venus, and the Sun, have negative magnitudes. The faintest objects detectable from ground-based telescopes are about magnitude 30.

These are *apparent magnitudes* – the brightness of those objects as seen from Earth. **Absolute magnitude** is the true brightness of an object, taking into account its distance from Earth. The difference between the apparent and absolute magnitudes is called the *distance modulus*. SEE ALSO **bolometric magnitude**, **photoelectric magnitude**, **photographic magnitude**, **photovisual magnitude**, **visual magnitude**.

main sequence The region on the **Hertzsprung–Russell diagram** where most stars lie, including the Sun. It runs diagonally from hot, bright stars at the upper left down to cool, faint stars at the bottom right. The position of a star on the main sequence depends on its mass, the most massive stars being the brightest. The upper limit is about 60 solar masses, at stars of spectral types O and B. The lower limit is at 0.08 solar masses, with stars of spectral type M. Stars spend most of their lives on the main sequence, producing energy by the fusion of hydrogen to helium in their cores. The *zero-age main sequence* is where stars lie when they first start to burn hydrogen. As stars use up their internal hydrogen they move away from the main sequence. The more massive a star, the sooner it evolves off the main sequence. All stars on the main sequence are termed **dwarfs**, irrespective of whether they are larger or smaller than the Sun.

major planet Any of the nine largest planets of the Solar System – Mercury, Venus, Earth, Mars, Jupiter, Saturn, Uranus, Neptune and Pluto – as opposed to the asteroids, which are also known as *minor planets*. Some astronomers feel that Pluto, because of its small size, does not merit being classified as a major planet.

Maksutov telescope A modification of the **Schmidt camera**, devised independently by Albert Bouwers (1940) and Dmitrii Maksutov (1944). The Schmidt's aspherical correcting plate is replaced by a concave lens which cancels out the **chromatic** and **spherical aberration** of the primary concave spherical mirror. A secondary mirror directs the light path through a hole in the primary to a Cassegrain focus, although various other configurations have been used. The Maksutov is a compact instrument once popular with amateur observers, but it has lost ground to the **Schmidt–Cassegrain telescope**. (After Dmitrii Dmitrievich Maksutov, 1896–1964.)

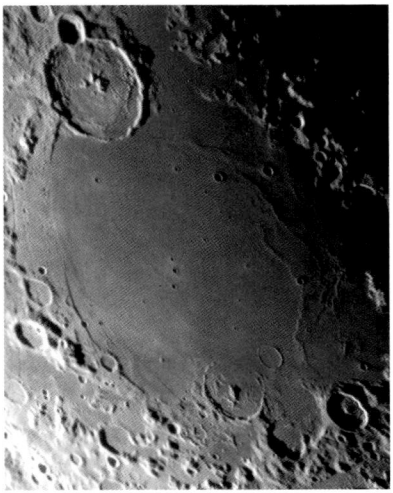

▲ **mare** Mare Humorum is a lava-filled basin on the Moon, some 370 km (230 miles) in diameter. It has fewer impact craters than the areas around it, indicating its relative youth.

mantle SEE **differentiation**.

many-body problem (*n*-body problem) The problem in celestial mechanics of finding how a number (*n*) of bodies move under the influence of their mutual gravitational attraction only. It is an extension of the **three-body problem**. The many-body problem has no general solutions, but with modern computers good approximate solutions can be obtained. Analysing how the planets of the Solar System perturb one another, and how space probes move between them, are examples of the many-body problem.

mare (plural **maria**) An extensive dark plain on the Moon, composed of basalt (solidified lava). Most of the maria were formed 3 to 4 billion years ago, after large meteorites struck the Moon and weakened the crust, and lava later erupted from below. They range from the vast Oceanus Procellarum, covering much of the northwestern quadrant of the Moon, down to areas less than 100 km (60 miles) across. Nearly all maria are on the Moon's nearside, where the crust is thinner than on the farside. 'Maria' is Latin for 'seas', which is what early observers thought they were. Similar plains on Mars were also known as maria, but since detailed mapping by spacecraft these areas are now known as *planitiae*.

Mariner A series of US space probes to the planets. Mariner 2 flew past Venus in 1962, discovering that the surface temperature was hot. Mariner 4 flew past Mars in July 1965 and photographed craters on its surface. Mariner 5 passed Venus in October 1967, making measurements of the planet's atmosphere. Mariners 6 and 7 obtained further photographs of Mars in 1969. Mariner 9 became the first probe to enter planetary orbit when it went into orbit around Mars in November 1971. Mariner 10, the last of the series, was the first two-planet mission, passing Venus in February 1974 and then encountering Mercury three times, in March and September 1974 and March 1975.

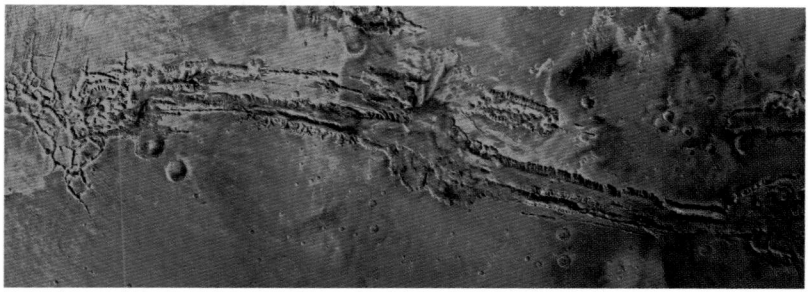

▲ **Mariner Valley** This composite Viking image shows the huge, sprawling canyon system of the Mariner Valley (Valles Marineris) on Mars. Ancient dry river channels run northwards from the chaotic terrain at the east (right).

Mariner Valley (Valles Marineris) A system of enormous, interconnected canyons on Mars, stretching for 4500 km (2800 miles) just south of the equator. Individual canyons extend for up to 200 km (125 miles) and are 7 km ($4\frac{1}{2}$ miles) deep. Mariner Valley is a fault structure whose formation is probably linked to that of the volcanic Tharsis region at its western end. The name commemorates the **Mariner** 9 probe, from which it was first imaged.

Markarian galaxy Any of the galaxies catalogued by the Armenian astronomer Benjamin Eghishe Markarian (1913–85), distinguished by their strong ultraviolet emission. They include **Seyfert galaxies** and **quasars**.

Mars The fourth major planet from the Sun, and the second-smallest of the terrestrial planets. Mars is known popularly as the Red Planet, from its naked-eye appearance. Through a telescope it shows a small orange-red disk with lighter and darker markings, and white patches often visible at one pole or the other. Because of its markedly eccentric orbit, its distance from Earth at opposition varies from 56 to 101 million km (35 to 63 million miles), with corresponding variations in apparent size and brightness of 14 to 25 arc seconds and magnitude −1.0 to −2.8. Favourable, *perihelic* oppositions occur at intervals of 15 to 17 years; that of 2003 was the closest for 60,000 years.

Mars shows a slightly **gibbous** phase at quadrature. Its axial tilt is similar to the Earth's, so it passes through a similar cycle of **seasons**, during which the polar caps vary in size, and the darker markings can vary in shape and extent as seasonal dust storms cover and uncover darker surface rocks. The planet was once imagined to be an abode of intelligent life, and some observers reported seeing linear surface markings which they believed were artificial

MARS: DATA

Globe	
Diameter (equatorial)	6792 km
Diameter (polar)	6752 km
Density	3.91 g/cm^3
Mass (Earth = 1)	0.107
Volume (Earth = 1)	0.152
Sidereal period of axial rotation	24h 37m 23s
Escape velocity	5.0 km/s
Albedo	0.15
Inclination of equator to orbit	25° 11′
Surface temperature (average)	220 K
Surface gravity (Earth = 1)	0.38
Orbit	
Semimajor axis	1.524 AU = 227.9 × 10^6 km
Eccentricity	0.0934
Inclination to ecliptic	1° 51′
Sidereal period of revolution	686.980d
Mean orbital velocity	24.13 km/s
Number of satellites	2

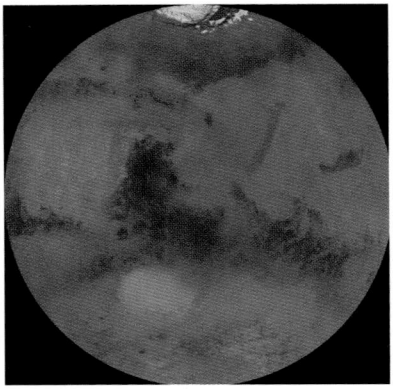

▲ **Mars** The Viking spacecraft gave unprecedentedly sharp images of Mars. The prominent dark area left of centre is Syrtis Major, and the bright area below it is Hellas Planitia.

canals. In 1996 it was announced that an **SNC meteorite** designated ALH 84001, recovered from Antarctica in 1984, appeared to contain the fossilized remains of primitive life forms, but most biologists remain sceptical. The main data for Mars are given in the table on page 141.

Although similar geological processes have operated on Earth and Mars, the results on Mars have been dramatically different, because of its smaller size and gravity and the absence of plate **tectonics**. Mars's northern hemisphere is covered by largely smooth, lowland volcanic plains, while the more ancient terrain of the south comprises heavily cratered uplands crossed by many channels. Two large impact basins, Hellas (1800 km; 1500 miles) and Argyre (900 km; 550 miles) dominate the southern hemisphere. There is also a prominent bulge, 5000 km (3000 miles) across, in the volcanic Tharsis region, which straddles the equator. The biggest volcanic structure on Mars is **Olympus Mons**. It is possible that Mars has been volcanically active as recently as 100 million years ago.

Stretching east from Tharsis is the network of giant canyons known collectively as the **Mariner Valley**. While some Martian canyons and similar features are geological faults, others are clearly channels in which water once flowed. Some features have even been interpreted as dried-up seabeds. At some time in the past Mars must have had a very different climate, with a denser atmosphere. In 2001–2 the Mars Odyssey spacecraft found evidence for a permafrost layer, less than a metre below the surface, into which the water may now be frozen.

The variable polar ice caps appear to be composed of solid carbon dioxide with underlying caps of water ice. The northern water cap is exposed in the summer when the carbon dioxide warms and becomes part of the atmosphere, but the southern cap always retains a carbon dioxide covering.

The atmosphere consists of 95% carbon dioxide, $2\frac{1}{2}$% nitrogen and $1\frac{1}{2}$% argon, with

▲ **Mars Exploration Rover** Images taken by the panoramic camera on the rover *Opportunity* were combined to produce this approximately true-colour picture of the crater 'Fram' in Meridiani Planum. The crater is about 8 m (26 feet) in diameter. The images were obtained on 23 April 2004.

smaller quantities of oxygen, carbon monoxide and water vapour. The atmospheric pressure at the surface is less than 1% of the Earth's. Winds of up to 300 km/h (200 mile/h) drive seasonal dust storms that may last for weeks. Water vapour forms into clouds and produces morning mists and ground frosts. The surface temperature on Mars varies between 130 and 290 K.

The crustal rocks of Mars are iron-rich, giving the planet its red colour. The minerals present include olivine, and the presence of haematite, jarosite and goethite, confirmed by the twin **Mars Exploration Rovers**, provides strong evidence for liquid water having had a role in modifying Mars's surface chemistry in the past. Much of the surface may be covered with a hard, ferrous, clay-like deposit. Basalts cover the volcanic plains. From the planet's density, its interior structure probably consists of a mantle overlying a core. However, the extremely weak magnetic field suggests that if there is an iron or nickel–iron core it is not hot and fluid.

Mars has two tiny **satellites** in very close orbits: Phobos and Deimos.

Mars Exploration Rover A NASA mission to land two roving vehicles on Mars. The first, *Spirit*, was launched on 10 June 2003 and landed in the crater Gusev, a possible lake-bed, on 4 January 2004. The second, *Opportunity*, was launched on 7 July 2003 and landed on 25 January 2004 in Meridiani Planum, an area rich in haematite – a mineral that forms under water. Both rovers were equipped with cameras and many mineralogical instruments. They were still operating in early 2005, well beyond their planned lifetimes of 90 days.

Mars Express A European Space Agency probe, with international participation, launched on 2 June 2003. The main craft entered a polar orbit in January 2004, from where it carried out high-resolution imaging to examine Martian geology. The UK's Beagle 2 lander was a failure.

Mars Global Surveyor An American probe to Mars, launched on 7 November 1996. It arrived in September 1997, but problems with aerobraking manoeuvres delayed its achieving the desired polar orbit until 1999. The probe obtained over 300,000 high-resolution images of the surface, some of which showed conclusively that water had once flowed there.

Mars Pathfinder A NASA probe to Mars, part of the **Discovery** programme, launched on 4 December 1996. It landed on 4 July 1997, using a combination of parachute and airbags which were deflated after touchdown. A miniature rover vehicle, Sojourner, analysed rocks in the vicinity of the lander. Lander and rover operated for nearly three months – well beyond expectations. Images of the terrain at the landing site, in Ares Valles, showed that water had once flowed across the surface.

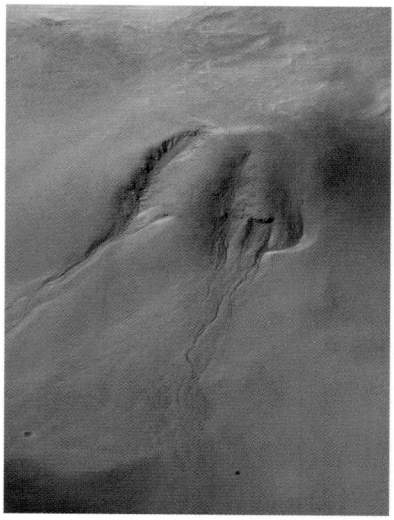

▲ **Mars Global Surveyor** The existence of water on Mars, whether as subsurface liquid or as ice, is vital for the question of whether there was, or is, life on Mars. Several outflow channels in crater walls, including this example in Noachis Terra, were imaged by Mars Global Surveyor, and appear to resemble gullies on Earth.

Mars Reconnaissance Orbiter A NASA craft, due for launch in August 2005, to enter Martian orbit in March 2006. It will carry a high-resolution camera that will be able to detect objects as small as 20 cm (8 inches) on the surface, searching for signs of water and locations for potential future landing sites. The craft will also be equipped with a subsurface sounding radar to search for underground water.

Mars Science Laboratory A long-range, long-duration NASA roving mobile laboratory that will collect and analyse Martian soil and rock samples, looking for organic compounds. Scheduled to be launched in 2009, it will be powered by a radioisotope thermoelectric generator, a more powerful version of the ones that provided electrical energy for the Viking landers in 1976.

mascon A high-density region below the Moon's surface; the word is derived from 'mass concentration'. Mascons cause small, localized increases in the Moon's gravitational field, which is how they were first detected by Lunar Orbiter 5 in 1967. The dozen or so mascons that are known have diameters of 50–200 km (30–125 miles) and lie about 50 km (30 miles) below the surface. They were probably formed when impacts from large meteorites blasted away the lighter crust, and the denser, underlying mantle material bulged upward.

maser A cloud of molecular **interstellar matter** in which **electromagnetic radiation** emitted by high-energy molecules at a particular wavelength in the microwave region stimulates other molecules to emit radiation at the same wavelength. The word is an acronym of 'microwave amplification by stimulated emission of radiation'. A maser therefore amplifies natural microwave emissions, just as a laser does for light.

◀ **Mathilde** Asteroid no. 253 was imaged by the NEAR Shoemaker probe. Its surface was seen to have several craters about 20 km (12 miles). It has a low albedo.

Maskelyne, Nevil (1732–1811) English astronomer. He was appointed the fifth Astronomer Royal in 1765. Building on a method of his for determining longitude at sea by observing the Moon, in 1767 he published the first *Nautical Almanac*, an annual book for the same purpose. In 1774 he measured the mass of the Scottish mountain of Schiehallion by seeing how far it caused a plumb-line to deviate, from which he calculated the average density of the Earth to be 4.7 g/cm^3 (the true value is 5.52 g/cm^3).

massive compact halo object SEE **MACHO**.

mass–luminosity relation A relationship between luminosity and mass for stars on the **main sequence**, derived theoretically by Arthur **Eddington** in 1924. It is expressed approximately as

$$L \propto M^x,$$

where L is the luminosity, M is the mass, and the exponent x varies between 2 and 5 according to the mass. The relationship is not obeyed by white dwarfs, or by giants or supergiants. The relationship enables the mass of a star to be determined if its absolute magnitude is known. The theoretical upper limit (SEE **Eddington limit**) of stellar mass is about 100 solar masses, but very few stars exceed 30 solar masses. The lower limit of a star's mass is 0.08 of a solar mass.

Mathilde Asteroid no. 253, discovered in 1885 by Johann Palisa. It was imaged in 1997 by the NEAR Shoemaker probe en route to Eros, and was found to have a diameter of 53 km (33 miles) and an unusually slow rotation period of 17.4 days. Mathilde is heavily cratered but with a very low density, about 1.3 g/cm^3, which suggests that it is a 'rubble pile' – many fragments from a past impact held together by gravity.

matter era The period, from about 40,000 years after the **big bang** to the present day, in which the energy density of matter in the Universe exceeds that of radiation. SEE ALSO **radiation era**.

Mauna Kea Observatory A major observatory in Hawaii, at an altitude of 4200 m (13,800 ft) – the world's highest and best observatory site for optical, infrared and millimetre-wave astronomy. Its main telescopes include the twin 10-m (396-inch) **Keck Telescopes**; the 3.6-m (142-inch) Canada–France–Hawaii Telescope; the 3.8-m (150-inch) UK Infrared Telescope; the 15-m (49-ft) James Clerk Maxwell Telescope, operated by the UK, Canada and the Netherlands and used for millimetre-wave astronomy; the Japanese 8.3-m (326-inch) **Subaru Telescope**; and one of the twin 8.1-m (319-inch) **Gemini Telescopes**.

Maunder minimum The period 1645–1715, when few sunspots were observed, as identified in the 1890s from historical records by British solar physicist E. Walter Maunder (1851–1928). This was not simply the result of a lack of data from those years: recent studies, especially of tree rings from the period, reveal a genuine lull in solar activity. Also, the Maunder minimum correlates well with the long succession of below-average annual temperatures known as the *Little Ice Age*. The Maunder minimum was caused by a number of much longer-term fluctuations acting together to damp down the 11-year **solar cycle**.

mean anomaly SEE **anomaly**.

mean solar time The timescale based on the motion of the **mean Sun**, which travels at a uniform rate along the celestial equator.

▲ **Mercury** Mariner 10 made three sets of observations of Mercury. This mosaic of images shows that the small planet is heavily cratered, indicating it has undergone no resurfacing for many millennia.

mean Sun A fictitious Sun that moves along the celestial equator at a constant rate equal to the average rate of the true Sun along the ecliptic. It completes its annual course with respect to the vernal equinox in exactly the same time as the true Sun. The true Sun travels at a variable rate along the ecliptic, and cannot therefore give a uniform measure of time.

Mensa A far-southern constellation, introduced by **Lacaille**, representing the Table Mountain at the Cape of Good Hope, where he worked. Mensa has no star above the 5th magnitude, but part of the Large Magellanic Cloud extends into it.

Mercury The innermost planet of the Solar System, and the smallest of the terrestrial planets. Mercury is a challenge to observe because it is never more than 28° from the Sun, and its small angular diameter (8 arc seconds at greatest elongation)

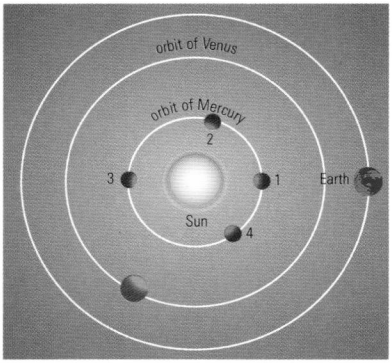

▲ **Mercury** Because Mercury is nearer to the Sun than the Earth, it displays phases. At (1) it is new, at (2) it is half-phase, at (3) full, and at (4) half-phase again.

makes telescopic identification of surface detail (light and dark areas) very difficult. As an inferior planet, Mercury exhibits phases. It reaches magnitude −1.3 at brightest. A **transit** of Mercury across the Sun's disk occurs at intervals of either 7 or 13 years. The planet has no satellite. The main data for Mercury are given in the table.

Mercury rotates three times on its axis for every two orbits around the Sun. As a result, the planet has a 'hot pole' in each hemisphere – areas that spend longer facing the Sun and get heated more. Also, the **synodic period** of 116 days is very close to twice the rotation period. Thus, at greatest elongation west, say, Mercury presents to Earth almost the same face as at the previous greatest elongation west. This led earlier observers into thinking that Mercury had **synchronous rotation**. The true rotation period was established in 1965 by **radar astronomy**.

Very little was known about Mercury's surface until the Mariner 10 probe (the only mission to Mercury so far, until **Messenger** arrives in 2008) made three close approaches to the planet in 1974 and 1975 and returned pictures of nearly half the surface. These showed a heavily cratered, lunar-like world. The largest impact feature, at 1300 km (800 miles) across, is the **Caloris Basin**, similar in size to the Moon's Imbrium Basin; the next-largest is the 625-km (390-mile) diameter crater Beethoven. Like the Moon, Mercury has **ray** craters. Unlike the Moon, it has *lobate scarps*: cliffs up to 3 km (2 miles) high and hundreds of kilometres long, and sometimes cutting across craters. They may be compression faults which formed when the planet's crust cooled and contracted. The *intercrater plains* appear to be volcanic, like the lunar maria, but unlike the maria they contain many small craters. There are also valleys and ridges.

The surface temperature at noon on the equator can reach over 800 K at perihelion, while on the night side it can fall to 90 K. Radar mapping of Mercury's polar regions in 1991 and 1992 revealed what may be metallic deposits on the floors of craters permanently in shadow, where the temperature is estimated never to rise above 112 K. There is a very tenuous atmosphere, mainly of helium and sodium, deriving from the **solar wind**, but also from particles sputtered (knocked off) from the surface by solar radiation, and possibly from material slowly outgassing from the interior.

MERCURY: DATA

Globe	
Diameter	4880 km
Density	5.43 g/cm^3
Mass (Earth = 1)	0.0553
Volume (Earth = 1)	0.0562
Sidereal period of axial rotation	58d 15h 30m
Escape velocity	4.4 km/s
Albedo	0.06
Inclination of equator to orbit	0°
Surface temperature	100–800 K
Surface gravity (Earth = 1)	0.38
Orbit	
Semimajor axis	0.387 AU = 57.91×10^6 km
Eccentricity	0.206
Inclination to ecliptic	7° 00′
Sidereal period of revolution	87.969d
Mean orbital velocity	47.89 km/s
Satellites	0

The presence of a significant magnetic field and the planet's high density for its size suggest that Mercury has a very large iron-rich core, proportionally far bigger than for any other Solar System body.

meridian SEE **celestial meridian**.

MERLIN (Multi-Element Radio-Linked Interferometer Network) An array of six radio telescopes linked to **Jodrell Bank**. Five of the telescopes are 25 m (82 ft) in diameter located at distances up to 127 km (79 miles) from Jodrell Bank; a sixth dish, of 32 m (105 ft) diameter at Cambridge, gives a maximum baseline of 217 km (135 miles).

Messenger A NASA Discovery series spacecraft launched on 3 August 2004 which will fly by Mercury and later orbit the planet in 2008–09. The first craft to orbit the innermost planet, Messenger will operate from a 200 km (125 miles) by 15,190 km (9430 miles), 80° inclination orbit monitoring the properties and environment of the planet.

Messier, Charles Joseph (1730–1817) French astronomer. He kept a record of various fuzzy extended objects that were in fixed positions in the northern sky so that he should not confuse them with comets, of which he was a successful hunter, discovering over twenty. But it is for this record, the catalogue of **Messier objects**, what we now know to be galaxies, clusters and nebulae, that he is remembered today.

Messier object Any of the nebulae, star clusters and galaxies, numbering over 100, which appear in the list begun by Charles **Messier**. Messier drew up the list so that he and other comet-hunters would not confuse these permanent, fuzzy-looking objects with comets. The first edition of the catalogue was published in 1774, with supplements in 1780 and 1781. Not all the objects were discovered by Messier himself; several were found by Pierre Méchain. The numbered objects in the catalogue are given the prefix M, and are still widely known by these *Messier numbers*. Some of the best-known Messier objects are listed in the table.

metals (heavy elements) In astronomy, all elements heavier than helium. Population I stars (SEE **stellar populations**), which are young, have relatively high metal contents (up to 2%), while Population II stars, which formed first, have low metal contents (as little as 0.01%). The metal content of a star affects both its structure and its evolution.

SOME OF THE BEST-KNOWN MESSIER OBJECTS

Messier number	Popular name	Description
1	Crab Nebula	Supernova remnant in Taurus
8	Lagoon Nebula	Emission nebula in Sagittarius
11	Wild Duck	Open cluster in Scutum
13	—	Globular cluster in Hercules
17	Omega, Swan or Horseshoe Nebula	Emission nebula in Sagittarius
20	Trifid Nebula	Emission nebula in Sagittarius
27	Dumbbell Nebula	Planetary nebula in Vulpecula
31	Andromeda Galaxy	Spiral galaxy in Andromeda
33	Pinwheel Galaxy	Spiral galaxy in Triangulum
42	Orion Nebula	Emission nebula in Orion
44	Praesepe or Beehive	Open cluster in Cancer
45	Pleiades	Open cluster in Taurus
51	Whirlpool Galaxy	Spiral galaxy in Canes Venatici
57	Ring Nebula	Planetary nebula in Lyra
64	Black Eye Galaxy	Spiral galaxy in Coma Berenices
97	Owl Nebula	Planetary nebula in Ursa Major
104	Sombrero Galaxy	Spiral galaxy in Virgo

meteor The brief streak of light in the night sky caused by a **meteoroid** entering the Earth's upper atmosphere at high speed from space. Meteors are popularly known as *shooting stars* or *falling stars*. They occur at altitudes of about 100 km (60 miles). The typical meteor lasts for a few tenths of a second to a second or two, depending on the meteoroid's speed of entry into the atmosphere, which can vary from about 11 to 70 km/s (7 to 45 mile/s).

Most meteor-producing meteoroids are low-density (about 0.3 g/cm^3) dust particles that originated in comets. They are rapidly eroded away by **ablation** and leave a trail of ionized and excited atmospheric atoms that is visible as the meteor. Most such meteors have a magnitude in the range 0–4. Larger, denser meteoroids that originated from asteroids produce brighter meteors, the brightest of which may rival the full Moon or, in rare cases, even the Sun (SEE **fireball**, **bolide**). Ionization trails are detectable by radar (SEE **radar astronomy**), enabling meteors to be studied in the daytime.

A few meteors per hour may be seen on any clear, moonless night at any time of year. But at certain times of the year there are **meteor showers**, when meteors are more numerous than usual. Showers occur when the Earth passes through a **meteor stream** – dust particles spread around the orbit of a comet. Most of the meteors appearing during the year are *sporadic meteors*, not associated with cometary orbits.

Meteor Crater (Arizona Meteor Crater, Barringer Crater) A circular crater, about 1200 m (4000 ft) in diameter and 180 m (600 ft) deep, near Flagstaff in the Arizona desert. It is the best-known of the Earth's impact craters, and the first to be identified as such. The impacting body is estimated to have been a nickel–iron meteorite, about 40 m (130 ft) across and weighing a quarter of a million tonnes, which fell around 50,000 years ago. Quantities of meteoritic nickel–iron have been recovered from the plain around the crater, but most of the meteorite was probably vaporized on impact. The crater was discovered in 1891.

meteorite The part of a large **meteoroid** that survives passage through the Earth's atmosphere and reaches the ground. Hundreds of tonnes of meteoroidal matter is swept up by the Earth each day. Most of it burns up in the atmosphere to produce **meteors**, but about 10% reaches the surface as meteorites and **micro-meteorites**.

Meteorites whose fall is observed are known as *falls*. Their atmospheric passage produces a **fireball** or **bolide**, and analysis of the trajectories of some falls reveals them to have been in orbits like those of **near-Earth asteroids**. Over 99% of meteorite falls are unobserved. Meteorites discovered on the ground are known as *finds*. Many recent finds come from Antarctica, recovered from the ice surface, and also from desert and semi-desert areas in, for example, Libya and Australia.

There are three main classes of meteorite: **stony meteorites** are composed mainly of silicates of iron, calcium, aluminium, magnesium and sodium; **iron meteorites** are composed of iron and nickel, and when polished and etched show characteristic **Widmanstätten patterns**; **stony-iron meteorites** have an intermediate composition. Stony meteorites, which make up 92% of all falls, are further subdivided into **chondrites** and **achondrites**. Iron meteorites account for just 7% of falls but over half of all finds, iron being much easier than stony material to pick out on the ground. Just 1% of falls are stony-irons. The exterior of all meteorites is a *fusion crust* of material, melted by friction during passage through the atmosphere and then resolidified.

The largest known single meteorite weighs about 60 tonnes and lies where it fell, at Hoba in Namibia. Stony bodies weighing several tonnes often fragment before impact; the fall at Allende, Mexico, in 1969 deposited 2 tonnes of material over 400 km^2 (150 mile^2). The area over which fragments are scattered is called a *strewnfield*. Meteoroids weighing more than 100 tonnes that do not break up are not decelerated as much as lighter bodies, and produce impact craters such as **Meteor Crater**.

Most meteorites are fragments of asteroids, but a few are of lunar origin, and the **SNC meteorites** are believed to be of Martian origin. Analysis of meteorites can shed light on the geological histories of the parent bodies.

meteoroid A particle or body less than 10 m (30 ft) across in orbit around the Sun. Those following an Earth-crossing orbit have the potential to become a **meteor** or **meteorite**. Meteoroids are of cometary or asteroidal origin. Cometary dust particles spread around their parent comet's orbit can produce **meteor showers**. Asteroidal meteoroids are fragments produced by collisions in the main **asteroid** belt (or small asteroids in their own right), and may survive passage through the Earth's atmosphere to land as **meteorites**.

meteor shower The appearance of **meteors** from a common source around the same time each year. The particles that cause meteors are travelling on parallel paths, but the effect of perspective makes them appear to radiate from the same point in the sky, known as the **radiant**. Nearly all showers are named after the constellation in which their radiant lies. There are a dozen or so major showers and many minor ones. Meteors that belong to a shower are called *shower meteors* to distinguish them from the ever-present random background of **sporadic meteors**. Major showers have typical durations of one to two weeks, with activity showing a gradual rise to maximum followed by a rapid decline. The peak **zenithal hourly rate** for a major shower may be between 10 and 100; some minor showers produce only one or two meteors per hour, even at maximum.

A shower occurs when the Earth passes through a **meteor stream**. In an *annual shower* the meteor rates vary little from one year to the next because the meteoroids in the stream are spread evenly around the orbit. In other, usually younger streams, the meteoroids may be concentrated mainly in one part of the orbit, leading to a *periodic shower* from which strong activity, possibly including a **meteor storm**, is seen only in some years.

MAJOR METEOR SHOWERS

Shower	Dates	ZHR*
Quadrantids	Jan 1–6	100
Lyrids	Apr 19–25	10–15
Eta Aquarids	Apr 24–May 20	50
Delta Aquarids	Jul 15–Aug 20	25
Perseids	Jul 23–Aug 20	80
Orionids	Oct 15–Nov 2	30
Taurids	Oct 20–Nov 30	10
Leonids	Nov 15–20	—
Geminids	Dec 7–15	100

**Approximate zenithal hourly rate*

meteor storm By loose but common definition, a period of activity in a **meteor shower** during which the **zenithal hourly rate** exceeds 1000. Meteor storms have been seen from the **Leonids**, **Giacobinids** and **Andromedids**, and are usually of short duration (less than an hour).

meteor stream A large quantity of **meteoroids** sharing a common orbit around the Sun. With the exception of the **Geminids** (associated with asteroid Phaethon), meteor streams are associated with **comets** and consist of particles shed by comets at perihelion passage over the course of their active lifetimes. A **meteor shower** occurs when the Earth intersects a meteor stream. In a young meteor stream, cometary debris may orbit the Sun in narrow, concentrated strands or *filaments*, and an encounter between the Earth and such a filament can lead to a **meteor storm**. Over time, the debris becomes spread more evenly around the orbit, the orbits of individual meteoroids moving farther away from the primary orbit to produce a torus-shaped stream. The showers to which such streams give rise are low-activity showers lasting a number of weeks. Eventually, meteoroids in a stream become so spread out by gravitational perturbations by the planets and solar radiation effects that they merge into the 'background' concentration of dust that pervades the inner Solar System (SEE **interplanetary matter**).

Metis One of the small inner **satellites** of Jupiter discovered in 1979 during the Voyager missions. It orbits within Jupiter's main ring.

Metonic cycle A 19-year cycle of **lunations**, after which the phases of the Moon begin to repeat themselves on the same days of the year. This is because 19 **tropical years** are equal to 235 **synodic months**, to within 2 hours. The discovery is often attributed to the 5th-century BC astronomer Meton of Athens. The cycle forms the basis of the Greek and Jewish calendars. SEE ALSO **saros**.

microlensing The action of a **gravitational lens** on a small scale. If **MACHOs** exist in the halo of our Galaxy they ought to make themselves apparent by microlensing the light of more distant stars. Such events would consist of temporary brightenings of the distant star lasting a few days. Stars of the Large Magellanic Cloud have been monitored since the mid-1990s, and a few hundred such brightenings have been observed.

micrometeorite A micrometre-sized **meteorite**. Micrometeorites are continually settling on the Earth's surface. The *micrometeoroids* that give rise to them are too small to be burnt up in the atmosphere; they are quickly decelerated and drift gently downward. The quantity that falls greatly exceeds the total of all other meteorites. There are three main types: spheres of high density, irregular compacted particles and fluffy, non-compacted particles of lower density.

▲ **Mimas** This Cassini image of Saturn's tiny satellite Mimas was taken from 213,000 km (132,000 miles) on 16 January 2005. It shows a heavily cratered world, with the giant crater Herschel prominent.

Microscopium A small southern constellation representing a microscope, introduced by **Lacaille**. Its two brightest stars are only of magnitude 4.7.

microwave background radiation SEE **cosmic microwave background**.

midnight Sun The name given to the Sun when it is visible at midnight from inside the Arctic or Antarctic Circle. At the North Pole the Sun is continuously visible for the 6-month period for which it is north of the celestial equator, and is continuously below the horizon for the 6 months it spends south of the celestial equator. A similar situation holds for the South Pole.

Milky Way The faint band of light visible on clear dark nights encircling the sky along the line of the galactic equator. It is the combined light of an enormous number of stars, in places obscured by clouds of interstellar gas and dust such as the **Coalsack**. It is in fact the disk of our **Galaxy** viewed from our vantage point within it. It is brightest in Sagittarius, towards the galactic centre.

millimetre-wave astronomy Observations of **electromagnetic radiation** at wavelengths from about 1 to 10 mm, at the shortest end of the radio spectrum. Millimetre waves are given out by molecules in dense clouds where stars are forming. Waves with a wavelength shorter than 1 mm are known as *submillimetre waves* and bridge the gap between radio and infrared astronomy (SEE **submillimetre-wave astronomy**).

millisecond pulsar A **pulsar** that flashes every few thousandths of a second,

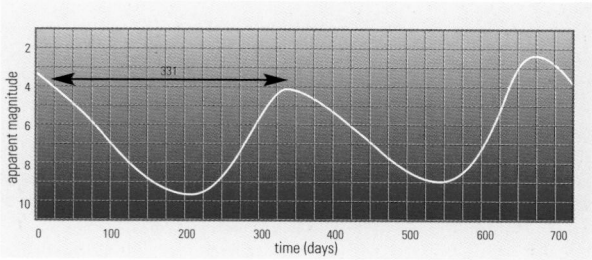

► **Mira** The light curve of Mira (o Ceti) shows a slow fall in brightness followed by a long minimum and then a steep climb up to a short, sharp maximum.

the first of which was discovered in 1982. Unlike normal pulsars, millisecond pulsars are not slowing down, and they have weak magnetic fields. Over 50 are known, about half of them occurring in binary systems.

Mimas The smallest of Saturn's regular **satellites**, with a diameter of 398 km (247 miles), and the innermost, orbiting near the inner edge of the tenuous E Ring. It was discovered by William Herschel in 1789. Mimas is heavily cratered; the largest crater, named Herschel, has a diameter of 130 km (80 miles).

minor planet An alternative term for an **asteroid** which tends to be preferred by professional astronomers.

minute of arc SEE **arc minute, second**.

Mir A Russian space station, the main module of which was launched on 20 February 1986. Mir served as a testbed for long-term human presence in space, and despite failures and accidents in the late 1990s was crewed almost continuously until 1999. With American participation from 1995, it was used to prepare the way for the **International Space Station**. It was destroyed on re-entry in 2001.

Mira The star Omicron Ceti, a red giant that is the prototype of the **long-period variables**, also known as *Mira stars*. It was the first star discovered to vary in a periodic manner. David Fabricius, a Dutch astronomer, first noticed Mira at 3rd magnitude in August 1596. A few months later it vanished, but he saw it again in February 1609.

Mira's mean period is 332 days, but this is subject to irregularities. Its mean range is magnitude 3.5 to 9.1, but maxima as bright as 2.0 and minima as faint as 10.1 have been observed; its light curve is shown in the diagram. Its spectrum varies from M5 to M9. Mira has a faint companion, VZ Ceti, a variable with a magnitude range of 9.5 to 12.0, possibly a white dwarf with an **accretion disk**. Mira lies 196 light years away.

Miranda The smallest and innermost of the five main **satellites** of Uranus, discovered in 1948 by Gerard **Kuiper**. Its diameter is 480 × 466 km (298 × 290 miles) and it has the lowest density of the five, at 1.35 g/cm^3. Miranda displays one of the

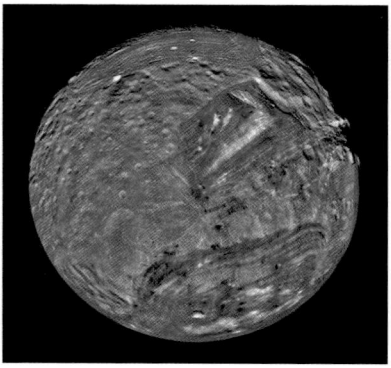

▲ **Miranda** The surface of Uranus's satellite Miranda is unlike anything else in the Solar System. Cliffs and craters can be seen.

most complex and baffling surfaces of any planetary body. A rolling, cratered terrain containing both old and new craters is interrupted by well-defined regions called *coronae* containing parallel or concentric light and dark bands. One of these, Inverness Corona, contains a bright, tick-shaped feature known as the Chevron, and faults running from it lead to cliffs 20 km (12 miles) high. These features are probably the result of extensive ice **volcanism** and geological faulting, or **differentiation** affecting only part of the surface.

mirror An optical component of an instrument (which may be plane, **spheroidal**, **paraboloidal** or **hyperboloidal**) that redirects an incoming beam of light, and may also modify it, so as to aid the formation of an image of the light source. Mirrors are used in many astronomical instruments, in particular the **reflecting telescope**. They are made of glass or quartz, ground and polished to achieve the required accuracy of curvature, and coated with a thin layer of a highly reflective material. Aluminium is the usual coating material as it is cheap and easily deposited on the glass. A variety of sophisticated and exotic techniques, such as honeycomb structures and the use of ceramic materials, are now being used to fashion primary mirrors with diameters of 10 m (33 ft) or more for professional reflecting telescopes.

missing mass Invisible or low-luminosity matter that surrounds individual galaxies and pervades clusters of galaxies. Analysis of the rotation of our Galaxy strongly suggests that the galactic halo is embedded in an enormous *corona* that is roughly 600,000 light years in diameter and contains at least 10^{12} solar masses of low-luminosity matter. Other galaxies have similar rotation curves, so they too are presumed to be surrounded by large, massive coronae. In addition, clusters of galaxies must contain a substantial amount of non-luminous matter – otherwise there would not be enough gravity to hold the clusters together. The matter deduced to exist from the motions of galaxies is called dynamical matter as opposed to the luminous or ordinary matter that we can see directly. It is also known as **dark matter**. Early results from the **Wilkinson microwave anisotropy probe** (WMAP) show that there is sufficient dynamical matter within the Universe to bring its mean density to about 23% of the **critical density**. Luminous matter adds about another 5% to this value.

The issue of the missing mass also arises on the cosmological scale. The measurements made by WMAP place the total density of the universe to within $\pm 2\%$ of the critical density, but luminous and dynamical matter account for only about 28% of this. The remainder may be in the form of **dark energy**.

Mizar A double star in Ursa Major, also known as Zeta Ursae Majoris, of magnitude 2.2. Its wide companion, Alcor, magnitude 4.0, can be distinguished with the naked eye. Mizar and Alcor are too far apart to be a genuine binary. However, with a small telescope a 4th-magnitude companion can be seen closer to Mizar, and this does form a very long-period binary with Mizar. In addition, Mizar, Alcor and Mizar's companion are all **spectroscopic binaries**.

MK system ABBREVIATION FOR **Morgan–Keenan classification**.

MMT Observatory An observatory on Mount Hopkins, near Amado, Arizona. It houses the 6.5-m (255-inch) MMT Telescope, which began operation in 2000. This instrument was originally opened in 1979 as the *Multiple-Mirror Telescope* (MMT), an innovative design in which six 1.8-m (72-inch) mirrors were mounted in a hexagonal array on an altazimuth mounting. The six images were collected at a common focus, giving the same light-gathering power as a 4.5-m (176-inch) telescope. The MMT closed for conversion to a single-mirror instrument in March 1998; the old initials have been retained.

mock Sun SEE **parhelion**.

molecular cloud An interstellar cloud in which the gas is in the form of molecules. Over a hundred different **interstellar molecules** have been identified in these clouds. Their temperature, about 20 K, is enough to make them visible to infrared telescopes (SEE **infrared astronomy**); optically they may show up as **dark nebulae**. They range in size from clouds of 1 solar mass to giant molecular clouds (GMCs) of 10^5 to 10^7 solar masses. GMCs are found in the spiral arms of galaxies, and are sites of star formation. The nearest one is part of the **Orion Nebula** complex.

Monoceros A constellation, representing a unicorn, which adjoins Orion. Its brightest star, Alpha, is only of magnitude 3.8, but it is crossed by the Milky Way and contains some famous clusters and nebulae, including the **Rosette Nebula** and the open cluster M50. Also in Monoceros is the massive binary known as **Plaskett's Star**.

monopole problem A problem with the standard **big bang** model, which predicts that enormous numbers of magnetic monopoles should have been formed – yet none are now observed. **Inflation** solves the problem because the enormous expansion involved has spread out the monopoles until they are now very widely separated.

month The period of revolution of the Moon around the Earth. Various months, defined according to the choice of reference point, are listed in the table. The *calendar months* are the 12 divisions of the Gregorian **calendar**, and are approximately equal in length to one synodic month. SEE ALSO the entries for each type of month listed in the table.

TYPES OF MONTH	
anomalistic month	27.55455 mean solar days
draconic month	27.21222 mean solar days
sidereal month	27.32166 mean solar days
synodic month	29.53059 mean solar days
tropical month	27.32158 mean solar days

▲ **Moon** When it is at its brightest, at full Moon, the differences between the bright lunar highlands and the dark, lava-filled basins become very apparent.

moon (with lower-case 'm') Another word for **satellite**.

Moon (with capital 'M') The Earth's only natural **satellite**, orbiting at a mean distance of only 384,000 km (239,000 miles). Its apparent diameter is about half a degree. With an actual diameter of 3476 km (2160 miles) and a density of 3.34 g/cm^3, the Moon has 0.0123 of the Earth's mass and 0.0203 of its volume. The Earth and Moon revolve around their common centre of gravity, the **barycentre**. Although its proximity makes it the brightest object in the sky apart from the Sun (the full Moon has a visual magnitude of −12.7), its surface rocks are dark, and the Moon's albedo is only 0.07.

As the Moon orbits the Earth, it is seen to go through a sequence of **phases** as the illuminated proportion of the visible hemisphere changes. One complete sequence, from one new Moon, say, to the next, is called a *lunation*. At the crescent phase, the dark side of the Moon is seen to be faintly illuminated by *earthshine* – sunlight reflected from the Earth. At new Moon or full Moon, **eclipses** can occur. An observer on Earth always sees the same side of the Moon (the *nearside*) because it has **synchronous rotation**: its orbital period (the

sidereal month) around the Earth is the same as its axial rotation period. **Libration** brings some of the *farside* into view at certain times.

The Moon has been studied by many space probes. The crewed **Apollo** missions and some of the later Luna probes returned samples of lunar material to the Earth for study. The surface features may be broadly divided into the darker maria (SEE **mare**), which are low-lying volcanic plains, and the brighter highland regions, which are found predominantly on the southern part of the nearside and over the entire farside. There are **impact features** of all sizes. The largest are called *basins*, produced during the early history of the Moon when bombardment by impacting objects was at its heaviest. On the nearside, where the crust is thinner, some basins were subsequently filled with upwelling lava to produce the maria, which in turn became cratered. In craters at the north and south poles, whose interiors are in permanent shadow, water ice exists just below the surface. The smaller basins are similar to the largest craters, which have flat floors and are surrounded by a ring of mountains. Other features are mountain peaks and ranges, valleys, elongated depressions known as **rilles**, **wrinkle ridges**, low hills called *domes* and bright **rays** of ejecta radiating from the sites of recent cratering impacts.

Early lunar cartographers assumed that the darker lunar features were expanses of water, and named them after fanciful oceans (oceanus), seas (mare), bays (sinus), lakes (lacus) and swamps (palus). Other features are named after famous people, principally astronomers and other scientists. The old names are still in use on lunar maps.

The most widely accepted theory for the Moon's origin, known as the *giant impactor* or *big splash* theory, is that a Mars-sized body collided with the newly formed Earth, and debris from the impact accreted (SEE **accretion**) to form the Moon. Subsequent large impacts produced the Moon's basins, and smaller impacts the craters. A very few lunar craters may have had a volcanic origin. Mild *moonquakes* occur at depths of roughly 700 km (450 miles), and **lunar transient phenomena** appear to be associated with them; otherwise, the Moon is now geologically inactive.

Seismic measurements at Apollo landing sites yielded some information on the Moon's internal structure. The lunar crust is about 100 km (60 miles) thick in the highland regions, but only a few tens of kilometres under the mare basins. Under the largest basins the mantle has bulged upward to form **mascons**. A small iron core at the Moon's centre accounts for about 4% of its mass, but there is no overall magnetic field.

Rocks brought back from the Moon have been found to be igneous. The maria have a basaltic composition, while the highlands consist largely of anorthosite. As a result of the constant bombardment, crustal rocks are in the form of **breccias**, perhaps to a depth of several kilometres, and the surface is covered with a **regolith**. The Moon has only the most tenuous of atmospheres. The atmosphere is probably made up of **solar wind** particles retained temporarily, and atoms sputtered (knocked off) from the surface by the solar wind. Consequently the surface temperature variation is extreme: from 100 to 400 K.

Morgan–Keenan classification The categorization of stellar spectra from their absorption features. The system was devised by William W. Morgan, Philip C. Keenan and Edith Kellman of Yerkes Observatory for their *An Atlas of Stellar Spectra*, published in 1943; it is also known as the *MK system*, *MKK system* or *Yerkes system*. It established formal rules for describing a star by spectral type (O, B, A, F, G, K, M, in sequence from hot to cool surface temperature) and luminosity class (I = supergiant, III = giant, V = dwarf; with interpolations Ia, Iab, Ib, II, and IV). SEE ALSO **spectral classification**.

Morning Star Name sometimes given to the planet Venus when it is visible in the east before sunrise.

Mount Graham International Observatory An observatory on Mount Graham, near Safford, Arizona. It has three main instruments: the Vatican Advanced Technology Telescope (also known as the Alice P. Lennon Telescope), a 1.8-m (70-inch) reflector; the 10-m (400-inch) Heinrich Hertz Submillimeter Telescope; and the **Large Binocular Telescope**.

mounting A means of supporting the weight of a telescope and aiming it at different points in the sky. Of necessity, all large professional instruments are on permanent mounts; most amateur telescopes and their mounts are portable. Permanent or portable, rigidity is a prime requirement.

To enable a telescope to be pointed at different parts of the sky, the mounting incorporates two axes, perpendicular to each other, about which the telescope can be rotated. In the **altazimuth** mount the axes are parallel and perpendicular to the horizon. The disadvantage of this mounting is that the telescope has to be moved about the two axes simultaneously in order to keep an object in view. This is overcome in **equatorial telescopes**, in which one axis (the polar axis) is parallel to the Earth's axis of rotation, and so points to the celestial pole. The telescope then only needs to be moved about the other axis, in the direction opposite to that of the Earth's rotation, to keep an object in the field of view.

Since the introduction of computer-controlled drive mechanisms, the altazimuth mount has become the choice for all large professional telescopes. A simple altazimuth platform is incorporated in the **Dobsonian telescope**, a large and portable amateur reflecting telescope.

Mount Palomar The location of the **Palomar Observatory**.

Mount Stromlo Observatory An observatory near Canberra, operated by the Australian National University. Founded in 1924 as a solar observatory, it changed its focus to stellar astronomy after World War II. Many of its facilities, including several historic telescopes, were destroyed in a bush fire in January 2003. It also runs telescopes at **Siding Spring Observatory**.

Mount Wilson Observatory An observatory in the San Gabriel Mountains near Pasadena, California. Its main instrument is the 2.5-m (100-inch) Hooker Telescope, opened in 1917, with which Edwin **Hubble** discovered the expansion of the Universe; it was renovated and reopened in 1993. The Mount Wilson Observatory was founded in 1904 by George Ellery **Hale**, originally for solar astronomy; a 1.5-m (60-inch) reflector was opened in 1908. There are also solar telescopes and interferometers, including the CHARA optical/infrared interferometer.

moving cluster A group of stars, often widely scattered, that share the same motion through the Galaxy. The stars in a moving cluster all follow nearly parallel paths. However, because of a perspective effect, they seem to be moving towards, or away from, a point known as the *convergent point*. This allows the distance of the

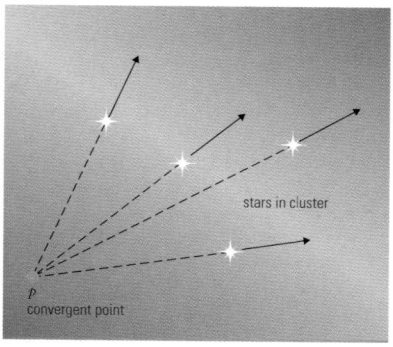

▲ **moving cluster** The velocity of stars in an open cluster can be determined if they are close enough for their parallax to be measurable over a short period. Their parallax is combined with their line-of-sight rate of motion towards or away from Earth to obtain their speed of movement at right angles to the line of sight. Combined with the measurement of the angle of movement, this allows astronomers to judge their distances.

cluster to be determined. First, the position of the convergent point must be established. Some clusters are very scattered, for example the Ursa Major moving cluster, which includes five stars of the Big Dipper (Beta, Gamma, Delta, Epsilon and Zeta). Several other bright stars show the same motion through the Galaxy as these five Ursa Major stars. They include Sirius, Delta Leonis, Beta Aurigae, Beta Eridani and Alpha Coronae Borealis. In total about 100 members of the group are known, scattered widely across the sky. Their separation in space is larger than in a normal cluster. Our Sun is moving through the outskirts of the group, which is why its members appear so widely spread across the sky.

M star A star of spectral type M, characterized by absorption bands of titanium oxide molecules in its spectrum. M-type dwarf stars, also known as red dwarfs, define the lower end of the **main sequence**. Proxima Centauri and Barnard's Star are examples. The surface temperatures of M-type dwarfs range from 3500 to 2500 K; their masses range from 0.08 to 0.5 solar masses, and their radii from 0.1 to 0.6 times that of the Sun. The most massive M-type dwarfs are nearly one-tenth as luminous as the Sun, while those at the bottom of the main sequence have less than one-thousandth of the Sun's luminosity. Most of their radiation is emitted in the infrared, so their visual luminosity is even fainter, from one twenty-fifth to one ten-thousandth that of the Sun. Many M-type dwarfs are **flare stars**; none are visible to the naked eye.

M-type giants and supergiants are all evolved stars with masses much greater than those of M-type dwarfs. Their surface temperatures are typically a few hundred degrees cooler than dwarfs of the same spectral type. The brightest M-type supergiants are up to a million times as luminous as the Sun. The largest of them are several thousand times the Sun's diameter – as large as the orbit of Saturn or Uranus. Even the coolest M-type giants are about 100 times the diameter of the Sun. Such distended stars are unstable and therefore variable; examples are **Betelgeuse** and **Mira**. The high luminosity and low surface gravity of M-type giants and supergiants give rise to a **stellar wind** of gas in which the mass loss can be as high as one solar mass every 100,000 years.

Mullard Radio Astronomy Observatory The radio astronomy observatory of the University of Cambridge, situated at Lords Bridge near Cambridge. It was there that the technique of **aperture synthesis** was developed. The Observatory's principal instrument is the Five-Kilometre Telescope (opened in 1972), consisting of four fixed and four movable dishes 13 m (43 ft) in diameter, arranged in an east–west line. Other radio interferometers on the site are the One-Mile Telescope (opened in 1964) and the Half-Mile Telescope (opened in 1968). The Cambridge Optical Aperture Synthesis Telescope (COAST) was opened in 1995.

Multi-Element Radio-Linked Interferometer Network SEE **MERLIN**.

Multiple-Mirror Telescope SEE **MMT Observatory**.

multiple star A gravitationally connected group of more than two stars. **Castor**, for example, has six components. In multiple systems there is a recognizable hierarchy or order. In Castor, for instance, the two main pairs rotate about a common centre of gravity, and a pair of cool red dwarfs rotates about the same centre but at a larger distance. There is no clear distinction between a multiple star with many members and a small cluster. A good example is the **Trapezium** in the Orion Nebula, which consists of four young stars enmeshed in nebulosity with a further five possible members nearby.

Musca A small but distinctive southern constellation representing a fly, adjoining Crux. Its brightest star, Alpha, is magnitude 2.7. Musca contains the globular clusters NGC 4372 and 4833.

N

nadir The point on the **celestial sphere** vertically below the observer. It is diametrically opposite the zenith.

Nagler eyepiece A telescope eyepiece with a very wide field of view, 80° or more, making it suitable for activities such as searching for comets. (After its designer, Albert Nagler, 1935– .)

Naiad One of the small inner **satellites** of Neptune discovered in 1989 during the Voyager 2 mission.

naked-eye A term applied to observations made by the eye alone, without the aid of any optical instrument, or to celestial objects thus visible.

NASA ABBREVIATION FOR **National Aeronautics and Space Administration**.

Nasmyth focus A focus of a reflecting telescope on an **altazimuth** mount, situated to one side on the altitude axis. The light beam is directed there by an extra flat mirror angled at 45°. Heavy instruments positioned at a Nasmyth focus do not have to be moved bodily as the telescope tracks an object. It is named after Scottish engineer James Nasmyth (1808–90), who incorporated it in his 6-inch (150-mm) reflector. Its use has been revived with the present generation of large, altazimuthally mounted professional instruments. SEE ALSO **coudé**.

National Aeronautics and Space Administration (NASA) The US government agency for the exploration of space, formed in 1958. Its headquarters are in Washington, DC, and it operates various field stations. They include the Goddard Space Flight Center at Greenbelt, Maryland, which handles space science research and Earth satellite tracking; the Jet Propulsion Laboratory in Pasadena, California, which is mission control for NASA space probes and manages NASA's deep-space tracking stations around the world; the Ames Research Center, San Francisco, for aeronautics and planetary science; the John F. Kennedy Space Center at Cape Canaveral, Florida, the main US launch site; and the Lyndon B. Johnson Space Center in Houston, Texas, mission control for crewed space missions.

National Radio Astronomy Observatory (NRAO) The major US radio observatory, with headquarters at Green Bank, West Virginia. Its instruments include a 42.7-m (140-ft) radio dish, and an interferometer consisting of three 26-m (85-ft) dishes plus one of 13.7 m (45 ft). The Green Bank Telescope was dedicated in 2000. With a dish measuring 100 × 110 m (330 × 360 ft), it is the world's largest fully steerable radio telescope.

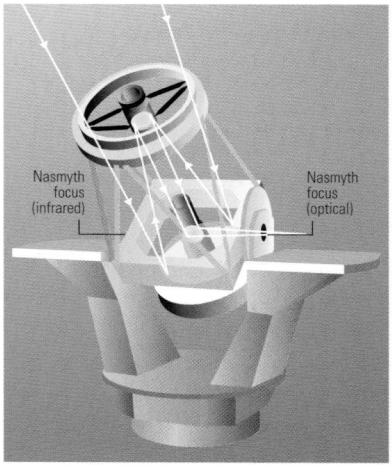

▲ **Nasmyth focus** A number of large, professional telescopes have instruments at their Nasmyth foci. In the William Herschel Telescope there are two foci, one used for infrared and the other for optical observations. The third mirror (the Nasmyth flat) can be moved to direct light to either of them.

NRAO also operates a 12-m (39-ft) dish for millimetre-wave studies at Kitt Peak, Arizona, the **Very Large Array** in New Mexico and the **Very Long Baseline Array**.

nautical twilight SEE **twilight**.

***n*-body problem** SEE **many-body problem**.

neap tide SEE **tides**.

near-Earth asteroid (NEA) A small **Amor asteroid**, **Apollo asteroid** or **Aten asteroid** whose orbit brings it relatively close to the Earth. The term *near-Earth object* is also used, to include close-approaching comets. *Potentially hazardous asteroids* (PHAs) are those whose orbits bring them particularly close to the Earth. Close approaches have been made by objects as small as 9 m (30 ft) across. Objects this size striking the Earth would burn up in the atmosphere; larger objects would penetrate further, causing a powerful low-level airburst as in the **Tunguska Event**, or producing a crater like **Meteor Crater**.

NEAR Shoemaker A NASA probe to the asteroid **Eros**, launched on 17 February 1996. The mission was originally called Near Earth Asteroid Rendezvous. After a flyby of the asteroid **Mathilde**, NEAR Shoemaker became the first probe to orbit an asteroid when it entered orbit around Eros in February 2000. After returning a wealth of images and geological data, the spacecraft made a successful soft landing on the surface of Eros on 12 February 2001, at the end of the mission.

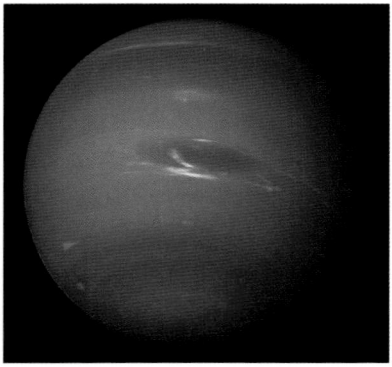

▲ **Neptune** The planet appears bluish because methane in its atmosphere absorbs red light. Ice particles in the upper cloud decks also scatter blue light.

nebula (plural **nebulae**) A region of interstellar gas and dust. The word nebula is Latin for 'cloud'. There are three main types of nebula.

Emission nebulae are bright diffuse nebulae which emit light and other radiation as a result of ionization (removal of electrons) and excitation of the gas atoms by ultraviolet radiation. The source of the ultraviolet is usually one or more hot stars. When gas ions recombine with free electrons, and forbidden transitions (SEE **forbidden lines**) occur in excited atoms, radiation is emitted, giving rise to emission lines in the spectrum. Examples of emission nebulae are **H II regions** such as the Orion Nebula, and **planetary nebulae** such as the Ring Nebula in Lyra. **Supernova remnants** are a form of emission nebula in which the gas is made to glow not by the ultraviolet radiation of the star within, but by frictional heating as it collides with surrounding interstellar gas; an example is the Cygnus Loop. The Crab Nebula is a type of supernova remnant that shines by **synchrotron radiation**.

In contrast, the brightness of **reflection nebulae** results from the scattering by dust particles of light from nearby stars.

Dark nebulae are not luminous: interstellar gas and dust absorb light from background stars, producing apparently dark patches in the sky. This third class includes a group known as **Bok globules**, which are dense absorption nebulae of nearly spherical shape.

Neptune The eighth major planet from the Sun, and the smallest and most distant of the four giant planets; it marks the inner edge of the **Kuiper Belt**. With a mean magnitude of 7.8, Neptune is invisible to the naked eye. Its average apparent size is 2.2 arc seconds. Through a telescope it appears as a small, featureless greenish-blue disk. In its size, mass, atmosphere and colour, Neptune resembles Uranus. The main data for Neptune are given in the table.

Neptune was first identified by Johann **Galle**, assisted by Heinrich D'Arrest, on 23 September 1846, close to a position predicted by Urbain **Le Verrier**. The flyby of the Voyager 2 probe in 1989 provided most of our current knowledge of the planet. The upper atmosphere is about 85% molecular hydrogen and 15% helium. Its predominant blue colour is due to a trace of methane, which strongly absorbs red light. Several different atmospheric features were visible at the time of the Voyager encounter. There were faint bands parallel to the equator, and spots, the most prominent being the Great Dark Spot, about 12,500 km (8000 miles) long and 7500 km (4500 miles) wide. It was about the same size relative to Neptune as the **Great Red Spot** (GRS) is to Jupiter, and like the GRS was a giant anticyclone. White, cirrus-type clouds of methane crystals cast shadows on the main cloud deck some 50 km (30 miles) below. Neptune's atmosphere shows **differential rotation**, some features taking 19 hours or more for one rotation. The atmosphere thus has **retrograde** (1) rotation with respect to the interior, which rotates in just over 16 hours. There are also the highest wind speeds recorded on any planet – over 2000 km/h (1250 mile/h) in places.

Neptune's dynamic and changing atmosphere, which since the Voyager encounter has been followed by the Hubble Space Telescope, probably has to do with its internal heat source – it radiates more than twice as much heat as it receives from the Sun. The amount of white cloud also varies with the **solar cycle**, which is hard to explain for such a distant planet.

Below the visible atmosphere the composition and structure are less certain. A chemical cycle, driven by ultraviolet radiation from the Sun, might operate in which methane is converted into other hydrocarbons and back again. At deeper levels there may be a mantle of water and ammonia about ten times the mass of the Earth, surrounding an iron–silicate core of about one Earth mass. Like Uranus, Neptune has a magnetic field in which the axis is markedly tilted (by 47°) to the axis of rotation and is displaced (by about 10,000 km/6000 miles) from the planet's centre. In Neptune's case it could be that the mantle is at least partially fluid and the ammonia–water mixture is ionized, and therefore electrically conducting, and that this is the source of the magnetic field.

Of Neptune's thirteen known **satellites**, **Triton** and **Nereid** were discovered before the Voyager 2 flyby, and another (**Larissa**) was suspected from an **occultation** measurement. Triton's orbital motion is **retrograde** (1), and Nereid's orbit is the most eccentric of any Solar System satellite. Six inner satellites, including the second-largest, **Proteus**, lie in close, regular orbits. Three small satellites orbit at around 21 million km and two at around 48 million km.

Neptune's ring system was detected from Earth during an **occultation** in 1984,

NEPTUNE: DATA

Globe	
Diameter (equatorial)	49,528 km
Diameter (polar)	48,686 km
Density	1.64 g/cm^3
Mass (Earth = 1)	17.14
Volume (Earth = 1)	57.67
Sidereal period of axial rotation	16h 07m
Escape velocity	23.5 km/s
Albedo	0.84
Inclination of equator to orbit	29° 34′
Temperature at cloud-tops	55 K
Surface gravity (Earth = 1)	0.98
Orbit	
Semimajor axis	30.06 AU = 4497×10^6 km
Eccentricity	0.0097
Inclination to ecliptic	1° 46′
Sidereal period of revolution	164.79y
Mean orbital velocity	5.43 km/s
Satellites	13

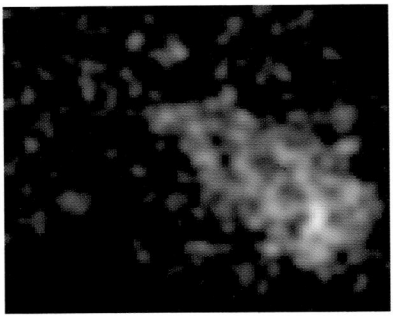

▲ **neutron star** A hot X-ray source within the supernova remnant IC443 is thought to be the neutron star created in the explosion.

but the measurements suggested a number of incomplete 'ring arcs'. Voyager revealed four continuous rings: two narrow ones, and two faint broad ones. The outermost and brightest, the Adams Ring, contains 'clumps' of material that appeared as arcs in the occultation data. SEE ALSO **ring, planetary**.

Nereid Neptune's third-largest **satellite**, discovered by Gerard **Kuiper** in 1949. It has a highly eccentric orbit, taking it from 1.4 million km (850,000 miles) out to 9.7 million km (6 million miles). It is almost certainly a captured body.

neutrino astronomy The detection of neutrinos from astronomical sources. Neutrinos are atomic particles with no charge and very low (possibly zero) mass. They therefore interact hardly at all with matter. To maximize the chance of detecting one, neutrino 'telescopes' consist of huge tanks filled with a suitable liquid. One type uses cleaning fluid, as an incoming neutrino has a chance of reacting with a chlorine atom to form a radioactive atom of argon, which can be detected. Another type contains water, as a neutrino may interact with an electron in a water molecule and emit a particular type of radiation. Neutrinos are emitted by nuclear reactions inside stars. Early attempts to detect neutrinos from the Sun found less than half the number that theorists expected. This *solar neutrino problem* was resolved when it became clear that neutrinos emitted by the Sun oscillated between different types en route to the Earth, and detectors were adapted to record the three types. Neutrinos are also produced by the complex reactions that take place in **supernovae**, and neutrinos from **Supernova 1987A** were detected. Neutrinos are also produced in the Earth's atmosphere by cosmic rays; these can be distinguished by their different energies.

neutron star An extremely small, dense star that consists mostly of neutrons. Neutron stars are observed as **pulsars** and are also thought to be one of the components in **X-ray binaries**. They are formed when a massive star explodes as a **supernova**, ejecting its outer layers and compressing the core so that its component protons and electrons merge into neutrons. Neutron stars have masses comparable to that of the Sun, but diameters of only about 20 km (12 miles) and average densities of about 10^{15} g/cm^3. A neutron star's gravity is so great that objects would weigh 10^{11} times more at its surface than on the surface of the Earth, and it also produces a significant **gravitational redshift** (z about 0.2). The maximum mass for a neutron star is about 3 solar masses, beyond which it would collapse further, into a **black hole**.

Being so small and having such strong gravity, neutron stars can spin rapidly without disintegrating – up to several hundred times per second (SEE **millisecond pulsar**). They retain the magnetic field of the original star but compressed into a much smaller volume, so the magnetic fields of neutron stars are about a million million times stronger than that of the Earth. A neutron star's structure is complicated because of the great variation in pressure from the surface to the centre. The crust is made of iron about 10,000 times denser than terrestrial iron and a million times stronger than steel. Beneath this is a liquid of neutrons, and at the very centre may be a core of exotic nuclear particles.

Newcomb, Simon (1835–1909) Canadian-born American mathematical astronomer. He calculated highly accurate values for astronomical constants, such as the solar **parallax**, and prepared extremely accurate tables of the motions of the Moon and the planets. It was Newcomb who calculated the discrepancy in the advance of **Mercury**'s perihelion that was to provide a proof of **Einstein**'s Special Theory of Relativity.

new Moon The **phase** of the Moon when at conjunction. The dark side of the Moon then faces the Earth, and the illuminated side is invisible to us.

New Technology Telescope A 3.6-m (142-inch) reflector at the **European Southern Observatory** in Chile, completed in 1989. It uses a system of **active optics** to keep the main mirror in shape.

Newton, Isaac (1642–1727) English physicist and mathematician. His achievements were so important for astronomy that the phrase 'Newtonian revolution' is often used. Most of his theories – on light, gravitation and calculus, for example – he developed in basic form while in his early twenties. In 1668 he built the first reflecting telescope in order to overcome the **chromatic aberration** inherent in the lenses used in refracting telescopes. In 1684, encouraged by Edmond **Halley**, he began work on his *Philosophiae naturalis principia mathematica*. Published in 1687, and usually referred to as the *Principia*, this was one of the most epoch-making books in the history of science. In it he put forward his law of **gravitation** and laws of motion (SEE **Newton's laws of motion**), and showed that the orbit of a body moving under an **inverse-square law** of gravitation would be an ellipse. He went on to demonstrate how gravitation explained the second and third of **Kepler's laws** and also many outstanding problems in astronomy. In his *Opticks* (1704) he set out his theories of light, including how white light is made up of the colours of the spectrum.

Newton The ESA X-Ray Multi-Mirror Mission satellite, launched on 10 December 1999. Newton carries X-ray telescopes with a large effective collecting area, enabling long uninterrupted exposures, and an optical monitor, the first to be flown on an X-ray observatory.

Newtonian telescope The simplest form of **reflecting telescope**. The first, built by Isaac **Newton** in 1668, had a primary concave spherical mirror of speculum metal (an alloy of copper and tin) which reflected the light up the tube of the telescope. A flat speculum mirror near the top of the tube, angled at 45°, directed the light to a focus – the *Newtonian focus* – to one side of the tube, where the image of the object being observed was viewed through a small plano-convex eyepiece. This is still the basic design, although speculum has given way to glass, and for all but the smallest apertures the primary mirror is now **paraboloidal**. The Newtonian has long been a favourite with amateur astronomers.

Newton's law of gravitation SEE **gravitation**.

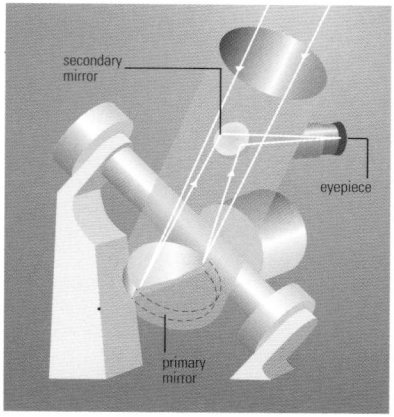

▲ **Newtonian telescope** In this form of reflecting telescope, light is reflected from a paraboloidal primary mirror to a flat secondary mirror placed diagonally within the tube, and directed to a side-mounted eyepiece. The resulting image is inverted.

Newton's laws of motion The fundamental laws of dynamics, set out by Isaac Newton in his *Principia*:

1. *Law of inertia*: Every body continues in its state of rest, or of uniform motion in a straight line, until acted upon by some outside force.

2. *Law of acceleration*: The acceleration of a body is directly proportional to the force acting upon it, and is in the direction in which the force is acting.

3. *Law of action and reaction*: To any action there is an equal and opposite reaction.

NGC Abbreviation for the *New General Catalogue of Nebulae and Clusters of Stars*, published in 1888 by J. L. E. **Dreyer**, and incorporating previous catalogues by William and John **Herschel**. The NGC contains nearly 8000 objects – including galaxies, although their real nature was not then known – and another 5000 are listed in two subsequent ***Index Catalogues*** (IC). Objects are still referred to by their NGC or IC numbers.

noctilucent clouds (NLC) Clouds that form at very high altitude (average 82 km/51 miles) in the Earth's atmosphere, sometimes visible in the summer in twilight hours (noctilucent means 'night-shining') from latitudes between 50° and 70°. NLCs have a pearly or blue colour and resemble delicately interwoven cirrus cloud. They are made up of water vapour that has been carried aloft and then condensed and frozen onto particles, probably of meteoric origin; ions produced by solar radiation and even products of industrial pollution have also been suggested as the condensation nuclei. NLCs have become brighter and more common in recent years.

node The point at which one orbit cuts another. Specifically, a node is one of the two points in the orbit of a planet or comet at which it cuts the ecliptic, or at which the orbit of a satellite cuts that of its primary. The *ascending node* is the node at which a celestial body passes from south to north; the *descending node* is the node at which a celestial body passes from north to south. The *line of nodes* joins the ascending and descending nodes. The *regression of the nodes* is the backward motion of the Moon's nodes around its orbit. They move slowly westward so that the nodal line makes one complete revolution of the orbit in 18.6 years. The regression is caused by the Sun's gravitational attraction.

non-gravitational force Any force acting on a celestial body that is not gravitational, in particular one that produces similar effects to gravity. The term is most commonly used in the context of comets which, on their approach to perihelion, lose volatile material through one or several vents in their crust, producing a jet or jets in a particular direction. This can alter their course in much the same way as the gravitational **perturbations** to which they are prone. The **Poynting–Robertson effect** results from another type of non-gravitational force. SEE ALSO **magnetohydrodynamics**.

Norma A small southern constellation introduced by **Lacaille** between Ara and Lupus, representing a surveyor's level. It contains no star above magnitude 4.0.

North America Nebula A large emission nebula about 3500 light years distant in the constellation Cygnus. It is also known as NGC 7000. An associated region of nebulosity, IC 5070, is the *Pelican Nebula*.

Northern Cross A popular name sometimes given to the cross-shaped arrangement of the brightest stars in the constellation Cygnus.

North Polar Sequence A list of stars within 2° of the north celestial pole with accurately measured magnitudes. It includes stars from magnitude 2 down to magnitude 20. Comparison with the North Polar Sequence enables the magnitudes of other stars to be determined.

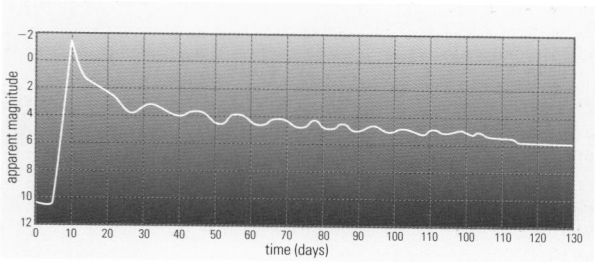

► **nova** The light curve of a nova typically shows a steep rise in brightness by a factor of many thousands followed by a slow, irregular decline. The rapid rise in brightness is caused by explosive hydrogen burning on the surface of the white dwarf.

nova (plural **novae**) A star, usually quite faint, whose brightness suddenly increases by 10 magnitudes or more before slowly returning to its prenova state. A typical nova rises to maximum in a few days, and thereafter declines in brightness by a factor of about ten in 40 days, although cases of both slower and faster declines are known. The diagram shows the light curve of the brightest nova (as opposed to **supernova**) of modern times, Nova Aquilae, in 1918. About 40 novae are estimated to explode in the Galaxy each year, although we see only a small percentage of them.

Novae occur in close binary systems where one component is a white dwarf and the other is a giant. Matter flows from the surface of the giant, where gravity is weak, to the surface of the white dwarf, where gravity is strong. When the infalling material reaches a sufficient density – after a long period of time – it suddenly ignites in a *thermonuclear runaway*, producing a violent explosion on the surface of the white dwarf. The energy of a typical explosion is about the same as the Sun emits in 10,000 years. Clouds of gas are ejected at speeds around 1500 km/s (900 mile/s) during the outburst. The total amount of material lost is estimated to be about one ten-thousandth the mass of the white dwarf. Some stars, known as **recurrent novae**, have been observed to undergo outbursts repeatedly. All novae probably recur over periods of thousands of years. SEE ALSO **dwarf nova**.

nucleosynthesis The production of chemical elements from other chemical elements via naturally occurring nuclear reactions. These reactions are of two types: *fusion*, in which heavier elements are built up from lighter ones, and *fission*, in which heavier elements are broken down into lighter ones. Fusion created most of the Universe's helium from hydrogen in the first few minutes following the **big bang**, and the **carbon–nitrogen cycle** and other fusion processes operate inside stars to create helium and heavier elements which are carried into interstellar space by stellar winds. A few light elements, such as lithium, are produced in the interstellar medium by cosmic rays. Both fusion and fission occur in the extreme conditions of novae and supernovae to produce many different elements which are then ejected into the surrounding space.

nucleus (1) The solid part of a **comet**.

nucleus (2) The central region of a **galaxy**.

nutation The periodic oscillation in the precessional motion of the Earth's axis of rotation (SEE **precession**). It was discovered by James **Bradley** in 1747. It is in effect a slight 'nodding' of the Earth's axis, caused by the combined effect of the gravitational attractions of the Sun and Moon upon the Earth. The effect is continuously varying as the relative positions of the three bodies are continuously changing. *Lunar nutation*, which causes the Earth's axis to trace out an ellipse, has a period of 18.6 years. *Solar nutation* has a period of 0.5 of a tropical year. The *fortnightly nutation* has a period of 15 days.

O

OB association A region of space in which recent star formation has produced high-mass O and B stars. Such associations are less tightly bound than a normal star cluster and are slowly dispersing. Since O and B stars are very luminous, OB associations are recognizable at great distances, even through intervening dust. Tens or hundreds of these massive stars can occur in a single association. Fainter, less massive stars may also be present, but are harder to see at great distances. The Orion Nebula is part of an OB association which is relatively close to us, so young, low-mass stars are also detectable within it. OB associations are a feature of the gas-rich spiral and irregular galaxies, but not of elliptical galaxies, in which no young stars are found. COMPARE **T association**.

Oberon One of the five main **satellites** of Uranus, discovered in 1787 by William Herschel. It is 1520 km (945 miles) in diameter and consists largely of a mixture of rock and water ice, with a density of 1.58 g/cm^3. Oberon has many craters, some with **rays** and some with dark floors, and mountains, one of which appears to be 20 km (12 miles) high. The high density of craters is evidence of a world unaltered by geological activity.

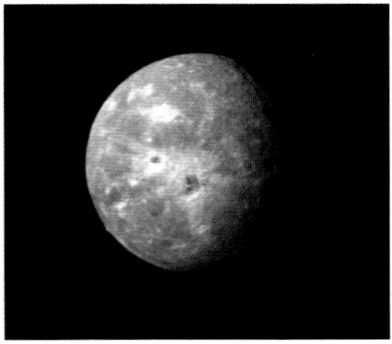

▲ **Oberon** This is the closest image obtained of Uranus's satellite Oberon during the Voyager 2 flyby of January 1986. Oberon's surface shows several impact craters surrounded by bright rays.

objective The main light-gathering element of an optical instrument. In a refracting telescope it is the lens or lens system (called the *object glass*, or OG, in older texts) that collects and focuses light from a celestial body to form an image of it.

objective prism A large, narrow-angle prism mounted across the aperture of a telescope so that the image of each star in the field of view is transformed into a low-resolution spectrum.

oblateness (ellipticity) A measure of how far a **spheroidal** body deviates from a true sphere, given by the difference between the equatorial and polar radii divided by the equatorial radius. The oblateness of a celestial body is an indication of the speed at which it is rotating.

obliquity of the ecliptic The angle between the plane of the ecliptic and the celestial equator. Alternatively, it may be defined as the angle between the axis of rotation of a planet and the pole of its orbit. The Earth's obliquity is currently about 23° 26′, and is decreasing by 0.47′ per year. The angle will start to increase again in about 1500 years; the value ranges between 21° 55′ and 28° 18′ in a period of 40,000 years. The obliquity is responsible for the seasons. It represents the greatest angular distance that the Sun can lie north and south of the equator.

observatory A structure built and equipped for making celestial observations and measurements, acting principally as a housing for a telescope. The best location for astronomical purposes is on a mountain or high plateau where there are many cloudless or near-cloudless nights. The most favourable locations include Hawaii,

Chile, western parts of the USA and the Canary Islands.

Professional optical and infrared telescopes are installed in buildings with a rotating roof, usually dome-shaped, although square buildings are now becoming common, and in some cases the entire building rotates. Domes have a slit, closed by sliding shutters when not in use, through which observations are made. In addition there are control rooms, laboratories, workshops, living accommodation for personnel and other facilities. There are over 200 professional optical observatories throughout the world.

Sites for radio telescopes are less critical than for optical observatories because radio waves are not affected so much by the atmosphere. Radio telescopes consisting of an array of antennas do, however, require a considerable area of land.

Observatories are also launched into Earth orbit, entirely above the absorbing and distorting effects of the Earth's atmosphere. In addition to optical and radio studies, these craft carry out measurements in the infrared, ultraviolet, X-ray and gamma-ray regions of the spectrum, which do not penetrate the Earth's atmosphere and so cannot be studied from ground-based observatories. SEE ALSO the names of individual observatories.

occultation The temporary cutting off of the light from one celestial body as another, nearer one passes in front of it. For example, the Moon may pass in front of a star; a planet may do the same and may also occult its own satellites, as does Jupiter; at certain times Jupiter's **Galilean satellites** may undergo mutual occultations. The rings of **Uranus** were first detected when the planet occulted a star in 1977. Strictly speaking, a solar eclipse is an occultation of the Sun by the Moon. Accurate timings of lunar occultations are valuable for accurately determining the Moon's position and can also provide information about the occulted body. *Grazing occultations*, in which the star appears to skim the Moon's limb, are of particular interest because they reveal the topological profile of the lunar limb.

Octans A faint constellation, notable only because it contains the south celestial pole. The nearest naked-eye star to the pole is Sigma Octantis, at magnitude 5.4 the official south pole star. Octans was introduced by **Lacaille** and represents an octant, the forerunner of the sextant. The brightest star in the constellation is Nu, magnitude 3.7.

ocular An older term for **eyepiece**.

Olbers, Heinrich Wilhelm Matthäus (1758–1840) German doctor and amateur astronomer. He is best known for having posed the question now known as **Olbers' paradox**. He made the second and fourth asteroid discoveries, of Pallas (1802) and Vesta (1807). Olbers invented a method for calculating cometary orbits, and discovered several comets. He also proposed the theory that the tails of comets are highly rarefied matter expelled from the head of the comet by pressure of some kind from the Sun, thus anticipating the concept of **radiation pressure**.

Olbers' paradox The paradox that arises in attempting to answer the question, 'Why is the sky dark at night?' If space is infinite and uniformly filled with galaxies, then an observer would end up looking at the surface of a star or galaxy at every point in the sky, so the entire night sky should be bright. In fact, space is not infinite. The Universe is believed to have been born in the **big bang** about 15,000 million years ago, and so light from the most distant objects has not yet reached us. In addition, the light from distant galaxies is weakened by the **redshift** caused by the expansion of the Universe. The paradox is named after Heinrich **Olbers**, who drew attention to it in 1823.

Olympus Mons An extinct volcano on Mars, and the largest known volcano in the Solar System, rising to 24 km (15 miles) above the planet's mean surface level. It is about 700 km (450 miles) across, including the surrounding area of eroded volcanic rock. The caldera (summit crater) is over 80 km (50 miles) across.

Omega Centauri The brightest globular cluster in the sky, visible to the naked eye as a fuzzy star of magnitude 3.7 in the constellation Centaurus. Designated NGC 5139, it is 17,000 light years away, making it one of the nearest globular clusters. It contains several hundred thousand stars in a volume 200 light years across.

Omega Nebula An emission nebula in the constellation Sagittarius, designated M17 or NGC 6618. Other names for it are the Horseshoe Nebula and the Swan Nebula. It lies 6800 light years away.

Oort, Jan Hendrik (1900–1992) Dutch astronomer. He carried out research on the structure and dynamics of stellar systems, especially the Galaxy, whose rotation he discovered in 1927. He was also a pioneer of radio astronomy, in particular with respect to the 21 cm radiation (SEE **twenty-one centimetre line**) of interstellar hydrogen, collaborating with Hendrik van de **Hulst** and Bart **Bok** to map the Galaxy at this wavelength. In 1950 he proposed the existence of what has come to be called the **Oort Cloud** of comets.

Oort Cloud (Oort–Öpik Cloud) The region of space surrounding the Solar System in which **comets** are thought to reside. It is spherical, extending from about 6000 AU from the Sun to halfway to the nearest star, with the greatest concentration of comets in a torus-shaped region about 40,000 AU from the Sun. The size and structure of the Oort Cloud have been worked out from statistical studies of comets' orbits, although there is no direct evidence for its existence. The idea was put forward by Ernst Öpik (1893–1985) in 1932, and first worked out in detail by Jan **Oort** in 1950. SEE ALSO **Kuiper Belt**.

open cluster A group, often irregularly shaped, of young stars in the spiral arms of our Galaxy, containing from a few tens of stars to a few thousand. Open clusters were formerly also known as *galactic clusters*. They contain young (Population I; SEE **stellar populations**) stars, are found in or near the plane of the Galaxy, and are typically several light years across. Examples are the **Hyades** and **Pleiades** in Taurus,

◀ **Omega Nebula** Also known as the Swan Nebula, M17 in Sagittarius is part of a huge complex of emission nebulosity. It contains about 800 solar masses of material.

and **Praesepe** in Cancer. Open clusters are susceptible to disruption by gravitational interactions between their individual stars, and also by tidal interaction with the Galaxy as a whole or encounters with interstellar clouds. Only the richest survive for more than 10^9 years, while the smallest and least tightly bound last no more than a few million years.

Related to open clusters are *associations*, loose groupings of stars of common origin. Often one or more open clusters are found in the central parts of associations. Because all stars in an open cluster are moving together, it is possible to find the distances of nearby clusters by the **moving cluster** method. SEE ALSO **star cluster**, **OB association**, **T association**.

open universe In cosmology, a mathematical model that describes the behaviour of a universe in which there is not enough matter to halt the expansion triggered in the **big bang**. The universe thus continues to expand for ever. In an open universe the **curvature of space** is negative, the **deceleration parameter** is less than 0.5 and the **density parameter** is less than 1 (i.e. the mean density of the Universe is less than the **critical density**). COMPARE **closed universe**, **flat universe**.

Ophelia One of the small inner **satellites** of Uranus discovered in 1986 during the Voyager 2 mission. It and Cordelia act as **shepherd moons** to the planet's Epsilon Ring.

Ophiuchus A large constellation spanning the celestial equator, commemorating Aesculapius, a great healer. Its brightest star, Alpha (Rasalhague), is magnitude 2.1 and lies near Alpha Herculis. RS Ophiuchi, a **recurrent nova**, flared up to naked-eye brightness in 1898, 1933, 1958, 1967 and 1985. Rho Ophiuchi is a celebrated multiple star, and 70 Ophiuchi is a well-known binary, of magnitudes 4.2 and 6.0, period 88 years. The constellation contains several fairly bright globular clusters – M9, M10, M12, M14, M19, M62 and M107 – and **Barnard's Star**.

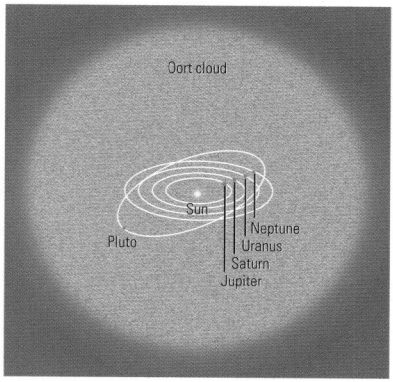

▲ **Oort cloud** The spherical Oort cloud of cometary nuclei surrounds the Solar System. Nuclei from this vast reservoir may be perturbed into orbits that carry them close to the Sun.

opposition The alignment of the Earth and Sun with another body in the Solar System so that the body lies exactly opposite the Sun in the sky, or the time when this alignment occurs. The term is used mostly for alignments with the major planets, although it is also applied to asteroids. Only a **superior planet** can come to opposition (SEE the diagram at **conjunction**). Opposition is the most favourable time to observe a body, for it is then closer to the Earth than at any other point in its orbit.

optical double Two stars that appear to be close together, but are not components of a binary system. They appear close together because they happen to be in almost the same line of sight, but are in fact separated by great distances.

optical interferometer An instrument for studying a celestial object by **interferometry** at optical wavelengths. The individual telescopes are linked either electronically or by laser beams, in conjunction with some form of **photometer**. In the *speckle interferometer*, an image enhancer or a CCD captures a succession of very-short-exposure, high-magnifica-

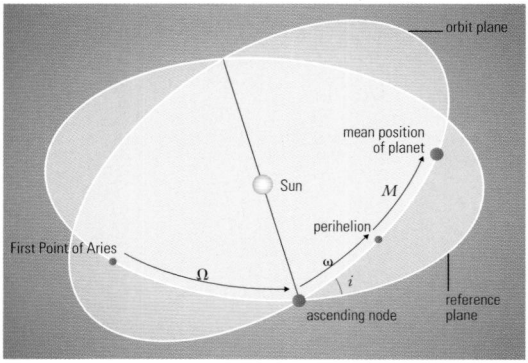

◀ **orbital elements** The orbit of a planet around the Sun can be defined by a number of characteristics, including the inclination (i) relative to the reference plane of the ecliptic, the longitude (Ω) of the ascending node measured in degrees from the First Point of Aries, and the argument of perihelion (ω) measured in degrees from the ascending node.

tion pictures. These individual pictures, which 'freeze' the effects of atmospheric turbulence, are made up of a large number of 'speckles'. Computer image-processing combines these pictures and yields an image from which the effects of the atmosphere are removed electronically. Other types, such as the **Fabry–Pérot interferometer**, exploit interference to capture high-resolution spectra. SEE ALSO **radio interferometer**.

orbit The path of a celestial body in a gravitational field. The path is usually a closed one about the focus of the system to which it belongs, as with those of the planets about the Sun, or the components of a binary system about their common centre of mass (SEE ALSO **ellipse**). Most celestial orbits are elliptical, although the eccentricity can vary greatly. It is rare for an orbit to be parabolic or hyperbolic. To define the size, shape and orientation of the orbit, seven quantities must be determined by observation. These are known as *orbital elements* (SEE the diagram). For a planetary orbit, only six elements are needed. They are:

- a the *semimajor axis*, in AU,
- e the *eccentricity*,
- i the *inclination* of the orbital plane to the plane of the ecliptic,
- Ω the *longitude of the ascending node*,
- ω the *longitude of perihelion*,
- T the *epoch* (the time of perihelion passage).

For the orbit of a binary star system a seventh element is needed, if the mass is not known – the *period* (P), or alternatively the *mean motion* (n).

In general, the same considerations apply when determining the orbit of a satellite as when determining the orbit of a planet, except that the inclination is usually referred to the equatorial plane of the primary planet, not to the plane of the ecliptic. SEE ALSO **parabola, hyperbola**.

orbital element One of the quantities needed to define an **orbit**, or the position of a body in its orbit at a given time.

Orion Perhaps the most splendid of all the constellations, representing a hunter. Its brightest stars are **Betelgeuse** and **Rigel**. Gamma (Bellatrix) is a giant of type B2, magnitude 1.6, 243 light years away. Orion is so distinctive as to be unmistakable. A line of three stars – Zeta (Alnitak), magnitude 1.8; Epsilon (Alnilam), 1.7; and Delta (Mintaka), 2.2 – make up *Orion's Belt*. The celestial equator passes closely north of Mintaka. From the belt hangs *Orion's Sword*, marked by the **Orion Nebula**. Within the nebula lies Theta, a multiple star known as the **Trapezium**. The **Horsehead Nebula** is a dark nebula near Zeta Orionis.

Orion Arm A spiral arm of the **Galaxy**, known also as the *local arm*. The Sun is located on the inner side of a part of this arm known as the *Orion Spur*.

Orionids A **meteor shower** active between 15 October and 2 November, with a broad peak over 20–22 October when the **ZHR** can reach 30. The shower has a diffuse, probably multiple radiant between the stars Betelgeuse and Gamma Geminorum. Like the Eta Aquarids, the Orionids are associated with Halley's Comet.

Orion Nebula A gaseous emission nebula, visible to the naked eye as a diffuse glow in Orion's Sword. It is known as M42, and is separated from a smaller nebula, M43, by a dark nebula called the Fish's Mouth. It lies about 1500 light years away and is 25–30 light years across. It is an **H II region** lit up by a group of young stars called the Trapezium, probably less than one million years old. The Trapezium stars have destroyed most of the volatile dust in the nebula, etching out a roughly spherical hollow that is still growing. The visible Orion Nebula is therefore a cavity in a much larger giant **molecular cloud**. Within the densest portions of the dark cloud, star formation is still under way. The most recent stars to be formed lie concealed within it.

orrery A mechanical model of the Solar System. Orreries varied from simple ones with just the Earth, Moon and Sun to highly complicated representations of the whole Solar System in which planets and their satellites not only revolve in their orbits but also rotate on their axes. The name is from Charles Boyle, fourth Earl of Orrery (1676–1731), who commissioned one from John Rowley in 1712, although this was not the first to be made.

orthoscopic eyepiece A standard telescope lens with good definition and **eye relief**. It is a two-element lens, one element being a cemented triple achromatic (SEE **achromatic lens**).

oscillating universe A variant of the **big bang** theory, in which the Universe passes through successive cycles of expansion and contraction (or collapse). At the end of the collapse phase, with the Universe packed into a small volume of great density, it is possible that a 'bounce' would occur, leading to another phase of expansion. The Universe could thus oscillate

► **Orion Nebula**
A colour composite of infrared images taken with the Very Large Telescope (VLT) at Cerro Paranal shows the central region of the Orion Nebula. The four bright, young Trapezium stars are prominent in the middle of this view.

between big bang and 'big crunch' episodes and so be infinite in age. However, for this to happen the density of the Universe would have to be above a certain value (the **critical density**), which at present is not thought to be the case.

osculating orbit The path that an orbiting body, such as a planet or comet, would follow if it were not subject to any perturbations, but were governed solely by the **inverse-square law** of attraction by the Sun or other central body. The orbital elements, in this case, are called *osculating elements*, and are used in calculating perturbations.

O star A star of spectral type O, characterized by absorption lines of ionized helium in its spectrum. O stars are the hottest and brightest stars on the main sequence, with surface temperatures in the range 30,000 to 50,000 K and luminosities from 30,000 up to 2 million times that of the Sun. They are also the heaviest stars, ranging from about 15 to 130 solar masses or even more. O stars are rare – there are only about 300,000 in the Galaxy at present – because they remain on the main sequence for less than a million years. However, there has been time for 10,000 generations of O stars during the estimated life of the Galaxy. Because they return most of their mass to space via stellar winds and supernova explosions, they have been an important source of heavy elements (SEE **metals**). Together with the hotter B stars, they form **OB associations**.

Most O and B stars form in binary systems, many with components of similar mass, an example being **Plaskett's Star**. When one star explodes as a supernova it may eject its companion from orbit at high velocity. As many as 20% of O stars may be such **runaway stars**. If the binary survives the supernova explosion intact, the result will be a neutron star or black hole orbiting through the stellar wind from its companion, producing an X-ray binary such as **Cygnus X-1**. Examples of O stars include Zeta Puppis, Delta Orionis and Zeta Orionis.

outer planets The planets Jupiter, Saturn, Uranus, Neptune and Pluto, which orbit beyond the main asteroid belt.

Overwhelmingly Large Telescope (OWL) The largest optical telescope yet proposed: a 100-m (330-ft) aperture instrument whose reflecting surface would be formed by 3048 hexagonal mirror segments, each 1.6 m (5.5 ft) across. **Adaptive optics** would give it a resolution of 1 milliarcsecond – enabling it to reveal surface features on nearby stars. OWL, at present a feasibility study by the **European Southern Observatory**, could be completed by 2021.

Owl Nebula A planetary nebula, designated M97, in Ursa Major, of 11th magnitude and apparent diameter 3 arc minutes. It takes its name from two dark patches, which look like eyes, that give it the appearance of an owl's face. It lies 2600 light years away.

► **Owl Nebula** The dark 'eyes' of the Owl Nebula, M97, are well seen in this image taken at the Burrell Schmidt Telescope at Kitt Peak. The Owl is a planetary nebula, produced as an aged star puffed off its outer layers.

P

PA ABBREVIATION FOR **position angle**.

Paaliaq A small outer **satellite** of Saturn, discovered in 2000.

Pallas Asteroid no. 2, the second to be discovered, by Heinrich **Olbers** in 1802. It is the second-largest, with a diameter of 580 × 470 km (360 × 290 miles).

Palomar Observatory An observatory located on Palomar Mountain, near San Diego, California, site of the 5-m (200-inch) Hale Telescope, a reflector opened in 1948, and the 1.2-m (48-inch) Oschin Telescope, a Schmidt camera. The Schmidt was used for the *Palomar Observatory Sky Survey*, an all-sky photographic survey. Palomar Observatory was founded by George Ellery **Hale** and is now run by the California Institute of Technology (Caltech).

Pan A small **satellite** of Saturn, identified in 1991 by Mark Showalter from images obtained during the Voyager missions. It orbits within the **Encke Division** in Saturn's rings.

Pandora A small inner **satellite** of Saturn discovered in 1980 during the Voyager missions. Pandora and Prometheus are the two **shepherd satellites** of Saturn's F Ring.

parabola An open curve. In mathematics the parabola is a type of conic section, so called because it is one of the intersections of a plane with a cone, and is formed when a plane cuts a cone parallel to its slope. The **eccentricity** of a parabola is 1. In astronomy, it is a type of open **orbit**, perhaps most familiar in the context of cometary orbits.

Half of all comets observed have an orbital eccentricity of approximately 1. Since these comets are observable over only a short arc of their orbits near perihelion, it is not possible to determine their orbital elements sufficiently accurately to distinguish between exactly parabolic orbits and extremely elongated elliptical orbits.

paraboloidal Having the form of the surface obtained by rotating a **parabola** about its axis. The paraboloid is the shape most often used for the primary mirror of reflecting telescopes, because a paraboloidal reflector, unlike a spherical one, brings an incoming beam of light to a single focus and is thus free from **spherical aberration**.

parallax (symbol π) The angular distance by which a celestial object appears to be displaced with respect to more distant objects, when viewed from opposite ends of a baseline, used as a measure of the object's distance. If the parallax is determined by direct means, using the principle of triangulation and a known baseline, it is known as **trigonometric parallax**. If the parallax is deduced by examining a star's spectrum, it is known as **spectroscopic parallax**.

In trigonometric parallax, the choice of baseline is made according to the remoteness of the object. For fairly close objects, such as members of the Solar System, the Earth's equatorial radius is used; this gives a *geocentric* (or *diurnal*) *parallax*. For more remote objects the radius of the Earth's orbit – the **astronomical unit** (AU) – is used; this gives the *annual* (or *heliocentric*) *parallax*. The parallax of an object is therefore the angular size of the radius of the Earth or of its orbit as it would appear from that object.

The *solar parallax* is the Sun's geocentric parallax, defined as the angular size of the Earth's equatorial radius from a distance of 1 AU. Its accurate determination has been historically very important in establishing the scale of the Solar System.

parhelion (mock Sun, sundog) A round patch of light at the same altitude as the Sun. Parhelia are images of the Sun produced when sunlight is refracted by ice crystals, most frequently in cirrus cloud. They often occur in pairs, one either side of the Sun, and may show elongated, spectrally coloured tails.

Parkes Observatory A radio astronomy observatory in New South Wales, part of the **Australia Telescope National Facility**. Its main instrument is a 64-m (210-ft) dish aerial, completed in 1961, which is now a component of the **Australia Telescope**.

parsec (symbol pc) The distance at which a star would have a **parallax** of 1 arc second, equivalent to 3.2616 light years, 206,265 astronomical units or 30.857×10^{12} km.

Parsons, William SEE **Rosse, Third Earl of**.

partial eclipse SEE **lunar eclipse**, **solar eclipse**.

Pasiphae The largest (60 km/40 miles in diameter) of Jupiter's outer **satellites**, discovered in 1908 by P. J. Melotte. It is in a **retrograde** (1) orbit and may be a captured asteroid.

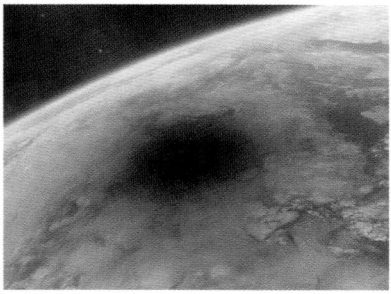

▲ **penumbra** (1) Mir cosmonauts took this photograph of the shadow cast by the Moon onto the Earth during the solar eclipse of 11 August 1999. The dark central umbra, under which the eclipse was total, is surrounded by a lighter outer penumbra.

patrol A systematic survey aimed at recording certain transient astronomical phenomena. In a meteor patrol, for example, networks of cameras are used to photograph the sky to maximize the chance of recording meteors. In a supernova patrol, images of selected galaxies are obtained on as many nights as possible in the hope of discovering a supernova.

Pavo A southern constellation, representing a peacock; its brightest star, Alpha, magnitude 1.9, is called Peacock. Kappa is a **Cepheid variable** with a magnitude range from 3.9 to 4.8 in a period of 9.1 days.

Pegasus A large constellation of the northern sky, representing the flying horse of Greek mythology. Its most prominent feature is the *Square of Pegasus*, made up of four stars, although one of these stars is actually part of Andromeda. Beta (Scheat) is a red giant irregular variable with a small magnitude range (2.3 to 2.7). Pegasus is not a rich constellation but it does contain a major globular cluster, M15, of magnitude 6.

Pelican Nebula SEE **North America Nebula**.

penumbra (1) The outer part of the shadow cast by a celestial body illuminated by an extended source such as the Sun. An observer in the penumbral region of a shadow sees only part of the source obscured, as, for example, in a partial **solar eclipse**. SEE ALSO **lunar eclipse**. COMPARE **umbra** (1).

penumbra (2) The lighter zone surrounding the dark central part of a sunspot. COMPARE **umbra** (2).

penumbral eclipse SEE **lunar eclipse**.

Penzias, Arno Allan (1933–) German-American physicist. In 1964 he was working with Robert **Wilson**, using a small radio telescope to measure noise that might interfere with satellite communications.

They detected a strange signal that, after consultation with Robert Dicke and James Peebles, they realized was the **cosmic microwave background** – radiation left over from the big bang, as had been predicted by George **Gamow** and others in the 1940s. Penzias and Wilson shared the 1978 Nobel Prize for Physics for what was the most important discovery in modern cosmology.

peri- A prefix referring to the point in an object's orbit at which it comes closest to its primary, as in *perigee* and *perihelion*.

periastron The point of nearest approach of the components of a binary star in their orbit.

perigee The point of nearest approach to the Earth by the Moon or an artificial satellite.

perihelion The point in the orbit of a planet or comet at which it is nearest to the Sun.

periodic comet A common term for a short-period **comet** – one whose period is less than 200 years.

period–luminosity law A relationship between the period of light variation and the mean absolute magnitude (luminosity) of a **Cepheid variable** star: the absolute magnitude increases as the period increases. Once the period of a Cepheid is known, its absolute magnitude can be deduced from the period–luminosity law. Furthermore, the absolute magnitude taken in conjunction with the apparent magnitude then enables the distance of the Cepheid to be calculated. The law was discovered by Henrietta **Leavitt** in 1912.

Perseids A **meteor shower** active between 23 July and 20 August peaking around 12 August, at which time its radiant lies in the north of Perseus. The Perseids are one of the strongest, best known and most long-lived of all the meteor showers, showing a peak **ZHR** in excess of 80. The shower is associated with Comet **Swift–Tuttle**, and showed enhanced activity near maximum in the early 1990s around the time of the comet's perihelion passage. In the late Middle Ages they were known as the *tears of St Lawrence*, who was martyred in AD 258, on 10 August, close to the shower's maximum.

Perseus A prominent northern constellation, representing the Greek hero who rescued Andromeda. Alpha Persei (Algenib or Mirfak) is magnitude 1.8. Beta (**Algol**) is the prototype eclipsing binary and the brightest in the sky; the second-brightest is Gamma, whose first eclipse was observed only in 1990 (components are magnitude 3.2 and 4.4, period 15 years). Rho is a red giant semiregular variable with a magnitude range from 3.3 to 4.0 and a very rough period of 50 days. Perseus is a rich constellation, crossed by the Milky Way; there are several open clusters, including the famous **Double Cluster** and M34.

personal equation A correction factor applied to visual observations made by a particular person. An individual observer may record timings of, for example, transits which are consistently early or late. Such systematic errors cannot be measured directly, but in any programme of work carried out by several observers a personal equation can be found by performing a statistical analysis of all the results. It can then be applied to improve the accuracy of future observations.

perturbation An irregularity in the orbital motion of a celestial body, brought about by the gravitational attraction of other bodies. *Periodic perturbations* are small oscillations with periods similar to the orbital period of the perturbed body. *Secular perturbations* are slow, continuous changes in orbital motion, for example the slow rotation of the orbit of a planet (SEE **perihelion**), and in the orbit's eccentricity and inclination, that arise from commensurabilities (SEE **commensurable**). It is such perturbations that produce

precession. Perturbations are responsible for dramatic changes in the orbits of **comets**, converting long-period orbits into short-period or hyperbolic orbits. SEE ALSO **non-gravitational force**.

Phaethon Asteroid no. 3200, discovered in 1983 by Simon Green and John Davies, using the **Infrared Astronomical Satellite**. Phaethon is a 5-km (2.5-mile) diameter **Apollo asteroid** with a perihelion within the orbit of Mercury. Its orbit coincides with the **Geminid** meteor shower.

phase The proportion of the illuminated hemisphere of a body in the Solar System (in particular the Moon or an inferior planet) as seen from the Earth. The phase of a body changes as it, the Sun and the Earth change their relative positions. All the phases of the Moon (SEE the diagram) – new, crescent, half, gibbous and full – are observable with the naked eye. The inferior planets, Mercury and Venus, show phases from a slender crescent to a fully illuminated disk when observed through a telescope. Of the superior planets, Mars can show quite a marked gibbous phase, and Jupiter a very slight gibbous phase, but with Saturn and the planets beyond no phase is discernible. Phase is sometimes expressed as the percentage of the visible disk that is illuminated.

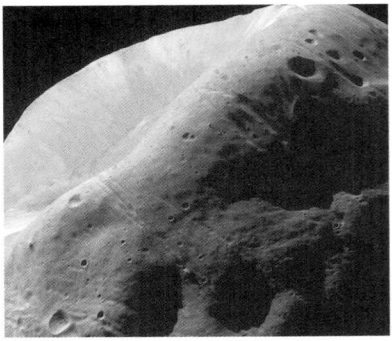

▲ **Phobos** Mars's larger moon, Phobos, is seen in this image obtained by the Mars Global Surveyor spacecraft in 1998. Believed to be a captured asteroid, Phobos is heavily cratered and shows linear structures resulting from impact shockwaves.

Phobos The larger of the two **satellites** of Mars, discovered in 1877 by Asaph **Hall**. It is a dark, irregular body measuring 26 × 21 × 18 km (16 × 13 × 11 miles), and may well be a captured asteroid. It has two large craters, named Stickney and Hall.

Phoebe The largest of Saturn's outer **satellites**, discovered in 1898 by William Pickering. Phoebe is quite small (diameter 230 × 210 km/145 × 130 miles), moves in a distant, inclined and **retrograde** (1) orbit, and has a non-**synchronous rota-**

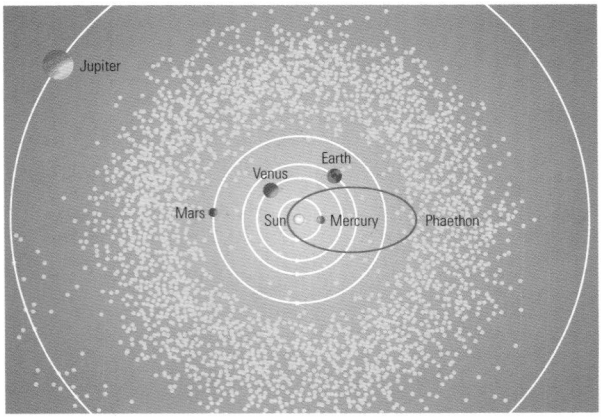

◀ **Phaethon** Phaethon's orbit (red) carries it from the inner part of the main asteroid belt at aphelion, to a perihelion closer to the Sun than Mercury. As an Apollo asteroid, Phaethon has an orbit that cuts across that of Earth.

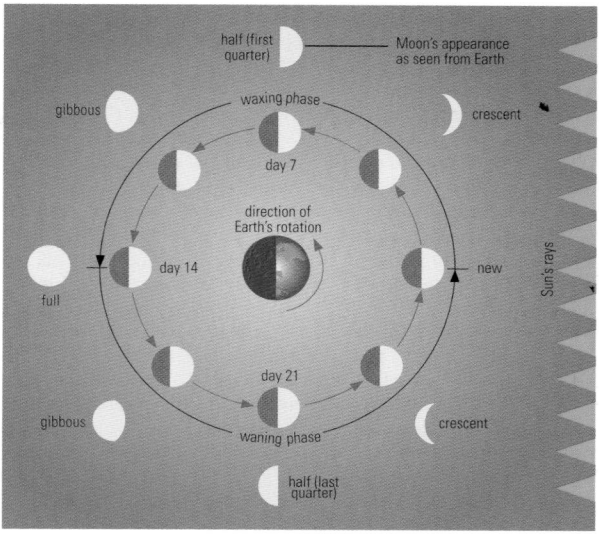

► **phase** The changing appearance of the Moon's illuminated disk results from its orbital motion around the Earth. At new (right), the Moon lies in line with the Sun, and the hemisphere presented to Earth is completely dark. At full, the Moon lies opposite the Sun in Earth's sky and is completely illuminated. Between these two extremes, varying degrees of illumination are seen as shown.

tion period of 9.4 hours, all of which suggests that it is a captured body, perhaps a comet nucleus or a **Centaur**. The **Cassini** spacecraft made a close flyby in 2004, imaging a heavily cratered body covered to a depth of several hundred metres with dark material.

Phoenix A southern constellation representing a phoenix. Alpha (Ankaa), magnitude 2.4, is the brightest star. Zeta is an eclipsing binary with a range from magnitude 3.9 to 4.4 in a period of 1.67 days.

photoelectric magnitude The **apparent magnitude** of an object measured by a **photoelectric photometer** attached to a telescope. The measurement of brightness is made through a number of coloured filters in turn. SEE ALSO **UBV system**.

photoelectric photometer An instrument that measures the brightness of a celestial object by generating an electrical signal that is directly proportional to the intensity of the light falling on it. There are various types of detector. A *photomultiplier tube* is a device in which electrons that are given off when light falls on a special screen liberate more electrons in a multi-stage cascade, so producing a measurable electric current. A **CCD** may be used instead (strictly speaking, it is then not a photoelectric device, but it performs the same function). The brightness thus measured

▲ **Phoebe** Two images taken by the Cassini spacecraft during its close flyby of Phoebe on 11 June 2004, were combined to produce the picture above. The bright craters are thought to be relatively young features.

is an object's **photoelectric magnitude**; an accuracy of 0.01 magnitude is possible. Photoelectric photometers are therefore used for accurate determination of stellar magnitudes, especially in the study of variable stars, and also for measuring the brightness variations of asteroids and planetary satellites, and thus their rotation periods. The light falling on the detector can be confined to a particular part of the visible spectrum, or infrared or ultraviolet wavelengths can be selected.

photographic magnitude (symbol m_{pg}) The **apparent magnitude** of an object as measured on a traditional photographic emulsion which is more sensitive to blue light than the human eye is. Hence the photographic magnitude differs significantly from the visual magnitude.

photographic zenith tube (PZT) A fixed telescope directed at the local zenith and used to image stars as they pass across its field of view. PZTs can be used to determine highly accurate values of the *clock error* – a very small non-uniformity in the Earth's rotation – and of the observatory's location.

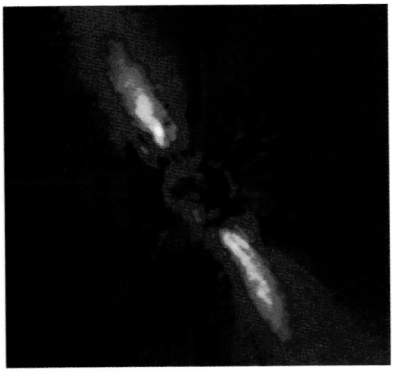

▲ **Pictor** This Very Large Telescope (VLT) false-colour image of the protoplanetary disk around Beta Pictoris, in Pictor, shows that it is not even in composition. The visible kink in the disk is about 20 AU from the star, which corresponds roughly to the distance of Uranus from the Sun.

photometer SEE **photoelectric photometer**.

photosphere ('sphere of light') The visible surface of the Sun. It is a layer of highly luminous gas about 500 km (300 miles) thick and with a temperature of about 6000 K, falling to 4000 K at its upper level, where it meets the **chromosphere**. Because this is the region of the Sun from which all of its output of visible light is emitted, the photosphere is the source of the Sun's visible spectrum. The lower, hotter gases produce the continuous emission spectrum, while the higher, cooler gases absorb certain wavelengths and so give rise to the dark **Fraunhofer lines**. This absorption contributes to **limb darkening**. Sunspots and other visible features of the Sun are situated in the photosphere.

photovisual magnitude (symbol m_{pv}) The **apparent magnitude** of an object measured photographically using emulsions and filters that mimic the spectral response of the human eye.

Piazzi, Giuseppe (1746–1826) Italian astronomer and monk. He discovered the first asteroid, Ceres, on 1 January 1801, during a lengthy programme of taking accurate measurements of star positions (published as a catalogue in two parts, in 1803 and 1814). Piazzi proposed the name 'planetoid', but 'asteroid' has prevailed.

Pickering, Edward Charles (1846–1919) American astrophysicist who was director of the Harvard College Observatory. His accurate stellar magnitudes, determined by photometry, were published in the *Revised Harvard Photometry*. He discovered spectroscopic binary systems, and invented a method of obtaining the spectra of several stars on one photographic plate. The spectroscopic work that Pickering instituted culminated in the *Harvard system* of **spectral classification** and the ***Henry Draper Catalogue*** of stellar spectra. This work was largely carried out by a team of women, originally hired as assistants but who grew to be able astronomers in their own right. They included Annie Jump **Cannon**, Williamina Fleming and Antonia C. Maury. Edward's younger brother, **William Henry Pickering** (1858–1938), instituted the search for **Pluto**; he also discovered Saturn's satellite Phoebe, in 1898.

Pictor An unremarkable constellation in the southern sky representing a painter's easel and introduced by **Lacaille**. Its leading star is Alpha, magnitude 3.3, and the only other star above fourth magnitude is *Beta Pictoris*, magnitude 3.8, which has a disk of gas and dust around it and is believed to be a planetary system in the process of formation.

Pinwheel Galaxy A large, face-on spiral galaxy in Ursa Major, also known as M101 and NGC 5457. It lies about 25 million light years away. The name is also sometimes used for the galaxies M33 in **Triangulum** and M99 in Coma Berenices, both also face-on spirals.

Pioneer A series of US space probes launched between 1958 and 1978. The first four were failed Moon probes. Pioneers 5 to 9 were put into orbit around the Sun to study the solar wind and solar flare activity. Pioneer 10, launched on 3 March 1972, was the first probe to travel through the asteroid belt; it passed Jupiter in December 1973, sending back photographs. Pioneer 11 followed it, flying past Jupiter in December 1974 and moving on to Saturn, which it reached in September 1979. Both probes are now on their way out of the Solar System; contact was maintained until 1997. The last two in the series were the **Pioneer Venus** probes.

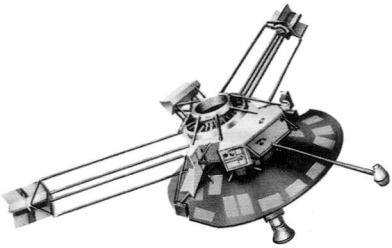

▲ **Pioneer** Pioneer 10, shown here in an artist's impression, was launched on 2 March 1972. It passed Jupiter on 3 December 1973, at a distance of 132,000 km (82,500 miles).

▲ **Pinwheel Galaxy** This face-on spiral galaxy, M33 in Triangulum, has loosely wound spiral arms surrounding a relatively small nucleus. M33 belongs to the Local Group, which also includes the Milky Way and the Andromeda Galaxy.

Pioneer Venus Two NASA space probes to the planet Venus, launched on 20 May and 8 August 1978. Pioneer Venus 1 went into orbit around the planet in December 1978, making a radar map of the surface and photographing the planet's clouds. Pioneer Venus 2 dropped five small probes into the atmosphere of Venus in December 1978, which sent back data on the planet's clouds and atmospheric conditions.

Pisces A constellation of the zodiac, representing a pair of fishes. It is very obscure, consisting of a chain of rather faint stars south of Pegasus. Its brightest star, Eta, is magnitude 3.6. Alpha (Alrescha) is a close binary, magnitudes 4.2 and 5.2, with a period of over 900 years.

Piscis Austrinus The constellation of the southern fish. First-magnitude **Fomalhaut** is its only star brighter than fourth magnitude.

plage A bright cloud of hot gas in the Sun's **chromosphere**. Plages are intimately connected with **faculae**, which occur below them in the photosphere, and are associated with active regions, where the local magnetic field is enhanced.

Planck A European Space Agency satellite, to be launched in 2007, to image the cosmic microwave background radiation with unprecedented sensitivity and angular resolution.

Planck era The first 10^{-43} seconds after the start of the **big bang**, during which time quantum fluctuations probably dominated the Universe and almost nothing can be understood about it with present-day physics. It is also called the *Planck time*. The size and density of the Universe at that instant (10^{-35} m and 10^{97} kg/m^3, respectively) are called the *Planck scale* and *Planck density*.

planet A large non-stellar body in orbit around a star, shining only by reflecting the star's light. In our **Solar System** there are nine **major planets**, as opposed to the thousands of small bodies known as **asteroids** or minor planets. SEE ALSO **extrasolar planet**, **Kuiper Belt**, **Planet X**.

planetarium A structure in which a representation of the stars and planets as visible in the night sky is projected onto the inside of a dome for an audience seated below. They vary in size and sophistication from small portable projectors used in inflatable 3-m (10-ft) domes to highly complex installations in permanent theatres seating hundreds of people. In all planetaria the positions and motions of celestial bodies can be simulated. The first planetarium, built by the Carl Zeiss company, began operation in Munich in 1923. From then until the early 1980s, planetarium projectors consisted of many individual optical projectors mounted on a complicated rotating frame. In the new generation of projectors, a computer-generated image is projected through a single, stationary fish-eye lens.

planetary nebula An emission nebula formed when a red giant or supergiant sheds its outer layers, leaving a hot core. The core's stellar wind compresses the gas, and its ultraviolet radiation ionizes it and makes it shine. Planetary nebulae range in size from about that of our Solar System to a light year or so across. They were so named because the first ones to be discovered gave the impression of planetary disks when viewed in small telescopes; the Ring Nebula (M57) in Lyra is a well-known example. However, only 10% of the 1000-plus now known have a circular shape, and about 70% have two lobes.

The central stars of planetary nebulae are hot and blue, with temperatures of 30,000 to over 100,000 K. They form a transitional stage at the end of a star's evolution between the giant stars and the white dwarfs. Ultraviolet light from the central star make planetary nebulae fluoresce, as in other emission nebulae. In addition to hydrogen and helium, planetary nebulae contain small proportions of oxygen, nitrogen and neon, which accounts for their various colours seen on images.

Planetary nebulae are thought to result from stellar winds blown out from red giants and supergiants, leaving the red giant's core as the central star (SEE **RV Tauri star**). Over time, this star cools to become a white dwarf. Because red giants are so big, the gravity at their surfaces is very low and flares or mass ejections on the surface can easily throw off material. A star below about 8 solar masses can lose most of its mass into space before it turns into a white dwarf of about 1.4 solar masses. The visible nebula is simply the central region of a much larger object, the outer parts of which are not lit up by the central star. All planetary nebulae are expanding, at speeds of around 20 km/s (12 mile/s).

The curious shapes of planetary nebulae are probably linked to the magnetic field and rotation of the stars. The magnetic field lines might control the outflow, channelling it in particular directions, and the rotation of the star could add spiral twists. The double-lobed structure of most planetary nebulae must be connected with

▲ **planetary nebula** This Hubble Space Telescope image shows the Ant Nebula (Mz 3), a planetary nebula in the southern constellation Norma. The twin-lobed structure has been produced by ejection of gas from the aged star at the nebula's centre. The Ant Nebula lies about 3000 light years away.

the star's direction of spin or magnetic axis (SEE **bipolar flow**). Some planetary nebulae, such as the Eskimo and Saturn nebulae, have a bright central nebula surrounded by fainter extensions, as though the central star has ejected a succession of shells. Because they are expanding as the central stars fade, planetary nebulae are relatively transitory objects with lifetimes of tens of thousands of years. Several new planetary nebulae form each year in our Galaxy.

planetary precession The smaller of the two components of **precession**. It results from the gravitational effects of other planets on the Earth and moves the equinoxes eastward by about 0.12 arc seconds per year.

planetesimal In theories of the origin of the Solar System (and other planetary systems), a body, between a few millimetres and a few kilometres in size, that condensed out of a cloud of gas and dust (the solar nebula), away from the centre where the Sun was forming. Once they had reached a few kilometres in size, the gravitational attraction of planetesimals would allow them to combine with one another, by a process of **accretion**, to form **protoplanets**.

Planet X The name given to a hypothetical tenth major planet orbiting the Sun beyond **Pluto**, the X representing both 'ten' and 'unknown'. It was originally the name given by Percival **Lowell** to the planet whose gravitational **perturbations** he supposed were causing irregularities in the orbits of Uranus and Neptune. Pluto's mass was calculated accurately in 1978 and found to be far too small to account for these perturbations, which are now almost certainly explained away by early errors in observing their positions. Several long-term searches, including one by Clyde **Tombaugh** that continued for 13 years after he discovered Pluto, have turned up no sizeable body. Some of the **Kuiper Belt** objects, such as **Varuna**, are quite large but nowhere near the size of a major planet. There is no longer any reason to invoke a Planet X, and it is highly unlikely that one exists.

planisphere A simple device for showing the stars on view from a certain latitude at any time and on any date. It consists of two disks held together by a pivot at their centres. On the lower disk is printed a map of the night sky, with the celestial pole at the pivot, and into the upper disk is cut an oval window. The planisphere is set by rotating one disk with respect to the other so that the date, marked on the rim of one, is aligned with the time, marked on the rim of the other. The part of the map visible through the window then shows the stars on view in the sky. There is some distortion of constellation shapes, particularly near the rim.

Plaskett's Star A massive binary star in the constellation Monoceros. In 1922 the Canadian astronomer John Plaskett (1865–1941) showed it to be a spectroscopic binary with a period of 14 days. Each component is a blue supergiant of about 50 solar masses. It is the second most massive binary known, exceeded only by the recently discovered WR 20a in Carina, where the two stars are each of about 80 solar masses.

Platonic year The time taken for the celestial pole to describe a circle around the pole of the ecliptic, as a result of **precession**. Its value is about 25,800 years. (After the Greek philosopher Plato, *c*.427–*c*.347 BC.)

Pleiades A young open cluster in the constellation Taurus, popularly called the *Seven Sisters*. Although only six or seven stars are visible to the naked eye, there are in fact over a thousand embedded in a reflection nebula. The brightest member is Alcyone, a B-type giant over 300 times as luminous as the Sun. Another member of the cluster, Pleione, is a slightly variable **shell star** that throws off shells of gas as a result of its fast rotation. The other named Pleiades are Maia, Atlas, Electra, Merope and Taygeta. The cluster lies 410 light years away.

plerion SEE **supernova remnant**.

Plössl eyepiece A telescope eyepiece with good **eye relief** and a wider field of view (about 40°) than an orthoscopic lens. The commonest form has two identical achromatic doublets (SEE **achromatic lens**), but some quite different designs have been called 'Plössls'. (After Austrian optician Simon Plössl, 1794–1868.)

Plough Popular name for the shape formed by the seven main stars of the constellation **Ursa Major**. In the USA, the stars are known as the *Big Dipper*.

◀ **Pleiades** The hot, young blue stars of the Pleiades illuminate a shroud of interstellar dust. Marking the shoulder of Taurus, the Bull, the Pleiades is a distinctive naked-eye object, prominent on winter evenings in the northern hemisphere.

► **Pluto** Sufficiently similar in size to be regarded as a 'double planet', Pluto (left) and its satellite Charon (right) are locked in synchronous rotation about the barycentre between them.

Pluto The smallest and outermost planet of the Solar System. It is in a much more inclined and eccentric orbit than are the other major planets, and for a 20-year period around perihelion (as, for example, from 1979 to 1999) it comes within the orbit of Neptune. However, a **resonance** between the two planets' orbits (their periods are in the ratio 3:2) means they never get close. Pluto's small size and great distance give it a maximum apparent diameter of 0.1 arc seconds, and it is never brighter than magnitude 14.0. The main data for Pluto are given in the table.

The planet was found by Clyde **Tombaugh** on photographic plates taken in January and February 1930 during a search for the **Planet X** originally mounted by Percival Lowell.

The first visit to Pluto by a space probe is scheduled for 2016, by **Pluto Kuiper Express**. Most of our current knowledge of it dates from the discovery in 1978 of its satellite, **Charon**, which allowed the planet's mass and unusual axial tilt to be determined. A series of mutual **occultations** of planet and moon from 1985 until 1990 gave information on their relative brightnesses and sizes, and a very rough brightness map of Pluto's surface. The Hubble Space Telescope has enabled these results to be refined further.

Charon has both a **synchronous orbit** and **synchronous rotation**, so it keeps the same face turned toward Pluto, and hangs motionless in Pluto's sky. This orbital coupling of Pluto and Charon could well have brought about sufficient tidal heating (SEE **tides**) to cause **differentiation** in both bodies, giving them rocky cores and icy mantles.

Pluto seems to have a mottled surface with light and dark regions, and signs of polar caps. The surface is covered with icy deposits consisting of 98% nitrogen, with traces of methane, and also probably water, carbon dioxide and carbon monoxide. For a 60-year period in each orbit, around perihelion, the surface warms to above 50 K, sufficient to release a thin, temporary atmosphere of nitrogen. Interaction with the **solar wind** probably draws some of this atmosphere into a comet-like tail.

PLUTO: DATA

Globe	
Diameter	2320 km
Density	2.0 g/cm^3
Mass (Earth = 1)	0.0021
Volume (Earth = 1)	0.0058
Sidereal period of axial rotation (retrograde)	6.387d
Escape velocity	1.1 km/s
Surface gravity (Earth = 1)	0.03
Albedo	0.9
Inclination of equator to orbit	122° 27′
Surface temperature (average)	45 K
Orbit	
Semimajor axis	39.54 AU = 5914×10^6 km
Eccentricity	0.249
Inclination to ecliptic	17° 09′
Sidereal period of revolution	248.54y
Mean orbital velocity	4.74 km/s
Satellites	1

Pluto is certainly not the massive planet whose supposed perturbations of Uranus and Neptune inspired the search for Planet X. Some astronomers have question its status as a major planet; Pluto and Charon may be alternatively regarded as the largest members of the **Kuiper Belt** so far detected.

Pluto Kuiper Express A NASA spacecraft to be launched no earlier than December 2006 to fly past the planet Pluto in 2016 at a distance of about 15,000 km (9500 miles) on the first exploration of the farthest planet and its moon, Charon. It will then proceed into the Kuiper Belt to explore some of the most distant objects in the Solar System.

Pointers The stars Alpha and Beta Ursae Majoris (Dubhe and Merak), which point towards Polaris – the north pole star and the brightest star in Ursa Minor.

polar caps Deposits of ice in the polar regions of a planet or satellite. The Earth's permanent polar caps are of water ice, and show only a little seasonal variation. Mars has caps of water ice and carbon dioxide that show pronounced seasonal variation because of the planet's eccentric orbit. The axial inclination of Neptune's moon **Triton** with respect to the Sun produces a migration of its nitrogen-ice cap from one pole to the other over the course of the planet's 165-year orbital period.

polar distance The angular distance of a celestial body from the celestial pole. It is the complement of **declination** (i.e. polar distance plus declination equals 90°).

Polaris The star Alpha Ursae Minoris, lying within 1° of the north celestial pole. **Precession** is bringing Polaris closer to the pole, and it will be at its closest around AD 2100. Polaris is a supergiant of spectral type F, magnitude 2.0, distance 431 light years, luminosity nearly 3700 times that of the Sun. It is a **Cepheid variable** whose amplitude steadily decreased during the 20th century, stabilizing at 0.03 magnitude in the mid-1990s. Its period is 3.97 days. It is an optical double, with an 8th-magnitude companion 18 arc seconds away.

pole The points on a sphere that are 90° from the equatorial plane. The *celestial poles*, the points where the Earth's axis of rotation intersects the celestial sphere, are 90° north and south of the celestial equator. Similarly, the *ecliptic poles* are 90° from the plane of the ecliptic, and the *galactic poles* are 90° from the galactic plane.

pole star The nearest naked-eye star to either celestial pole. In the northern hemisphere it is **Polaris**, magnitude 2.0; in the south it is fifth-magnitude Sigma Octantis. Because of **precession** the positions of the celestial poles, and hence of the pole stars, are continually changing. In 3000 BC the north pole star was Thuban, in Draco; by AD 10,000 it will be Deneb, and by AD 14,000 it will be Vega.

Pollux The star Beta Geminorum, magnitude 1.16, distance 34 light years, luminosity 40 times that of the Sun. It is a giant of type K0.

Population I, Population II SEE **stellar populations**.

pore A very small and short-lived **sunspot**.

Portia One of the small inner **satellites** of Uranus discovered in 1986 during the Voyager 2 mission.

position angle (PA) The orientation in the sky of one celestial body with respect to another, measured from 0 to 360° from north via east. For a binary system it is the angle between the direction of the celestial north pole and the direction of the line joining the stars.

Poynting–Robertson effect A **non-gravitational force** produced by the action of solar radiation on small particles in the Solar System which causes them to spiral inward towards the Sun. The effect is most marked for particles between about a micrometre and a centimetre in size. A similar effect makes particles in planetary ring systems slowly spiral inward. The effect is named after John Poynting (1852–1914), who first described it, and Howard Robertson (1903–61), who proved its reality.

Praesepe A large open cluster in the constellation Cancer, known also as the *Beehive* or *Manger*, and designated M44. It was known to the Greeks as a hazy naked-eye patch; binoculars and telescopes show about 60 stars. The cluster is believed to be about 700 million years old, and lies 525 light years away.

preceding (abbreviation *p*) Describing the leading edge, feature or member of an astronomical object or group of objects. For example, the *preceding limb* of the Moon is the edge facing its direction of motion; and the *preceding spot* (or *p*-spot) of a group of sunspots is the first of the group to be brought into view by the Sun's rotation.

precession The circular motion of the celestial poles around the poles of the ecliptic, and the associated westward movement of the equinoxes with respect to the background stars. The main component of precession is caused by the gravitational pull of the Sun and Moon on the Earth's equatorial bulge. This is called *lunisolar precession*. A smaller, similar effect caused by the gravitational influence of the other planets is *planetary precession*, but it acts in the opposite direction. The total effect, *general precession*, amounts to about 50.3 arc seconds per year. The celestial pole therefore describes a circle on the sky of radius $23\frac{1}{2}°$ (the inclination of the Earth's axis), centred on the pole of the ecliptic, in about 25,800 years. Because of precession, the observed positions of all stars (including, of course, the pole star) are continuously changing. Precession also has the effect of moving the **equinoxes** around the ecliptic, and is therefore sometimes called the *precession of the equinoxes*. The First

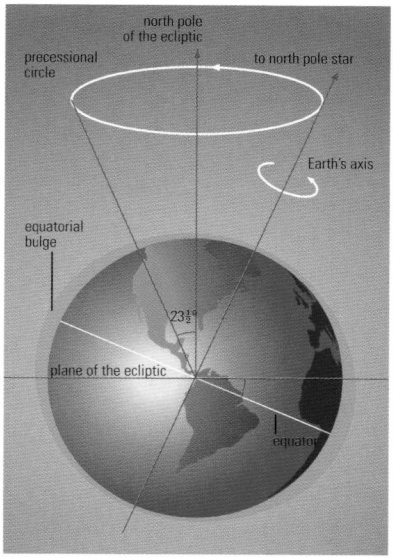

▲ **precession** Over the course of a 25,800-year cycle, the direction in which Earth's axis of rotation points describes a 47° diameter circle relative to the star background. One result of this is that different stars will, over time, become Pole Star.

Point of Aries, for example, is now in Pisces. Star catalogues give positions for a given **epoch**, e.g. 2000.0, so these positions must be corrected for precession to the date of observation.

primary (1) The largest of a system of celestial bodies, around which the others orbit. The term is used for the more massive component of a binary or multiple star, and for a planet with respect to its satellites.

primary (2) **(primary mirror)** The largest mirror in a **reflecting telescope** which collects light from the object being observed.

prime focus The point at which the objective lens or primary mirror of a telescope brings light to a focus when there is no other optical component in the light path, such as a secondary mirror. In large telescopes, special instruments are sometimes mounted at the prime focus. In the **Schmidt camera** a photographic plate or detector is placed at the prime focus.

In *prime focus photography*, a technique which can be used with small telescopes, the lens of a camera is removed and the camera back attached to the telescope via an adapter, in place of the eyepiece; photography – or CCD imaging – does not, despite the name, actually take place at the prime focus.

primordial black hole SEE **black hole**.

Procyon The star Alpha Canis Minoris, the eighth-brightest star in the sky at magnitude 0.40, 11.4 light years away and seven times as luminous as the Sun. It is a main-sequence star of type F5 and has a white dwarf companion of 11th magnitude which orbits it in 41 years.

prograde SEE **direct motion** (1).

Prometheus One of the small inner **satellites** of Saturn discovered in 1980 during the Voyager missions. Prometheus and Pandora are the two **shepherd moons** of Saturn's F Ring.

prominence A cloud of matter extending outwards from the Sun's chromosphere, into the corona. Prominences are regions of higher density and lower temperature than the surrounding corona. Those near the Sun's limb are visible, during total eclipses or with the aid of an instrument such as a **spectroheliograph** or a special filter, as bright protrusions into the corona. Away from the limb, they are silhouetted against the bright lower chromosphere as dark **filaments** when viewed in the light of hydrogen-alpha or calcium. *Quiescent prominences*, which extend for tens of thousands of kilometres, can last for weeks or months. *Active prominences* are short-lived, high-speed, flame-like eruptions that can reach heights of up to 700,000 km (450,000 miles) in just 1 hour. All prominences have spectra showing lines of neutral hydrogen, helium and ionized calcium. Their forms and behaviour are very varied, and their direction of motion is controlled by the Sun's magnetic field.

proper motion (symbol μ) The apparent motion of a star on the celestial sphere, as a result of its movement relative to the Sun. About 300 stars are known to have

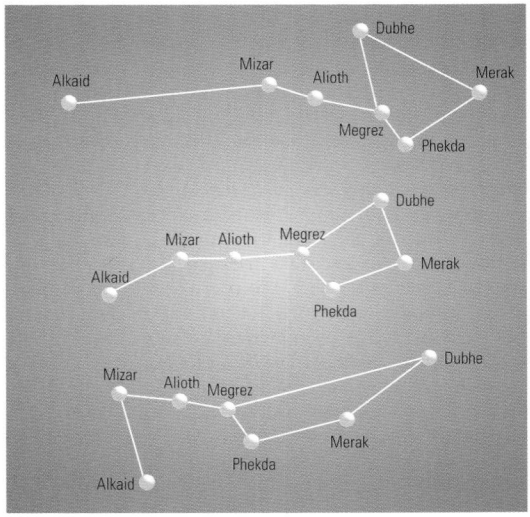

► **proper motion** Five of the stars making up the familiar pattern of the Plough are members of a cluster showing a common proper motion relative to the more distant stellar background. The two non-members, Alkaid and Dubhe, have different proper motions, and over the course of hundreds of thousands of years, the pattern will become distorted. The view in the centre shows the stars' relative positions today, with an earlier epoch represented at top, and the configuration in the distant future at bottom.

proper motions greater than 1 arc second per year, but most annual proper motions are smaller than 0.1 arc second. High proper motion is an indicator that a star lies in the Sun's neighbourhood. **Barnard's Star** has the largest known proper motion, 10.3 arc seconds per year. Proper motion is determined by comparing the star's position on images obtained on widely separated occasions, usually many years or decades apart, with the aid of a **comparator** or similar device. Accurate proper motions for over 100,000 stars were obtained by the position-measuring satellite **Hipparcos**.

Prospero A small outer **satellite** of Uranus discovered in 1999.

Proteus Neptune's second-largest **satellite**, discovered in 1989 during the Voyager 2 mission. It is a squarish, dark body, heavily cratered, and its most prominent feature is the large crater Pharos, 250 km (150 miles) across.

proton–proton reaction The nuclear reaction thought to be the main source of energy for main-sequence stars with masses equal to or less than the Sun's. In more massive stars the main energy source is the **carbon–nitrogen cycle**. The proton–proton reaction is a three-stage process in which hydrogen is converted into helium with the liberation of an enormous amount of energy. In the first stage, two protons (hydrogen nuclei) combine to form a deuterium nucleus, releasing a positron, a neutrino and radiation. In the second stage, the deuterium nucleus combines with a proton to form an isotope of helium (^{3}He), again releasing radiation. In the third stage, two ^{3}He nuclei combine to form the normal helium nucleus (^{4}He), releasing two protons and radiation. Overall, four hydrogen nuclei have combined to form one helium nucleus and give out radiation. The reaction requires a temperature of about 10 million K.

protoplanet In theories of the origin of the Solar System (and other planetary systems), a body formed by the accretion of **planetesimals**. Once the protoplanets attained the size of, say, the Moon, they could grow no further by accretion. Instead, collisions between them caused some to grow by accumulating fragments. The major planets of our Solar System are the result of this second stage of growth, while the collisional fragments that were not captured survive as bodies ranging in size from asteroids down to microscopic particles.

▲ **Proteus** Discovered by Voyager 2 during its 1989 flyby, Proteus is the largest of the inner satellites of Neptune. Proteus is notably dark, with an albedo of only 0.07.

protoplanetary disk A disk of dust and gas around a young star which is thought to represent the initial phase of planet formation. Such disks are thought to form around **protostars** as they grow by the gravitational collapse of interstellar matter in molecular clouds. As the disks spread out and become gravitationally unstable, clumps of matter within them form the progenitors of planets. More than a hundred protoplanetary disks have been observed in the **Orion Nebula**.

protostar An embryonic star formed when dense clumps of gas and dust collapse within molecular clouds. Protostars are the earliest, optically invisible stage of star formation. They are very red and can be observed at infrared wavelengths. The

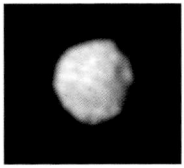

◀ **Puck** This distant view of Uranus's inner satellite Puck was taken from Voyager 2 during its flyby of January 1986. Puck has a diameter of about 154 km (96 miles).

Sun's protostellar stage lasted about 0.1 to 1 million years. SEE ALSO **stellar evolution**.

Proxima Centauri The nearest star to the Sun, at 4.22 light years slightly closer than **Alpha Centauri**. Proxima Centauri is an M-type dwarf, of visual magnitude 11.0, and is also a **flare star**. It was once thought to be part of the Alpha Centauri system, but is now known to be an unrelated star making a close approach on a hyperbolic orbit.

Ptolemy (Claudius Ptolemaeus) (2nd century AD) Egyptian astronomer and geographer. His chief astronomical work, known by its title in Arabic translation, the *Almagest*, is largely a compendium of contemporary knowledge, including a star catalogue, drawing heavily on the work of **Hipparchus of Nicaea**. His *Ptolemaic system* is based on the geocentric universe of the ancient Greeks, with the Earth fixed at the centre, and the Moon, Mercury, Venus, the Sun, Mars, Jupiter and Saturn revolving about it. Beyond these planets lies the sphere of fixed stars. Each of the bodies moves around a small circle called the *epicycle*, the centre of which in turn revolves around the Earth on a larger circle called the *deferent*. Ptolemy added to each orbit two more points, the *eccentric* and the *equant*, equally spaced either side of the Earth. He made the epicycle revolve around the eccentric rather than the Earth, and the planet have uniform motion with respect to the equant. The resulting model reproduced the apparent motions of the planets, including **retrograde** (2) loops, so well that it remained unchallenged until the revival of the heliocentric theory by Nicholas **Copernicus** in the 16th century.

Puck The largest of the small inner **satellites** of Uranus discovered during the Voyager 2 mission at the end of 1985, shortly before Voyager's encounter with the planet in January 1986. It is cratered and approximately spherical.

pulsar An object emitting radio waves in pulses of great regularity. They were first noticed in 1967 by Jocelyn **Bell**, working at the Mullard Radio Astronomy Observatory, Cambridge. The first pulsar to be discovered pulsed every 1.337 seconds.

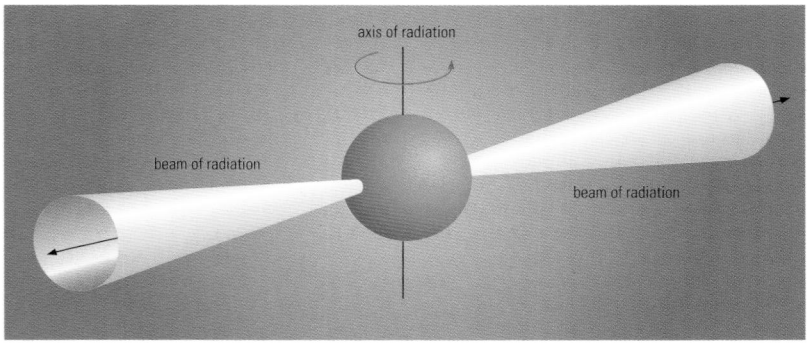

▲ **pulsar** This schematic illustration of a pulsar shows the two beams of radiation directed from the collapsed star along its magnetic poles, which need not coincide with the axis of rotation. If the rotational axis is so aligned that the radiation beams sweep across our line of sight, the rapidly pulsing radio signal characteristic of a pulsar will be detected.

Pulsars are rapidly rotating **neutron stars**, produced either as the result of a **supernova** explosion or by the collapse of a white dwarf that has accreted material from a companion star in a binary system and exceeded the **Chandrasekhar limit**. Acceleration of electrons by the powerful rotating magnetic field generates a conical beam of radio waves that sweeps across the Earth and is received in the form of pulses, in the same way that a lighthouse is seen to flash. With the Crab Pulsar, in the **Crab Nebula**, two pulses are observed for each rotation, which suggests that the magnetic axis is pointed almost directly at us so that we receive a beam from each pole as the star spins. The Crab Pulsar was the first pulsar to be seen flashing optically, in 1969. In some pulsars, the pulse amplitude and shape can change with time, and the pulsation can fade out temporarily.

Some highly energetic radio pulsars are now known to emit X-ray pulses. Of these, a few, the *anomalous X-ray pulsars* (AXPs) appear to be solitary neutron stars with exceptionally strong magnetic fields (known as **magnetars**). The magnetic field, rather than the star's rotation, provides the energy for the radiation.

Pulsars are gradually slowing down as their energy is dissipated by their radiation, but some, such as the Vela Pulsar, occasionally increase their spin rate abruptly. Such an event is called a *glitch* and is believed to be caused by a small, sudden adjustment in the internal structure of the star, producing a contraction of the stellar surface and increasing the rate of rotation. Over a thousand pulsars have been found, flashing at rates from 1.5 milliseconds (SEE **millisecond pulsar**) to nearly 5 seconds. The commonest period is just under 1 second. Pulsars are concentrated towards the disk of our Galaxy, which is where supernovae most commonly occur. Most pulsars are thought to be over a million years old, so that the supernova remnants that once surrounded them would have long since dispersed and faded from view. SEE ALSO **binary pulsar**.

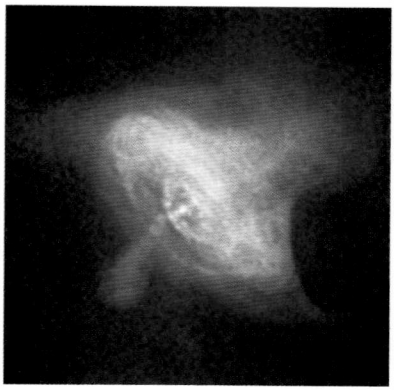

▲ **pulsar** This Chandra X-ray image shows the heart of the Crab Nebula, the site of a pulsar remnant from the supernova explosion of 1054. A thin jet almost 20 light years long emerges from the pulsar's south pole towards lower left.

pulsating star A type of variable star whose fluctuating brightness arises from a regular cycle of expansions and contractions. **Cepheid variables**, **RR Lyrae stars** and **long-period variables** are important classes of pulsating stars. Numerous **semiregular variables** and giant and supergiant **irregular variables** are also of this type.

Puppis A large southern constellation, once part of the larger grouping of Argo Navis, the ship Argo, and representing the ship's stern. Its brightest star is Zeta (Naos), magnitude 2.2. L^2 is a red giant semiregular variable with a range from 2.6 to 6.2, and a period of 140 days or so. Puppis is a rich constellation, with several bright open clusters including M46, M47, M93, NGC 2451 and NGC 2477.

Pyxis A southern constellation adjoining Vela and Puppis, representing a ship's compass, introduced by **Lacaille**. There is no star above magnitude 3.7, but there is one object of particular interest, the **recurrent nova** T Pyxidis.

Q

Quadrantids A **meteor shower** active between 1 and 6 January. The shower is one of the most prolific, reaching a narrow (less than 6 hours) peak **ZHR** in excess of 100 on 3 January. Its radiant lies in the constellation Boötes, near the borders with Hercules and Draco. The shower gets its name from a constellation called Quadrans Muralis which once occupied this region, but is no longer officially recognized.

quadrature The position of the Moon or a planet when it makes an angle of 90° with the Sun (SEE the diagram at **conjunction**). The Moon is at quadrature at first and last quarter.

Quaoar A Kuiper Belt object discovered by Chad Trujillo and Mike Brown on 4 June 2002. Quaoar has an estimated diameter of 1200 km (750 miles) and takes 288 years to orbit the Sun at a mean distance of 43 astronomical units.

quasar A compact and extremely luminous source of energy, radiated over a wide range of wavelengths from X-rays through optical to radio wavelengths. The name is derived from the term *quasi-stellar object*, which is what they were first called because of their starlike appearance on photographs. Quasars were discovered in 1963 when astronomers identified optical counterparts of certain strong radio sources. However, only a small proportion of quasars are actually radio sources. The most powerful of these are now known as *quasi-stellar sources* (QSSs), the original term 'quasi-stellar object' (QSO) being reserved for radio-quiet objects.

Every quasar shows a very large **redshift** which, if it results from the expansion of the Universe, indicates that quasars are the oldest and most distant objects in the Universe. In some cases, galaxies that lie between us and a quasar cause **gravitational lens** effects. Quasars inhabit the

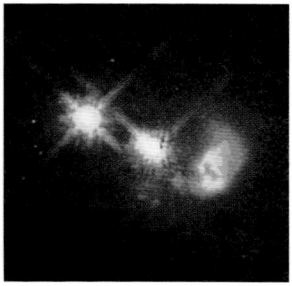

◀ **quasar** This set of quasar images shows the bright nuclear activity that lies at the heart of these galaxies. Each of these quasars is probably fuelled by material falling into a massive central black hole.

centres of very remote galaxies. Because they may vary in brightness on timescales of a few days, they can be no larger than a few hundred astronomical units wide. A surrounding galaxy has been seen around the nearest quasars, almost lost in the glare from the brilliant centre. Quasars are thus extreme examples of **active galactic nuclei**.

Quasar activity is thought to be caused by a **black hole** accreting material at the centre of the host galaxy. The most luminous quasars would need to be powered by black holes of 100 million solar masses, swallowing stars at a rate of about one every year. Quasars are bright ultraviolet sources, and this hot radiation seems to come from a swirling **accretion disk**. Also associated with the inner regions of quasars are emission lines of hydrogen, helium, and iron that indicate motions as rapid as 5000 km/s, as would be expected near a massive black hole.

Quasar-like activity may be possible within every galaxy, but at any given time the black hole is accreting matter in only a small proportion of galaxies. Activity in the gas-free elliptical galaxies may produce a **radio galaxy** or **BL Lacertae object**. In galaxies such as our own, where the black hole is small, even at its most active it is swamped by the light of surrounding stars.

Because quasars are the most luminous objects in the Universe, they can be seen to greater distances, and hence further back in time, than anything else. At redshifts near $z = 4$ we see the Universe as it was about one-tenth of its present age. Therefore quasars give us a way of examining the Universe in its youth.

quintessence SEE **dark energy**.

R

RA ABBREVIATION FOR **right ascension**.

radar astronomy The use of radar techniques to study bodies in space, by transmitting continuous waves or pulses and observing the reflections. Because of the restrictions of signal strength, radar astronomy is limited to objects within the Solar System.

Radar astronomy has been used to study ionized **meteor** trails in the Earth's atmosphere, revealing the existence of daytime meteor showers. Radar reflections from Venus provided an accurate value of the solar **parallax** and hence of the scale of the Solar System. The giant dish at **Arecibo Observatory** was used to establish the true rotation rates of Venus and Mercury, mapped the surface of cloud-covered **Venus**, and revealed what could be metallic deposits or perhaps water ice in craters near the poles of Mercury.

Radar altimeters have been carried aboard space probes to map Venus and, with the **Cassini** mission, the hidden surface of Titan. Other objects in the Solar System studied by radar include asteroids, Saturn's rings, planetary satellites and the nuclei of comets.

radial velocity The velocity of a celestial body along our line of sight, either towards us (positive) or away from us (negative). The radial velocity is determined from the **Doppler effect** on lines in the object's spectrum. SEE ALSO **redshift**.

radiant The point on the celestial sphere from which a **meteor shower** seems to radiate. Most meteor showers are named after the constellation in which their radiant is situated.

radiation belts (radiation zones) Doughnut-shaped regions of a planetary **magnetosphere** in which charged particles become trapped by the planet's magnetic field. The Earth's radiation belts are the **Van Allen Belts**. All the giant planets have radiation belts; Jupiter's are huge in extent and 10,000 times as intense as those of the Earth.

radiation era The period from about 10^{-32} seconds after the **big bang** to about 40,000 years later when the energy density of radiation in the Universe exceeded that of matter. It is subdivided into the *hadron era*, when most of the protons and neutrons now in the Universe were produced, and the *lepton era*, when most of the electrons were formed. SEE ALSO **matter era**.

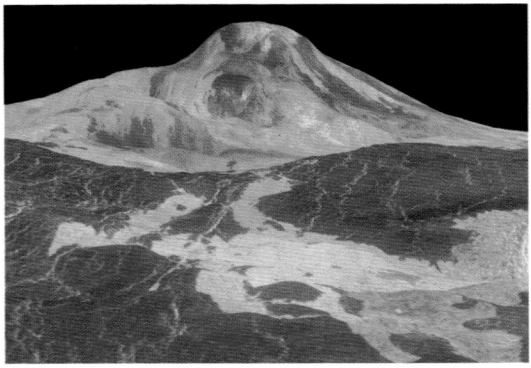

◀ **radar astronomy** This Magellan orbiter radar image shows Maat Mons on Venus. Radar observations have allowed planetary astronomers to penetrate Venus's dense atmosphere and obtain an idea of the planet's surface topography.

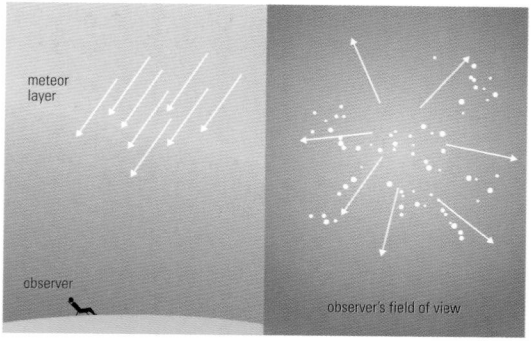

► **radiant** Meteors from a common source, occurring during a shower such as the Perseids, enter the atmosphere along parallel trajectories, becoming luminous at altitudes of around 80–100 km (50–60 miles). If an observer on the ground plots the apparent positions of meteors on a chart of the background stars, they will appear to diverge from a single area of sky, the radiant. The radiant effect is a result of perspective.

radiation pressure The pressure exerted by **electromagnetic radiation**. A beam of electromagnetic radiation can also be considered as a beam of particles called photons. When a beam of photons strikes a surface, the photons transfer their momentum to the surface and so exert a pressure. It is radiation pressure that keeps a star's outer layers from collapsing inwards under its own gravity. In interplanetary space, small particles are subjected to considerable radiation pressure from the **solar wind**. It is what causes the tail of a comet to point away from the Sun.

radio astronomy The study of radio emissions – **electromagnetic radiation** with wavelengths from about 1 mm to many metres – reaching the Earth from space. Observations down to about 2 cm can be made at sea level, but at shorter wavelengths it is necessary to observe from high mountains to avoid atmospheric absorption, mainly by water vapour. At the long-wavelength end, from 10 to 20 m, radio waves are absorbed and scattered in the Earth's **ionosphere**. Between these extremes, interference from terrestrial communications is a problem. International agreements have set aside certain wavebands for the sole use of radio astronomy.

Radio noise from the Milky Way was discovered in 1931, by Karl **Jansky**. The first radio telescope purposely built to study such emission was constructed in the USA by Grote Reber, who made a radio map of the sky. In 1942, James Stanley Hey in England found that the Sun emitted radio waves. The subject grew rapidly after World War II once it was realized that significant information about the Universe could be obtained in this region of the spectrum.

Radio emission has been detected from an extremely wide range of sources: the Sun; some of the planets (particularly those, such as Jupiter, with powerful magnetic fields); radio flares from certain stars; general **synchrotron radiation** from high-speed electrons in the Galaxy's magnetic field; and interstellar molecular clouds. Many discrete radio sources have been discovered, including **pulsars**, **supernova remnants**, **radio galaxies** and **quasars**. In addition, the **cosmic microwave background** is strong evidence in favour of the **big bang** theory of the origin of the Universe.

Individual spectral lines can also be detected at radio wavelengths, similar to optical spectral lines. In 1951 the emission from hydrogen at 21 cm (SEE **twenty-one centimetre line**) was detected, and it has proved to be an especially useful means of studying the structure of the Galaxy. The first known **interstellar molecule**, hydroxyl (OH), was discovered in 1965 from radio observations. Spectral lines have now been observed from over a hundred interstellar molecules. The study of these molecular lines has stimulated the extension of radio observations into the centimetre and millimetre range.

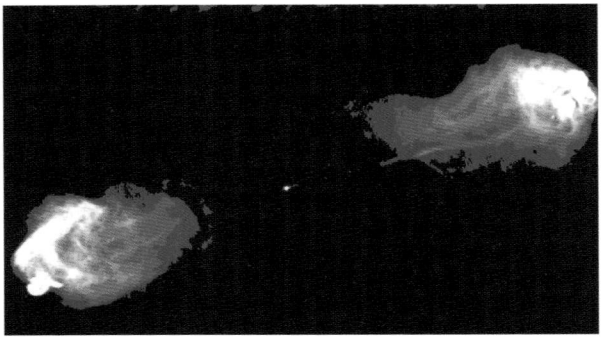

◀ **radio galaxy** Jets emerging from the active nucleus of Cygnus A produce two lobes of strong radio emission to either side of the galaxy, as seen in this image from the Very Large Array. Many radio galaxies show similar patterns of emission as material they eject impacts on the surrounding intergalactic medium.

radio galaxy A galaxy that is an abnormally high emitter of radio waves, around a million times stronger than the weak emissions from galaxies such as our own. Radio galaxies are usually giant ellipticals, such as M87 in the constellation Virgo. The emission from a typical radio galaxy is concentrated in two *lobes* well outside the galaxy, often located symmetrically either side, which appear to have been ejected explosively from the galaxy. The biggest radio galaxies have lobes around 15 million light years across, comparable in size to a typical cluster of galaxies. The extended radio lobes are almost invariably connected to the main galaxy by jets which seem to originate from the galactic nucleus. Strong, point-like radio sources exist in the nuclei of many of these objects.

It is not understood why some elliptical galaxies emit such enormous amounts of radio energy. However, it does seem clear that the source of the activity lies in the galaxy's nucleus, and in this respect radio galaxies are similar to radio **quasars** and **BL Lacertae objects**. A radio galaxy's enormous energy output comes from **synchrotron radiation**. The central 'engine' produces vast quantities of energetic electrons travelling near the speed of light. When these electrons encounter a magnetic field, they spiral around and, in so doing, lose energy. It is this energy that we detect as radio waves.

radio interferometer A type of radio telescope that operates on the same principle as an **optical interferometer**, achieving far higher resolution than is possible with a single dish. Single-dish radio telescopes typically have beam-widths of around one arc minute, which means in effect that they can only resolve angular sizes larger than this and measure positions to a few tens of arc seconds. A radio interferometer consists of two or more antennas which are spaced along a baseline and receive radio waves from the same source. The output signals from pairs of antennas are fed to a common radio receiver. The length of the baseline can be changed by varying the distance between the pairs of antennas. A small dish can be very effective as an interferometer in conjunction with a second dish of large collecting area. In **very long baseline interferometry** the radio telescopes can be on different continents, thereby providing a baseline the size of the Earth and achieving resolutions of about 1 milliarcsecond.

radio telescope An instrument used to collect and record radio waves from space. The basic design is a parabolic reflector, normally in the form of a large single dish similar to an optical reflector but much larger – 100 m (330 ft) or more in diameter; some instruments employ a parabolic 'trough' or a section of a much larger parabolic surface. The smoothness and accuracy of the reflecting surface determine the shortest wavelength that can be detected, because surface irregularities must be no more than a small fraction of this wavelength. However, for longer wavelengths, such as around 21 cm, it is adequate to use open wire mesh for the main reflector. Radio waves are reflected by the main reflector to a focus, sometimes via a secondary reflector, then converted into electrical signals that are amplified and recorded. A single-dish telescope does not produce a radio 'photograph' of a source: it only collects radio energy in a narrow beam along its axis. To build up an image of a radio source, the dish must be scanned across it, both in altitude and azimuth. In some cases, the azimuth scan is carried out by utilizing the rotation of the Earth.

A radio telescope should have high *sensitivity* (the ability to detect faint signals) and high *resolution* (the ability to detect fine detail). The largest individual dishes, such as the **Effelsberg Radio Telescope** and those at **Jodrell Bank** and **Parkes Observatory**, are very sensitive. They collect radio energy as weak as one hundred-million-millionth of a watt (10^{-14} W). Resolution depends on wavelength and aperture. Metre wavelengths are about a million times longer than the wavelengths of light, and resolution is correspondingly poorer. At 1-m wavelength, large radio telescopes have a resolving power of around 1°, about 20 times inferior to that of the naked eye. At a few centimetres their resolution equals that of the eye – still much poorer than optical telescopes.

Better resolution can be obtained by linking two or more dishes to form a **radio interferometer**, or by combining many of them to form an **aperture synthesis** array, which gives resolutions of a few hundredths of an arc second, 10 to 100 times better than that of the largest optical telescopes on Earth. The sensitivity remains that of the individual dishes added together.

radius vector The imaginary straight line between an orbiting celestial body and its primary, such as the line connecting a planet and the Sun. SEE **Kepler's laws**.

Ramsden eyepiece A basic telescope eyepiece consisting of two simple elements. It suffers from **chromatic aberration**, so the **Kellner eyepiece** is usually preferred. (After English optician Jesse Ramsden, 1735–1800.)

Ranger The first series of US Moon probes. After several failures, Rangers 7, 8 and 9, in 1964 and 1965, transmitted thousands of photographs of the lunar surface before they impacted. The photographs showed details of boulders and craters as small as 1 m (3 ft), a much higher resolution than was possible with Earth-based telescopes.

rays Bright streaks radiating from a crater. They consist of *ejecta* – material ejected by the impact that produced the crater. They are bright because the shock wave accompanying the impact is sufficient to vaporize rock, which then resolidifies in a glassy form. Ray craters are numerous on the Moon. The most prominent is Tycho, whose rays are seen at full Moon to extend across much of the Moon's disk. Ray craters are also found on other bodies in the Solar System, such as Jupiter's moon **Ganymede**. Solar radiation gradually

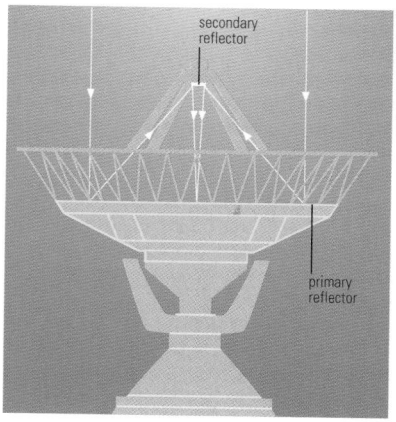

▲ **radio telescope** Just like an optical reflector, a radio telescope depends on collection of electromagnetic radiation by a large aperture, parabolic reflector, often in the form of a dish. Radio waves are brought to a focus on the detector. Most radio telescopes can be steered in altitude and azimuth.

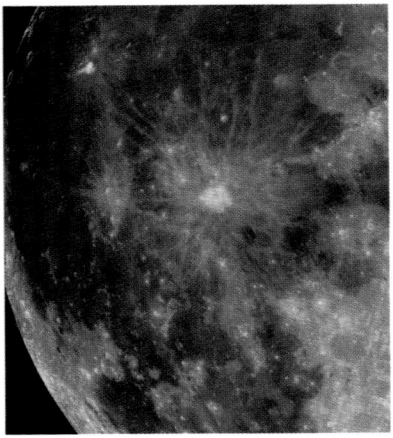

▲ **rays** This Galileo image of the Moon's northeast quadrant was taken as the spacecraft sped past Earth en route to Jupiter. A bright ray is seen crossing the oval Mare Serenitatis at left.

darkens rays, so the presence of a bright ray system around a crater indicates relatively recent formation.

R Coronae Borealis star A member of a class of variable stars that are subject to sudden, unpredictable fades. They normally spend most of their time at maximum, sometimes pulsating slowly by about half a magnitude over five to seven weeks. The fades can be anywhere between one and nine magnitudes and occur completely at random; the resultant deep minima can last for a few weeks or months, or for more than a year.

The type star, R Coronae Borealis, was first found to be variable in 1795. As shown in the diagram, it is normally at 6th magnitude, where it may remain for as long as ten years before suddenly fading. The minimum may be anywhere between 7th and 15th magnitude and can last for a few weeks or several years.

Only about three dozen stars of this sort are known. They are thought to be stars of 1 solar mass which have evolved past the red-giant stage and have ejected their hydrogen-rich outer layers to expose a core which is rich in carbon but depleted in hydrogen. Their declines are due to obscuration by clouds of carbon particles that have condensed in their outer atmosphere. Each cloud covers only a few per cent of the star's surface, and we see a fade when a cloud lies in our line of sight.

recombination era SEE **decoupling era**.

recurrent nova A **nova** that is seen to erupt more than once. Recurrent novae are binary stars in which the primary is a late-type (G, K or M) giant filling its **Roche lobe** and transferring material to a white dwarf secondary. At intervals of years to tens of years this material, which has accumulated on the surface of the white dwarf, explodes in a thermonuclear reaction, causing the brightness of the star to increase by 7 to 10 magnitudes for about 100 days. The longer the time between outbursts, the brighter the outbursts are. Examples are T Coronae Borealis, which underwent major outbursts in 1866 and 1946, and T Pyxidis (1890, 1902, 1922, 1944, and 1966). Only six are known, but all novae probably recur, over periods of thousands of years.

reddening A phenomenon exhibited by starlight as it is selectively absorbed and re-radiated on passing through clouds of interstellar dust and the Earth's atmosphere. Light is re-radiated at longer wavelengths than the wavelengths of the incident light, and is thus redder. A correction factor for interstellar reddening must therefore be introduced when distances of remote stars are estimated from their spectra. This reddening is not to be confused with **redshift**.

red dwarf A star at the lower end of the main sequence, of spectral type K or M. Red dwarfs have masses of between 0.8 and 0.08 of a solar mass. They are of small diameter, relatively low surface temperature (between 2500 and 5000 K) and low absolute magnitude. **Barnard's Star** and **Proxima Centauri** are examples.

red giant A giant star of spectral type K or M, having a surface temperature of less than 4700 K, 10 to 100 times the Sun's diameter and 100 to 10,000 times as luminous. A red giant is a star in a late stage of evolution, having exhausted the hydrogen fuel in its core. Red giants lie at the upper right of the **Hertzsprung–Russell** diagram. Examples are Arcturus and Aldebaran.

Red Planet Popular name for the planet Mars, which has a distinctly red colour.

redshift (symbol z) A lengthening of the wavelength of light or other electromagnetic radiation from a source, caused either by the source moving away (the **Doppler effect**) or by the expansion of the Universe. It is defined as the change in the wavelength of a particular spectral line, divided by the unshifted, or rest, wavelength of that line. For example, a redshift of $z = 0.1$ means that all the wavelengths of a source are lengthened by 10%; for $z = 0.2$ they are lengthened by 20%; and so on. The speed (v) at which a source is receding is found from its redshift by the formula $v = z \times c$, where c is the speed of light. Redshifts greater than 1 do not imply speeds greater than c; instead, for speeds greater than about $\frac{1}{3}c$, **Einstein**'s Special Theory of Relativity has to be used. An object with a redshift of $z = 2$, for example, is not receding from us at twice the speed of light. Instead, a redshift of 2 corresponds to a speed of 80% of the velocity of light, and a redshift of 4 corresponds to 92% of the velocity of light. By early 2004, quasars had been detected out to a redshift of $z = 6.6$, and lensed galaxies (SEE **gravitational lens**) out to at least $z = 7.0$.

Redshifts caused by the expansion of the Universe, called *cosmological redshifts*, have nothing to do with the Doppler effect. They are caused by the expansion of space itself, which stretches the wavelengths of light travelling towards us. The longer the travel time of light, the more its wavelength is stretched, as embodied in the **Hubble law**. The most highly redshifted photons are those in the **cosmic microwave background** radiation. These photons have been travelling towards us since matter and radiation decoupled (SEE **decoupling era**) from each other some 300,000 years after the **big bang**. They exhibit a redshift of about $z = 1000$, meaning that their wavelengths have been stretched by over a thousandfold.

Gravitational redshift is a phenomenon predicted by Einstein's General Theory of Relativity. Light emitted by a star has to do work to overcome the star's gravitational field. As a result there is a slight loss of energy and a consequent increase in wavelength, so that all the spectral lines are shifted towards the red.

redshift survey A mapping of the three-dimensional distribution of galaxies in space, with the aim of determining the large-scale structure of the Universe. Current redshift surveys, such as the **Sloan Digital Sky Survey**, employ bundles of optical fibres, each precisely aimed at a different galaxy, to enable the spectra – and hence the redshifts – of several hundred galaxies to be obtained at one pointing of the telescope.

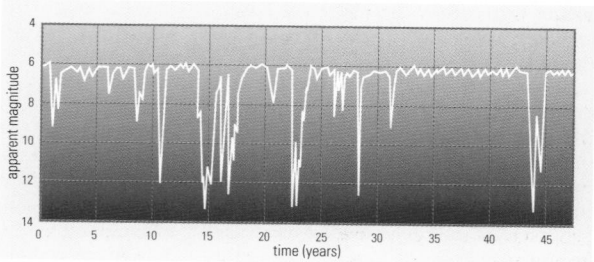

► R Coronae Borealis star The light curve of a typical R Coronae Borealis star shows long intervals spent at maximum brightness, punctuated by irregular episodes of abrupt dimming caused by condensation of carbon in the stellar atmosphere.

▲ **reflection nebula** The nebulosity surrounding the Pleiades star cluster is a well-known example of a reflection nebula. A small part of the nebula can be seen in this Hubble Space Telescope image. The star Merope is just out of frame at top right.

reflecting telescope (reflector) A telescope that forms an image by reflecting light from a primary concave mirror (usually parabolic) at the back of the tube, via a secondary mirror, to a focus whose position depends on the design of the instrument. The chief optical designs for smaller, amateur reflectors are the **Newtonian** and **Schmidt–Cassegrain**. Other configurations used for larger, professional instruments include the **Cassegrain**, **coudé** and **Schmidt camera**. Many large professional reflectors incorporate two or more different optical configurations, and with some it is possible to site instruments at the **prime focus**. Reflectors have the advantage of being free from **chromatic aberration**.

A reflector's mirrors are usually made of low-expansion glass or a ceramic material and coated with a thin layer of aluminium, which is highly reflective. Reflectors can be made larger than refractors because the primary mirror can be supported at the back, whereas a lens can be supported only around its rim, and is consequently liable to distort under its own weight. Also, new ways of building and controlling mirrors, as in the segmented mirrors of the **Keck Telescopes**, and the technique of **active optics**, have made possible further increases in the size and performance of big reflectors.

The Scottish mathematician James Gregory was the first to describe a reflecting telescope, in 1663; the first to build one was Isaac **Newton**, in 1668. In the late 18th century William **Herschel** demonstrated the instrument's potential by making large, accurately figured mirrors. SEE ALSO **Dobsonian telescope**, **Maksutov telescope**, **Ritchey–Chrétien telescope**.

reflection nebula A nebula in which light from a nearby star is reflected off the dust particles mixed in with the gas. Reflection nebulae generally appear blue in images; they are always bluer than the star that illuminates them because blue light is scattered more efficiently than red light. A good example is the Merope Nebula in the Pleiades.

refracting telescope (refractor) A telescope that forms an image by the refraction of light by a **lens** at the front of the tube. This lens, called the *objective*, is a compound lens, of perhaps more than one type of high-quality glass in order to minimize **chromatic aberration**, and it may be coated to reduce reflection and so increase the amount of light transmitted to the eyepiece. The tube is long so as to reduce **spherical aberration**. Refractors are still the first choice of some amateur observers for targets such as the planets, but the difficulties of making a large, optically perfect lens and supporting its weight mean that all large professional instruments are now **reflecting telescopes**.

The origins of the refracting telescope are uncertain. The Dutch spectacle-maker Hans Lippershey applied for a patent on one in 1608, although others were making similar instruments around the same time, and telescopes of some sort may have even been in use at the end of the 15th century.

Galileo was among the first to use the refractor in astronomy. Improvements were made by a succession of opticians, including John **Dollond** and Joseph von **Fraunhofer**, who devised means of reducing aberrations, and Christiaan **Huygens** and Jesse Ramsden, who devised better eyepieces. The largest refractor, at the Yerkes Observatory, has an aperture of 1 m (40 inches) and was built in 1897.

refraction SEE **atmospheric refraction, refracting telescope**.

regolith The dust and debris covering the surface of the Moon and other bodies in the Solar System. It is produced by the disintegration of crustal rocks caused by the impact of meteorites.

regression of the nodes The slow westward motion of the **nodes** of the Moon's orbit, caused by the gravitational pull of the Sun. A full circuit takes 18.6 years.

Regulus The star Alpha Leonis, magnitude 1.36, distance 77 light years, luminosity 135 times that of the Sun. It is a main-sequence star of type B7.

relative sunspot number (symbol R) An index of sunspot activity which reduces the sunspot counts of different observers to a common, statistical basis. For a particular day, it is given by

$$R = k\,(f + 10g),$$

where g is the number of groups of sunspots – irrespective of the number of spots each contains – and f is the total number of spots in all the groups; k is a factor based on the estimated efficiency of observer and telescope. The work of recording sunspots in this way was begun by Swiss astronomer Rudolf Wolf at the Zurich Federal Observatory, and the terms *Wolf number* and *Zurich relative sunspot number* were both formerly used.

relativity SEE **Einstein, Albert**.

residuals Differences between predicted and observed values of astronomical quantities. Residuals in, say, observations of the position of a planet, if collected over a sufficiently long period of time and then analysed statistically, may reveal some long-term trend. Residuals may also result from inaccurate observations.

resolution (1) SEE **resolving power**.

resolution (2) The size of the smallest detail distinguishable by an imaging system at a particular distance, or of the smallest detail visible in an image (of, for example, an image obtained by a space probe, often expressed in kilometres per pixel).

resolving power (resolution) The ability of an optical system, such as a telescope or the human eye, to distinguish close but separate objects such as the components of a close binary star, or to image a single small object. It is measured in

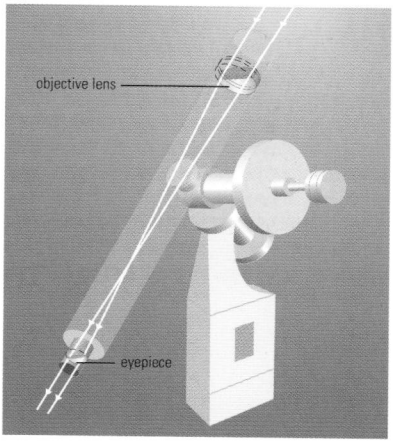

▲ **refracting telescope** The optical layout of a refractor is shown in this schematic cutaway. Light is collected by the main, objective lens at the front end of the telescope and brought to a focus. The observer views a magnified image of the focused light through a smaller lens, or set of lenses, making up the eyepiece. The telescope is shown on a German equatorial mount.

angular units. The resolving power of a telescope in arc seconds is found by dividing 116 by its aperture in millimetres (or 4.56 by the aperture in inches). This measure is called the *Dawes limit*, after the English amateur astronomer William Rutter Dawes (1799–1868). The resolving power of the human eye is about one arc minute. SEE ALSO **radio telescope**.

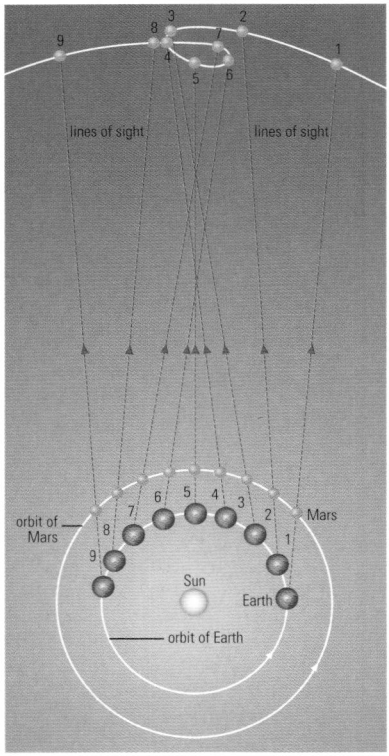

▲ **retrograde motion** (2) The cause of the apparent reversal of superior planets in their motion along the ecliptic close to the time of opposition is explained in the diagram. Here, Earth (inner orbit) is seen catching up Mars (outer orbit). Before opposition, Mars appears to move eastward (top). As Earth, moving more rapidly on its smaller orbit, catches up, Mars's eastward motion apparently slows, then reverses relative to the background stars as Earth overtakes. After opposition, the apparent eastward motion of Mars resumes.

resonance An effect produced by the gravitational interaction between two bodies orbiting the same primary. Resonance between bodies of similar size can lock bodies into **commensurable** orbits, as has happened with Jupiter's **Galilean satellites**. Between bodies of highly dissimilar size, however, it can prevent small bodies from occupying commensurable orbits. This is what keeps clear the **Kirkwood gaps** in the asteroid belt and the various gaps in the rings of **Saturn**.

Reticulum A small but distinctive southern constellation, introduced by **Lacaille**, representing a reticle used for measuring star positions. Its brightest star, Alpha, magnitude 3.3, forms a compact pattern with Beta, Gamma, Delta and Epsilon.

retrograde (1) Describing orbital or rotational motion in the opposite sense to the motion of the Earth: clockwise as viewed from above the Sun's north pole. Some satellites, such as the outermost ones of Jupiter, have a retrograde orbital motion, and so do certain comets, notably Halley's Comet. Venus, Uranus and Pluto all have retrograde rotation. Orbital or rotational motion is retrograde if the orbital or axial inclination is more than 90°. COMPARE **direct** (1).

retrograde (2) Describing the temporary east-to-west motion traced out by a superior planet or asteroid on the celestial sphere as the Earth catches up and overtakes it. The curve traced out by this reversal is called a *retrograde loop*. The places at which the planet changes between easterly and westerly motion are termed *stationary points* (points 4 and 6 in the diagram). COMPARE **direct** (2).

revolution The movement of a planet or other celestial object around its orbit, as distinct from the rotation of the object on its axis.

Rhea Saturn's second-largest **satellite**, diameter 1528 km (950 miles), discovered by Giovanni Cassini in 1672. It is a typical heavily cratered icy world, but with few large craters, the largest being the 250-km (150-mile) diameter Izanagi. Some resurfacing seems to have taken place in the past, but the surface is mostly ancient. The **Cassini** spacecraft also imaged a more recent, bright ray crater.

rich-field telescope (richest-field telescope) (RFT) A low-power telescope equipped with a wide-angle eyepiece (such as a **Nagler eyepiece**) designed to show a wide field of view at a relatively low magnification. Such an instrument is ideal for studying starfields and hunting for novae.

Rigel The star Beta Orionis, the brightest star in the constellation Orion, and marking Orion's left leg. It is of magnitude 0.18 (the seventh-brightest star in the sky), and a supergiant of spectral type B8. Its distance is 773 light years, and its luminosity 40,000 times that of the Sun.

right ascension (RA, symbol α) A coordinate used to define the position of a celestial object. It is the angular distance measured eastwards from the vernal equinox (also known as the First Point of Aries) to the point where the hour circle of an object meets the celestial equator. RA is usually expressed in hours, minutes and seconds, although it may be expressed as an angle, 1 hour of RA being equivalent to 15°. SEE ALSO **celestial sphere**.

Rigil Kentaurus Another name for the star **Alpha Centauri**.

rille A cleft on the surface of the Moon. Rilles that are curved or straight are geological faults. The winding *sinuous rilles* are most likely collapsed lava tubes. Lava was able to flow on the Moon for hundreds of kilometres. The top and sides of a flow can solidify, insulating the molten lava inside which continues to flow. When such a tube has been drained, the roof may collapse, leaving what we see today as a sinuous rille.

▲ **Rhea** This Cassini image of Saturn's satellite Rhea was obtained from a distance of 496,500 km (308,600 miles) on 16 January 2005. It gives a natural-colour view of Rhea's trailing hemisphere.

ring, planetary A narrow, circular filament or thin disk of large or small particles, or both, orbiting a planet. **Jupiter**, **Saturn**, **Uranus** and **Neptune** all have systems of rings. Saturn's rings, by far the most prominent and extensive, were discovered telescopically in the 17th century. Jupiter's were found in 1979 by Voyager 1; those of Uranus and Neptune were first observed during **occultations**, and were confirmed by Voyager 2 in the 1980s. Saturn's rings were found in 1849 to lie inside the **Roche limit**; any sizeable, rigid satellite inside this limit would be broken up by gravitational forces, and any smaller bodies would be prevented from accreting into a larger one. In 1857 the Scottish physicist James Clerk Maxwell showed that on theoretical grounds the rings had to be made up of many separate particles.

The ring systems around the four giant planets show similarities and dissimilarities. For example, Saturn's main rings are made up principally of bright, icy particles, whereas the other three planets have dark rings. Some rings are very narrow and distinct, others are continuous sheets, but still a kilometre or less thick, and confined to the equatorial plane, while still others are large, tenuous, doughnut-shapes haloes. Small **shepherd moons** keep some rings

PLANETARY RINGS

Ring or gap	Distance from centre (thousand km)	Width (km)
Jupiter		
Halo	100–122.8	22800
Main	122.8–129.2	6400
Gossamer	129.2–214.2	85000
Saturn		
D	67–74.5	7500
C (Crepe)	74.5–92	17500
Maxwell Gap	88	270
B	92–117.5	25500
Cassini Division	117.5–122.2	4700
A	122.2–136.8	14600
Encke Gap	133.6	325
Keeler Gap	136.5	35
S/2004 1R	138	300
F	140.2	30–500
G	165.8–173.8	8000
E	180–480	300000
Uranus		
U/1986 2R	37–39.5	2500
6	41.8	1–3
5	42.2	2–3
4	42.6	2–3
Alpha	44.7	7–12
Beta	45.7	7–12
Eta	47.2	0–2
Gamma	47.6	1–4
Delta	48.3	3–9
U/1986 1R	50	1–2
Epsilon	51.1	20–100
Neptune		
Galle	41.9–43.6	1700
Le Verrier	53.2	15
Lassell/Arago	53.2–59.1	5900
Adams	62.9	50

in check (in Neptune's case, creating clumps of material – *ring arcs*), while larger satellites create concentric gaps. Particles range from micrometre-sized in Jupiter's main ring to as much as 10 m (30 ft) in Saturn's B Ring.

ring galaxy A galaxy having the form of an elliptical ring around a nucleus, like the rim of a wheel. Ring galaxies are thought to form from a collision in which a small galaxy passes through a larger one.

Ring Nebula The planetary nebula M57, 4100 light years away in the constellation Lyra, also known as NGC 6720. It appears as a hollow ellipse; in three dimensions it is shaped not like a spherical shell but like a toroid, or ring doughnut. The nebula, centred on a blue star of 0.2 solar masses, is expanding at 19 km/s (12 mile/s) and is estimated to be about 5500 years old.

Ritchey–Chrétien telescope A modified form of **Cassegrain telescope**, designed originally by George Ritchey (1864–1945) and Henri Chrétien (1879–1956). The primary and secondary mirrors are **hyperboloidal**, although there are variants with near-hyperboloidal, **paraboloidal** or **spheroidal** components. The system is free from **coma** (1) over a wide field, though there is some astigmatism and the field is curved.

robotic telescope A telescope designed to operate automatically under control from a remote site via an Internet connection. Examples are the 400-mm (18-inch) *Bradford Robotic Telescope* in West Yorkshire (the world's first), and the two 2-m (6.6-ft) *Faulkes Telescopes*, one on Hawaii and the other at the Siding Spring Observatory in Australia, due to open in 2005.

Roche limit (symbol R) The minimum distance at which a planetary satellite can remain in orbit without being torn apart by gravitational forces; alternatively, the minimum distance at which smaller bodies in orbit round a planet can combine by **accretion** to form a larger one. A planet's Roche limit is about 2.5 times its radius (r). It depends, though not strongly, on the densities of the planet (d_p) and the satellite (d_s):

$$R = 2.44\, r_3 \sqrt{(d_p/d_s)}.$$

Actually the Roche limit applies only to 'fluid' satellites. Rocky bodies with a high structural integrity, such as Neptune's innermost satellites, can exist inside the Roche limit. All planetary ring systems lie inside their planet's Roche limit. The concept was originated by the French astronomer and mathematician Edouard Roche (1820–83).

Roche lobe A surface defining the maximum size of a star in a binary system. If one star's surface overflows this size as it evolves, the companion's gravity will pull matter off it.

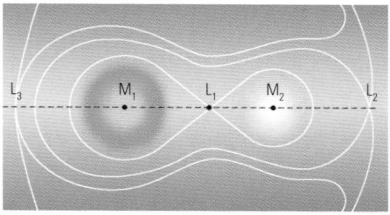

▲ **Roche lobe** In a binary star system, the stars (here marked M_1 and M_2) have their own gravitational sphere of influence, defined by the Roche lobe. If one component fills its Roche lobe, material can escape through the inner lagrangian point, L_1, onto its companion. If both stars fill their Roche lobes, material can escape entirely from the system through the outer Lagrangian points, L_2.

Römer, Ole Christensen (1644–1710) Danish astronomer. He was the first to measure the speed of light. Working at Paris in 1675, he found that the times of occultations of the **Galilean satellites** predicted by Giovanni Domenico **Cassini** did not match the observed times: intervals between successive occultations decreased as the orbital motions of the Earth and Jupiter brought them closer together, and increased as they moved farther apart. He deduced that the differences were caused by light taking time to travel the intervening distance. His timings gave the speed of light as 200,000 km (120,000 miles) per second, two-thirds of the true value. Römer also invented the first **transit instrument**.

Roque de los Muchachos Observatory A Spanish observatory on La Palma in the Canary Islands, the home of telescopes owned by several European nations, foremost among them the 10.4-m **Gran Telescopio Canarias**. Other major instruments include the 3.5-m Italian **Galileo National Telescope**, the UK's **William Herschel Telescope** and 2.5-m (100-inch) Isaac Newton Telescope, moved from England in 1984, and the 2.5-m Nordic Optical Telescope, opened in 1989.

Rosalind One of the small inner **satellites** of Uranus discovered in 1986 during the Voyager 2 mission.

Rosat A German satellite containing an X-ray telescope and a British wide-field camera that made the first survey of the sky at extreme ultraviolet wavelengths. It was launched on 1 June 1990, and operated until 1998.

► **Ring galaxy** Catalogued as AM 0644-741, this ring galaxy lies 300 million light years away in the direction of the constellation of Volans. Ring galaxies are thought to form when one galaxy passes through the disk of another. In this case the 'intruder' galaxy is out of shot.

Rosetta A European Space Agency probe, launched on 2 March 2004, to rendezvous with Comet Churyumov–Gerasimenko in 2014 and deposit a small lander on its surface. En route there are planned close flybys of asteroids 21 Lutetia and 2867 Steins.

Rosette Nebula A diffuse nebula 4900 light years away in the constellation Monoceros, surrounding the open cluster NGC 2244. The stars of the cluster were born within the past half million years from the Rosette Nebula and now illuminate it. Radiation pressure from the stars has created a cavity about 30 light years across in the central part of the nebula. Star formation continues in the nebula, which contains many **Bok globules**.

Rosse, Third Earl of (William Parsons) (1800–1867) Irish amateur astronomer. In the grounds of Birr Castle, near Parsonstown in Ireland, he built a 72-inch (1.83-m) reflector. Completed in 1845, it was the largest telescope in the world for the rest of its lifetime. Rosse studied in particular the hazy cloud-like objects then known collectively as nebulae. In 1845 he discovered the spiral structure of what is now called the **Whirlpool Galaxy** and other 'spiral nebulae' (recognized as other galaxies in the 1920s) and was able to resolve certain nebulae into groups of stars. He also discovered the ring structure of planetary nebulae.

rotation The turning of a celestial body on its axis, as distinct from its orbital revolution.

rotating variable A star whose variations in magnitude arise either because of its non-spherical shape, as in *ellipsoidal variables*, where a binary system is not eclipsing but presents a changing apparent surface area throughout its orbit, or because a star has a non-uniform surface brightness. In the latter case, the variation may be caused by surface spots, chromospheric activity, or alterations in the surface brightness related to fluctuations in the magnetic field, arising either because of varying field strength or because the star's magnetic axis does not coincide with the rotational axis.

Royal Greenwich Observatory (RGO) The former UK national astronomical observatory, founded at Greenwich, London, in 1675. After World War II the observatory moved to Herstmonceux in Sussex, and in 1990 it moved again, to Cambridge. When it closed in 1998, some of its functions were transferred to the Royal Observatory, Edinburgh.

Royal Observatory, Edinburgh (ROE) An administrative centre in Edinburgh which took on some of the functions of the **Royal Greenwich Observatory** when the RGO closed in 1998. Its two main departments are the Institute for Astronomy, run by the University of Edinburgh, and the UK Astronomy Technology Centre, which plans, designs and manufactures telescopes and instrumentation.

RR Lyrae star A pulsating variable star of a class formerly known as *short-period Cepheids* or *cluster variables* (because the first were found in globular clusters). RR

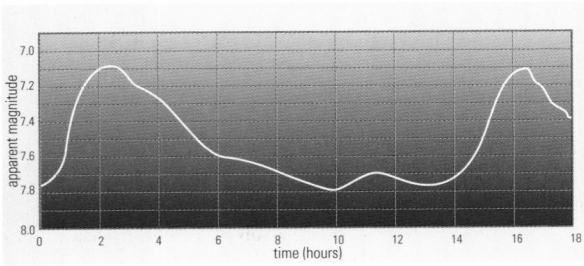

◀ RR Lyrae variable
This is the light curve of a typical RR Lyrae star. These stars show a rapid rise to peak light followed by a slower decline. They are commonly found in globular clusters.

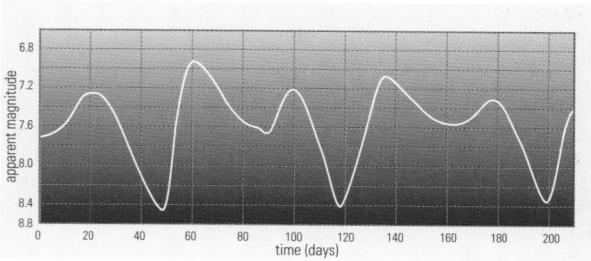

► **RV Tauri star** The light curve of a typical RV Tauri star shows alternating deep and shallow minima. The deep minima result from different modes of pulsation in the star's outer layers coming into phase with each other.

Lyrae, the type star, was the first to be found outside a globular cluster. RR Lyrae stars are Population II giants (SEE **stellar populations**) of spectral type between A and F. Their periods range from 0.2 to 1.2 days and their amplitudes from 0.2 to 2.0 magnitudes, although most have periods of between 9 and 17 hours. Their light curves differ from those of the so-called classical Cepheids; that of RR Lyrae itself is shown in the diagram. Most rise quickly to maximum in no more than a tenth of their total period. Their minima are comparatively long, so that for a few hours their light remains roughly constant. The absolute magnitude of all RR Lyrae stars is about +0.5, so they can be used as 'standard candles' for distance finding. However, they are too faint to be seen beyond the nearest galaxies. SEE ALSO **Cepheid variable**.

runaway star A star of spectral type O or B that has an unusually high space velocity. Such stars are former members of massive binaries, one member of which has exploded as a supernova, flinging off the other like a slingshot. Three runaway stars with a common origin in Orion – 53 Arietis, AE Aurigae and Mu Columbae – may have belonged to a quadruple star system in which the fourth star exploded as a supernova 3 million years ago. As many as 20% of O stars may be runaways.

Russell, Henry Norris (1877–1957) American astronomer. In 1910 he produced a diagram plotting the absolute magnitude of stars against their spectral type which showed the division between main-sequence stars and giant stars. Unknown to him, Ejnar **Hertzsprung** had done the same some years previously. The diagram, now called the **Hertzsprung–Russell diagram**, is the starting-point for all modern theories of stellar evolution, although Russell first thought, erroneously, that stars evolved along the main sequence. In 1928 he determined the composition of the Sun's atmosphere from a study of its spectrum.

RV Tauri star A member of a class of pulsating variable stars with alternating primary and secondary minima. The depth of the minima also vary, such that the primary and secondary minima occasionally exchange places. RV Tauri stars are luminous supergiants of spectral type between F and M, with periods ranging from 30 to 150 days and amplitudes of up to 4 magnitudes. The diagram shows the light curve of AC Herculis, which has a constant mean magnitude, but with some other stars of this class, such as RV Tauri itself and DF Cygni, the mean varies by up to 2 magnitudes with a secondary periodicity of 600 to 1500 days. Apart from their pulsations with multiple periods, RV Tauri stars are undergoing pulsations of more than one period, and appear to be giving off material in a strong **stellar wind** in what may be the early stages of formation of a **planetary nebula**.

Ryle, Martin (1918–84) English radio astronomer. In 1946 he discovered **Cygnus A**, the first radio galaxy to be found. In 1955 he built the first **radio interferometer**, and in 1960 developed the technique of **aperture synthesis**. For his pioneering work in radio astronomy he received a share of the 1974 Nobel Prize for Physics.

S

Sagan, Carl Edward (1934–96) American astronomer. He studied many aspects of the Solar System, including the physics and chemistry of planetary atmospheres and surfaces, particularly of Venus and Mars. He investigated the origins of terrestrial life (discovering in 1963 that ATP, an important biochemical, is formed when chemicals thought to be present on the early Earth are mixed together) and the possibility of life existing elsewhere. Sagan did much to popularize astronomy and science in general.

Sagitta A small constellation representing an arrow, adjoining Aquila. Its brightest star is Gamma, magnitude 3.5. The arrow shape is completed by the 4th-magnitude stars Alpha, Beta and Delta.

Sagittarius The southernmost of the zodiacal constellations, representing a centaur with a bow, and notable for containing the dense star clouds in the direction of the centre of our Galaxy. The outline of its main stars is sometimes likened to a teapot. Sagittarius's brightest star is Epsilon (Kaus Australis), magnitude 1.8, followed by Sigma (Nunki), 2.0. There are plenty of variable stars, including the Cepheids X (range 4.3–5.1) and W Sagittarii (range 4.2–4.9). The Milky Way is at its richest in Sagittarius, and there are no fewer than 15 Messier objects – globular clusters, open clusters and gaseous nebulae, including the **Omega Nebula**, the **Lagoon Nebula** and the **Trifid Nebula**.

Sagittarius A A strong radio source at the centre of our Galaxy. The exact centre of the Galaxy is marked by a small part of this source, referred to as *Sagittarius A**, less than 4 AU across. It is most likely a black hole surrounded by an **accretion disk**. Estimates of the hole's mass range from as high as 5 million solar masses to as low as 100 solar masses.

Saha, Meghnad (1894–1956) Indian astrophysicist and nuclear physicist. His theoretical work on spectra, the solar corona, radiation, and ionization resulted in the 1920s in an equation, now known as the *Saha equation* which relates the degree of ionization of atoms (i.e. how many electrons they have lost) in a star's atmosphere to the temperature. It provides an important insight into the differences in the spectra of stars of different spectral types, and allows their temperatures to be deduced from the lines present in their spectra.

Salpeter process SEE **triple-alpha process**.

Sandage, Allan Rex (1926–) American astronomer. Initially an assistant to Edwin **Hubble**, he continued to work towards establishing the value of the **Hubble constant** and thus fixing the age of the Universe. In 1960 Sandage made the first identification at optical wavelengths of what would prove to be a **quasar**, and five years later he found the first radio-quiet quasar.

saros A cycle of lunar and solar eclipses lasting just over 18 years. A saros consists of the same sequence of eclipses repeated at nearly the same time intervals. This is because 223 **lunations** (6585.32 days) have almost the same duration as 19 **eclipse years** (6585.78 days), after which time the Sun, the Moon and the Moon's **nodes** return to very nearly the same positions relative to one another. This period, known to several ancient peoples, enabled astronomers to predict eclipses, since, if an eclipse of the Sun or the Moon is observed, then after a complete saros there will be a similar eclipse. The difference of 0.46 of a day between the number of lunations and eclipse years in one saros means that each eclipse in the sequence will recur about 165° further west. In predicting a date, allowance must be made for leap days. SEE ALSO **Metonic cycle**.

satellite (moon) A celestial body in orbit around a planet. Of the major planets in the Solar System, Mercury and Venus have no known satellite. The Earth has one satellite, Mars two, Jupiter sixty-three, Saturn thirty-three, Uranus twenty-seven, Neptune thirteen and Pluto one (as of early 2005). There are almost certainly further small satellites of the giant planets awaiting discovery. Some asteroids, such as **Ida**, are known to have a satellite.

The satellites of the Solar System vary enormously in their size, orbit, surface features and supposed origin. One, Jupiter's Ganymede, is bigger than Mercury, and seven are bigger than Pluto, while the smallest are tiny bodies less than a kilometre across. Their orbital periods vary just as much, from Phobos, which circles Mars's equator more than three times a day, to the remotest satellites of Neptune, which takes over 25 Earth years to complete one orbit. In their surface features they are remarkably diverse, ranging from dead, densely cratered worlds to ones that show a fascinating variety of geological processes. Indeed, Jupiter's Io is the most geologically active body known. As to their origin, some, such as Jupiter's **Galilean satellites**, may have condensed out of the same part of the solar nebula as their parent planet, while several small ones are believed to be captured asteroids of cometary nuclei. The **Moon** is currently thought to have formed from debris flung out when a Mars-sized body collided with the Earth.

Data for planetary satellites are given in the table on pages 206–207; each satellite in the table has an entry of its own. SEE ALSO **artificial satellite**.

Saturn The sixth major planet from the Sun, and the second-largest of the four giant planets. Saturn was the remotest planet known in the pretelescopic era. Viewed through a telescope it appears as a flattened golden yellow disk encircled by white rings. Because of Saturn's axial inclination, the angle at which the ring system is presented to us depends on the planet's position in its orbit, and this has a considerable effect on its brightness. At perihelic opposition with the rings fully open, Saturn is magnitude −0.3; at aphelic opposition with the rings edge-on, it is magnitude 0.8. The disk's maximum apparent size is 21 arc seconds (and the rings span up to 48 arc seconds). The main data for Saturn are given in the table on page 208.

Saturn's magnificent system of rings was first identified as such by Christiaan Huygens in 1656. As seen from the Earth, the ring system consists of a few principal components, separated by a number of gaps. The three main rings, starting from the outside, are Ring A, which is greyish-white,

▲ **Saturn** This detailed, natural-colour image mosaic of Saturn is made up of 126 images obtained by the Cassini spacecraft on 6 October 2004. The smallest features visible are 38 km (24 miles) across.

PRINCIPAL SATELLITES OF THE MAJOR PLANETS

Satellite	Diameter (km)	Distance from centre of planet (thousand km)	Orbital period (days)	Mean opposition magnitude
Earth				
Moon	3,475	384	27.32	−12.7
Mars				
Phobos	22	9.4	0.32	11.4
Deimos	12	23.5	1.26	12.5
Jupiter				
Metis	43	128	0.29	17.5
Adrastea	16	129	0.30	18.7
Amalthea	167	181	0.50	14.1
Thebe	99	222	0.67	16.0
Io	3,643	422	1.77	5.0
Europa	3,122	671	3.55	5.3
Ganymede	5,262	1,070	7.15	4.6
Callisto	4,821	1,883	16.69	5.7
Leda	20	11,165	240.9	19.5
Himalia	170	11,461	250.6	14.6
Lysithea	36	11,717	259.2	18.3
Elara	86	11,741	259.6	16.3
Ananke	28	21,276	629.8 *R*	18.8
Carme	46	23,404	734.2 *R*	17.6
Pasiphae	60	23,624	743.6 *R*	17.0
Sinope	38	23,939	758.9 *R*	18.1
Saturn				
Pan	25	134	0.58	19.4
Atlas	32	138	0.60	19.0
Prometheus	100	139	0.61	15.8
Pandora	84	140	0.63	16.4
Epimetheus	116	151	0.69	15.6
Janus	178	152	0.69	14.4
Mimas	397	186	0.94	12.8
Enceladus	499	238	1.37	11.8
Tethys	1,060	295	1.89	10.2
Telesto	24	295	1.89	18.5
Calypso	19	295	1.89	18.7
Dione	1,118	377	2.74	10.4
Helene	32	377	2.74	18.4
Rhea	1,528	527	4.52	9.6
Titan	5,150	1,222	15.95	8.4
Hyperion	283	1,464	21.28	14.4
Iapetus	1,436	3,561	79.33	11.0
Phoebe	220	12,944	548.2 *R*	16.4
Paaliaq	19	15,199	686.9	21.3
Albiorix	26	16,404	783.4	20.5
Siarnaq	32	18,160	893.1	20.1
Ymir	16	23,096	1,312 *R*	21.7

R indicates retrograde orbit

continued

PRINCIPAL SATELLITES OF THE MAJOR PLANETS (continued)

Satellite	Diameter (km)	Distance from centre of planet (thousand km)	Orbital period (days)	Mean opposition magnitude
Uranus				
Cordelia	40	50	0.34	23.6
Ophelia	23	54	0.38	23.3
Bianca	51	59	0.43	22.5
Cressida	80	62	0.46	21.6
Desdemona	64	63	0.47	22.0
Juliet	94	64	0.49	21.1
Portia	135	66	0.51	20.4
Rosalind	72	70	0.56	21.8
Belinda	81	75	0.62	21.5
S/1986 U10	40	76	0.64	23.6
Puck	162	86	0.76	19.8
Miranda	472	130	1.41	15.8
Ariel	1,158	191	2.52	13.7
Umbriel	1,169	266	4.14	14.5
Titania	1,578	436	8.71	13.5
Oberon	1,523	584	13.46	13.7
Caliban	89	7,231	579.5 *R*	22.4
Stephano	20	8,004	677.4 *R*	24.1
Sycorax	190	12,179	1,288 *R*	20.8
Prospero	30	16,243	1,977 *R*	20.8
Setebos	30	17,501	2,234 *R*	23.3
Neptune				
Naiad	58	48	0.29	24.6
Thalassa	80	50	0.31	23.9
Despina	148	53	0.33	22.5
Galatea	158	62	0.43	22.4
Larissa	192	74	0.55	22.0
Proteus	416	118	1.12	20.3
Triton	2,707	355	5.88 *R*	13.5
Nereid	340	5,514	360.1	19.7
S/2002 N1	54	16,600	1,874 *R*	24.2
S/2002 N4	43	48,600	9,412 *R*	24.7
Pluto				
Charon	1,186	19	6.39 *R*	17.3

R indicates retrograde orbit

There are many minor satellites of the giant planets; only those with diameters greater than 10 km (Jupiter), 15 km (Saturn), (20 km) Uranus and (40 km) Neptune are included here.

Ring B, which is bright white, and Ring C, which is fainter and blue-grey. Ring D, the innermost, was discovered in 1969; Rings E, F and G, which lie outside Ring A, were discovered by space probes. The most prominent gap is the Cassini Division, between Rings A and B; the Encke Gap in Ring A is also quite prominent. These gaps are caused by the perturbing effect of Saturn's inner satellites. The rings are made up of components ranging from sub-centimetre-size particles to objects many metres across, all in individual orbits around Saturn. The main rings are only a kilometre or so thick. Ring A's diameter is about 275,000 km (170,000 miles).

SATURN: DATA	
Globe	
Diameter (equatorial)	120,000 km
Diameter (polar)	107,100 km
Density	0.69 g/cm^3
Mass (Earth = 1)	95.2
Volume (Earth = 1)	764
Sidereal period of axial rotation (equatorial)	10h 14m
Escape velocity	35.5 km/s
Albedo	0.70
Inclination of equator to orbit	26° 43′
Temperature at cloud-tops	95 K
Surface gravity (Earth = 1)	1.19
Orbit	
Semimajor axis	9.539 AU = 1427×10^6 km
Eccentricity	0.056
Inclination to ecliptic	2° 29′
Sidereal period of revolution	29.46y
Mean orbital velocity	9.65 km/s
Satellites	49

Following Pioneer 11 in 1979, the two Voyager probes flew past Saturn, Voyager 1 in 1980 and Voyager 2 in 1981. They revealed the ring system to be made up of thousands of separate ringlets. The gaps were not empty, but simply regions with a lower ring density. Some rings were eccentric, not circular. Ring F has a braided structure, being made up of intertwined ringlets and controlled by **shepherd moons**. *Spokes*, dark radial features in Ring B, apparently mark the trajectories of ring particles moving outwards along magnetic field lines. The **Cassini** probe, which reached Saturn in 2004, confirmed these findings, and suggested that only small changes had occurred in the rings since the Voyager probes' exploration.

Like Jupiter's disk, Saturn's shows a pattern of darker *belts* and lighter *zones* parallel to the equator. These atmospheric features appear rather muted in comparison with Jupiter's because of the presence of a high-altitude haze layer. Storms and associated lightning have been detected, but there is no equivalent of Jupiter's Great Red Spot, although short-lived white spots occasionally erupt near the equator. The most recent, in 1990, grew until it was spread around much of the equator, and was referred to as the Great White Spot. There is **differential rotation** on Saturn: features at the equator rotate in about $10\frac{1}{4}$ hours, and those near the poles in nearly $10\frac{3}{4}$ hours (the rotation period of the interior). Winds at the equator can reach speeds of 1800 km/h (1100 mile/h).

Saturn has an internal heat source, which probably drives its weather. Its density, 0.7 g/cm^3, is much lower than for any other planet (and its oblateness is higher than for any other). Saturn is thought to be composed predominantly of hydrogen, and to have an iron–silicate core about five times the Earth's mass, surrounded by an ice mantle. Near the mantle the hydrogen should be in metallic form (so compressed it behaves as a conducting metal), above this existing as liquid molecular hydrogen. The upper atmosphere contains 97% hydrogen and 3% helium, with traces of other gases including methane, ethane, ammonia and phosphine.

The magnetic field of Saturn is 500 to 1000 times the strength of the Earth's, and its magnetic axis is nearly coincident with the axis of rotation. The magnetic field generates a **magnetosphere** intermediate between those of the Earth and Jupiter in its shape and extent and in the intensity of its radiation belts of trapped particles. Aurorae have been detected around Saturn's polar regions.

As of early 2005, Saturn has 49 known **satellites**.

Saturn Nebula The planetary nebula NGC 7009 in Aquarius, just 160 light years away. Handle-like protuberances give it the appearance of Saturn. The nebula is small, of 8th magnitude, and its appearance probably indicates an internal structure consisting of a series of shells ejected by the central star.

scattering The deflection of light or other types of electromagnetic wave by particles. For particles very much larger than the wavelength of the light, the scattering is a mixture of reflection and **diffraction**, and the amount of scattering depends very little on wavelength. For particles very much smaller than the wavelength, the amount of scattering (d) is inversely proportional to the fourth power of the wavelength (λ): $d \propto 1/\lambda^4$. Thus blue light is scattered by small particles ten times as much as red light. Scattering of sunlight by atoms and molecules in the atmosphere is what makes the sky blue. It is the main cause of **atmospheric extinction**, making the Sun appear red at sunrise and sunset by preferentially scattering blue light out of the line of sight.

Schiaparelli, Giovanni Virginio (1835–1910) Italian astronomer. He was an assiduous observer of the inner planets, Mars in particular. He prepared a map of the surface of Mars, introducing a nomenclature for the various features, that remained standard until the planet was mapped by space probes. His use of the term *canali* (SEE **canals**) was the cause of much subsequent controversy. Schiaparelli discovered the connection between **meteor streams** and comets, and also studied double stars.

Schmidt, Maarten (1929–) Dutch-American astronomer. He is known for his work on **quasars**, in particular for his discovery in 1963 of the large **redshift** of the lines in the spectrum of the quasar designated 3C 273. This discovery, and his subsequent finding that the number of quasars increases with distance, provided important support for the big bang theory. Schmidt also ascertained the spiral structure of the Galaxy and the distribution of mass within it.

Schmidt camera (Schmidt telescope) A wide-field **catadioptric** telescope used mainly for photography, developed in 1930 by the Estonian instrument-maker and astronomer Bernhard Schmidt (1879–1935). Schmidt telescopes use a spherical primary mirror with a correcting plate, placed at the mirror's centre of curvature, figured so as to minimize optical aberrations in the image. The image is formed on a curved focal surface in front of the mirror (necessitating the use of a special plate-holder to curve the photographic plate or film) and is sharp over a diameter of 10° or more.

The Schmidt camera and its numerous variations enabled whole-sky photographic surveys to be carried out, and significant advances to be made in photographic astronomy. Modifications of the Schmidt camera design that have made popular amateur telescopes have been the **Maksutov telescope** and, more recently, the **Schmidt–Cassegrain telescope**.

Schmidt–Cassegrain telescope (SCT) A short-focus telescope combining features of the **Schmidt camera** and the **Cassegrain telescope**. The SCT has the Schmidt's spherical primary mirror and specially figured correcting plate, but light is reflected back down the tube by a convex secondary mirror mounted behind the plate and then through a hole in the primary to a Cassegrain focus. This makes for a highly compact and portable telescope which, since the 1970s, has become very popular with amateur astronomers.

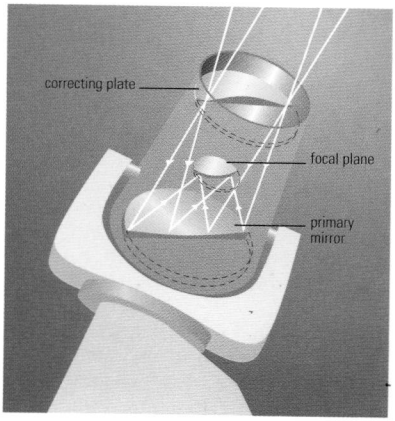

▲ **Schmidt camera** This type of reflecting telescope gives a wide field of view. They are used only for astrophotography and are useful in survey work.

Schmidt telescope SEE **Schmidt camera**.

Schröter effect A difference between the theoretical and observed dates of dichotomy for Venus. At western (morning) elongation dichotomy comes a day or two late, while at eastern elongation it is a day or two early. The effect, the cause of which is unknown, was first observed in the 1790s by the German astronomer Johann Schröter (1745–1816).

Schwarzschild, Karl (1873–1916) German astronomer. He worked on methods of measuring stellar magnitudes from photographs, and studied stellar atmospheres and motions. In the last year of his life he solved equations in **Einstein**'s General Theory of Relativity that dealt with the gravitational field of a point mass, from which he developed the concepts of what are now called the **Schwarzschild radius** and the **black hole**. His son, **Martin Schwarzschild** (1912–97), became a naturalized American and an astrophysicist, working on stellar evolution.

Schwarzschild radius The critical radius at which a very massive body under the influence of its own gravitation becomes a **black hole**. It is the radius of the event horizon of a black hole, from which nothing can escape, not even light. The radius is given by the expression

$$R = 2GM/c^2,$$

where G is the gravitational constant, M is the mass of the body and c is the speed of light. The Schwarzschild radius for the Sun is about 3 km, and for the Earth it is about 1 cm. (After Karl Schwarzschild.)

scintillation The twinkling of the stars – rapid changes in brightness and colour, particularly noticeable in stars low in the sky; through a telescope, the positions of stars are seen to undergo short, rapid changes. The effect is caused by the non-uniform density of the Earth's atmosphere, producing uneven refraction of starlight. In general, planets do not exhibit this phenomenon because the scintillations from different points on their surface are not in phase, and the fluctuations are lost in the general illumination. The inconvenience of the effects of scintillation is reduced by siting observatories at high altitudes and by the use of **adaptive optics**. It is almost entirely avoided by using telescopes carried by balloons, rockets or artificial satellites.

Radio sources also scintillate: the non-uniformity of the refractive indices of the interstellar medium, the interplanetary medium and the ionosphere causes the strength of radio waves to fluctuate.

SEE ALSO **seeing**.

Scorpius A constellation of the zodiac, in mythology representing the scorpion that killed Orion. Its long line of bright stars has the appearance of a scorpion and its sting, with its heart marked by first-magnitude **Antares**, the constellation's brightest star. The 'sting' is Lambda (Shaula, magnitude 1.6). Both Mu and Zeta are optical doubles. In addition to being immersed in a very rich part of the Milky Way, Scorpius contains some magnificent clusters, notably the globular cluster M4 near Antares, and M6 and M7, open clusters both visible with the naked eye.

Scorpius X-1 The first cosmic X-ray source, discovered in 1962, and the brightest known apart from occasional transients (SEE **X-ray burster**). It is a low-mass X-ray binary with an orbital period of 19.2 hours; one component is a neutron star and the nature of the other is unknown. X-rays are thought to originate from the neutron star and from a thin **accretion disk** around it.

SCT ABBREVIATION FOR **Schmidt–Cassegrain telescope**.

Sculptor A barren southern constellation representing a sculptor's studio, introduced by **Lacaille**. It contains no star above magnitude 4.3, but is rich in faint galaxies, including a dwarf member of our Local Group.

Scutum A small constellation adjoining Aquila, representing a shield. Alpha, magnitude 3.9, is the only star above magnitude 4, but Scutum contains the splendid open cluster M11, nicknamed the **Wild Duck Cluster**, and is crossed by the Milky Way. Delta Scuti is the prototype of a class of variable stars that pulsate in periods of 0.01 to 0.2 days with very small amplitudes, usually only several hundredths of a magnitude.

season One of a number of periods making up a cycle of changes in the surface conditions of a planet or satellite (average temperature and 'weather') as its axial tilt gives different parts of its surface longer or shorter periods of sunlight over the course of one revolution around the Sun. The Earth's seasons are defined by the different positions of the Sun with respect to the ecliptic. In the northern hemisphere, spring is reckoned from the vernal equinox (21 March) to the summer solstice (21 June), summer from the summer solstice to the autumnal equinox (23 September), autumn from the autumnal equinox to the winter solstice (22 December) and winter from the winter solstice to the vernal equinox. Spring in the northern hemisphere corresponds to autumn in the southern hemisphere, summer to winter, autumn to spring, and winter to summer.

Of the other planets, Mars has a similar axial inclination to the Earth's (25° 11′ compared to 23° 27′), and shows seasonal variations such as changes in its **polar caps**.

Secchi, (Pietro) Angelo (1818–78) Italian astronomer and priest. He was a pioneer of spectroscopy, which he used to study the Sun and the stars. In the 1860s he carried out the first spectroscopic survey of stars, cataloguing the spectra of over 4000 and dividing them into four classes according to their colour and spectral characteristics (SEE **spectral classification**).

secondary (1) Any of the smaller members of a system of celestial bodies that orbit around the largest, the **primary** (1).

secondary (2) **(secondary mirror)** The small mirror in a **reflecting telescope** that diverts the converging beam of light from the primary mirror towards the eyepiece.

second contact In a total **solar eclipse**, the moment when the leading edge of the Moon's disk covers the last visible part of the Sun's disk, and totality begins. In a **lunar eclipse** it is the moment when the trailing edge of the Moon enters the umbra of the Earth's shadow. The term is also used for the corresponding stage in eclipses involving other bodies.

second of arc SEE **arc minute, second**.

secular acceleration A gradual long-term change in the orbital motion of one body around another. For example, tidal friction (SEE **tides**) between the Earth and

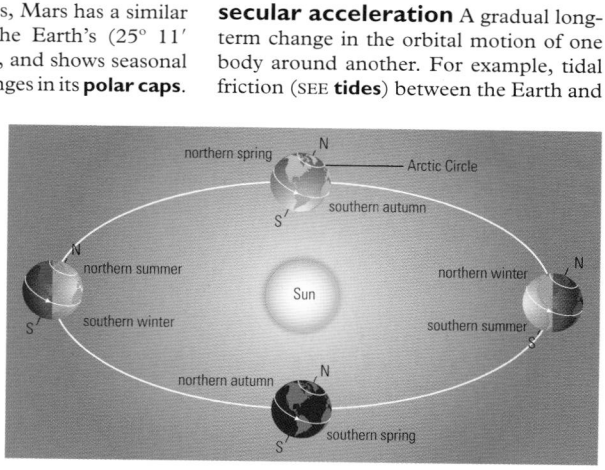

► **season** Because the Earth's axis is inclined relative to its plane of orbit, the Earth experiences seasons. The side that is tilted towards the Sun gets more light and has longer days and so experiences summer.

the Moon is slowing the Earth's rotation and transferring angular momentum from the Earth to the Moon.

secular parallax The apparent displacement of a celestial body in the sky over time as a result of the Sun's motion in space.

Sedna A distant Solar System object discovered by Mike Brown and colleagues in November 2003. Sedna follows a highly elliptical, 10,200-year orbit with perihelion 76 AU from the Sun, and aphelion at 860 AU. Sedna's distance is greater than that of known **Kuiper Belt** objects, and it may belong to an intermediate class of body gravitationally perturbed inward from the **Oort Cloud**. Sedna has an estimated diameter of 1800 km (1100 miles), rivalling that of **Pluto**.

seeing The quality of the image produced by a telescope, as affected by the steadiness of the air. Seeing depends on the calmness of the atmosphere immediately overhead, and also on air movements at ground level, and even in the telescope tube itself. The evaluation of seeing is largely subjective, though values assigned according to the **Antoniadi scale** can help achieve a degree of uniformity in the reporting of conditions under which observations are made.

seleno- Prefix referring to the Moon, as in selenography, the description and mapping of the Moon's surface features. (From *Selēnē*, Greek for Moon.)

semimajor axis (symbol *a*) Half the longest diameter of an **ellipse**. The semimajor axis of an elliptical **orbit** is the average distance of an object from its primary, and is one of the elements used when defining an orbit.

semiregular variable A pulsating giant or supergiant star, usually of late spectral type (typically M or C), with a period between 20 and 2000 days or more and an amplitude from a few hundredths of a magnitude to 1–2 magnitudes. Some semiregular variables show a definite periodicity, interrupted at times by irregularities. Others differ little from long-period variables except in having smaller amplitudes (less than 2.5 magnitudes). The various subtypes exhibit periodicities that cover a wide range from about 20 days to several thousand days.

separation The angular distance between two celestial objects, in particular the members of a visual binary or multiple star system.

Serpens A constellation in two parts, Caput (the head) and Cauda (the body), separated by the serpent-bearer Ophiuchus; Serpens represents a snake coiled around Ophiuchus. Alpha (Unukalhai), its brightest star, is magnitude 2.6. Theta is a wide, easy pair with components of magnitudes 4.6 and 5.0. Delta is also an easy telescopic double, magnitudes 4.2 and 5.2. The Mira-type variable R Serpentis can reach magnitude 5.2 (minimum 14.4, period 356 days). There are two bright Messier objects: M5, a globular cluster, and M16, an open cluster embedded in the **Eagle Nebula**.

Setebos A small outer **satellite** of Uranus, discovered in 1999.

SETI Abbreviation for 'search for extraterrestrial intelligence', a term used for a variety of projects seeking signs of intelligent life elsewhere in the Universe. The first was Project Ozma, which in 1960 began to search for radio signals. None has met with success.

Seven Sisters Popular name for the **Pleiades**, an open star cluster in Taurus.

Sextans An obscure constellation between Leo and Hydra representing a sextant. Its brightest star is only magnitude 4.5.

Seyfert galaxy A class of galaxies that have extremely bright, compact **active galactic nuclei** and whose spectra show strong emission lines. About 1% of all galaxies are Seyferts; M77 (NGC 1068) is one of the brightest. Seyferts emit strongly at ultraviolet and infrared wavelengths, and exhibit a degree of short-term variability. Some are strong X-ray sources but few are particularly strong radio emitters. Most are spirals. The gas clouds in Seyfert galaxies move at several thousand kilometres per second, probably because of the gravitational influence of a massive black hole at the nucleus. They are named after American astrophysicist Carl Seyfert (1911–60), who first studied them.

▲ **Seyfert galaxy** M77 (NGC 1068) is the prototype type 2 Seyfert galaxy. Its spectrum indicates that there is high-velocity gas in its inner regions. Visually, Seyfert galaxies are characterized by their bright, small nuclei.

shadow bands Grey ripples seen moving across light-coloured surfaces in the minutes either side of a total eclipse of the Sun. Close to totality the bands are more ordered and show higher contrast. They are caused by turbulence in the atmosphere affecting light from the very thin crescent of the Sun, in the same way that it affects starlight to produce **scintillation**.

Shapley, Harlow (1885–1972) American astronomer. In 1914 he explained how the variability of **Cepheid variables** is caused by their pulsations. In the years that followed he calibrated the period–luminosity law discovered by Henrietta **Leavitt**, using Cepheid variables in globular clusters, which enabled him to make the first accurate estimate of the size of the Galaxy and the Sun's position in it. He later catalogued many thousands of galaxies, and discovered that galaxies are arranged in clusters.

shell star A hot B-type star with prominent absorption lines in its spectrum, from a shell of material surrounding the star. Some shell stars also exhibit an equatorial disk thrown out by the star's rapid rotation. Strong shell spectra, which include emission lines from the surrounding disk, are seen in B0e to B3e giants and dwarfs. Examples are Gamma Cassiopeiae, BU Tauri (Pleione) and 48 Librae (FX Librae).

shepherd moon A minor moon that through its gravitational influence keeps in check the particles in a planetary ring. Shepherd moons often act in pairs. Examples are Prometheus and Pandora, orbiting either side of Saturn's F Ring, and Cordelia and Ophelia, orbiting either side of Uranus's Epsilon Ring.

Shoemaker–Levy 9, Comet A **comet** discovered in 1993 jointly by Eugene and Carolyn Shoemaker, David Levy and Philippe Bendjoya. It was in a temporary orbit around Jupiter, and a close approach to the planet in 1992 had disrupted its nucleus into over 20 separate fragments. Over a six-day period in July 1994 these fragments collided with Jupiter, creating fireballs and leaving temporary dark scars at the impact sites. This was a unique event in the history of observational astronomy, observed worldwide by amateur and professional astronomers.

shooting star A popular (chiefly British) name for a **meteor**.

short-period comet A **comet** whose apparitions are sufficiently frequent to permit correlation of orbital data; formally

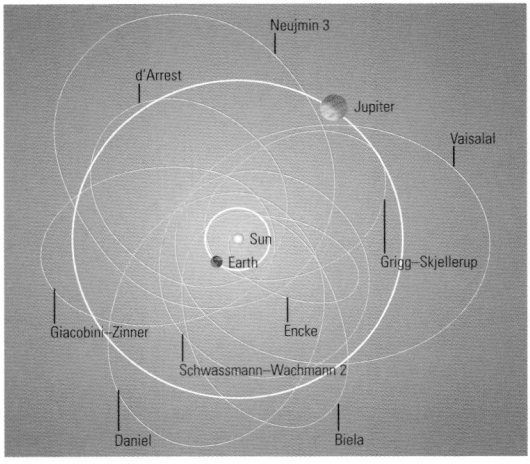

◀ **short-period comet** Most short-period comets are thought to have originated in the Kuiper Belt and to have been perturbed into the inner Solar System by the gravitational influence of the gas giants. Encke has an orbital period of only 3.3 years, Grigg–Skjellerup 5.1 years and Giacobini–Zinner 6.5 years. Biela's Comet suffered the probable fate of all short-period comets: on one orbit, its nucleus was seen to have broken into at least two pieces and it never reappeared.

defined as one whose period is less than 200 years.

Siarnaq A small outer **satellite** of Saturn, discovered in 2000.

sidereal Of or pertaining to the stars; measured or determined with reference to the stars.

sidereal day The interval between two successive passages of a given star, or the vernal equinox, across the observer's meridian. It is equal to 23h 56m 4.091s of mean solar time.

sidereal month The time taken for one revolution of the Moon around the Earth relative to a fixed star. The sidereal month is equal to 27.32166 mean solar days.

sidereal period The orbital period of a planet or other celestial body with respect to a background star. It is the true orbital period. COMPARE **synodic period**.

sidereal time Local time reckoned according to the rotation of the Earth with respect to the stars. The time is 0h when the vernal equinox crosses the observer's meridian. The sidereal day is 23h 56m 4s of mean solar time – nearly 4 minutes shorter than the mean solar day. Sidereal time is equal to the **right ascension** of an object on the observer's meridian.

sidereal year The time required for the Earth to complete one revolution around the Sun relative to the stars. It is equal to 365.25636 mean solar days.

siderostat A mirror mounted equatorially and driven so as to counteract the apparent movement of a star or other celestial object and direct its light into a fixed telescope. A more sophisticated version is the **coelostat**.

Siding Spring Observatory An astronomical observatory in the Warrumbungle mountains of New South Wales, founded in 1962 by the Australian National University, which operates several telescopes there including the 2.3-m (90-inch) Advanced Technology Telescope opened in 1984. The site is shared by the **Anglo-Australian Observatory**.

singularity A point at which the known laws of physics break down. For example, the singularity at the centre of a **black hole** is a point at which matter is compressed into an infinitely small volume and so has an infinitely large density. In the theory of the **big bang**, the Universe was born from a singularity of infinite density and temperature.

Sinope An outer **satellite** of Jupiter discovered in 1914 by Seth Nicholson. It is in a **retrograde** (1) orbit and may be a captured asteroid.

sinuous rille SEE **rille**.

Sirius The star Alpha Canis Majoris. It is the brightest star in the sky, magnitude −1.44, distance 8.6 light years, luminosity 22 times that of the Sun. Sirius is a main-sequence star of spectral type A0, and is the sixth-closest star system to us. It has a binary companion, Sirius B, whose existence was deduced by Friedrich **Bessel** in 1844 from its gravitational perturbations on Sirius, which affected the proper motion of Sirius. The companion, magnitude 8.4, was discovered optically by Alvan G. Clark in 1862. Sirius B is a **white dwarf** of 1.03 solar masses, but with a diameter only 0.92 times that of the Earth. Its orbital period is 50 years. It was the first star to be recognized as a white dwarf.

Sixty-one Cygni (61 Cygni) The first star to have its parallax measured (by Friedrich **Bessel** in 1838). It lies 11.4 light years away and is a binary consisting of two K-type main-sequence stars, magnitudes 5.2 and 6.0, with a period of 660 years.

Slipher, Vesto Melvin (1875–1969) American astronomer. On the death of Percival **Lowell** he became director of the Lowell Observatory. He was a specialist in astronomical spectroscopy, especially of galaxies. In 1912 he became the first to obtain the spectrum of another galaxy, the Andromeda Galaxy, and in 1920 he discovered the **redshifts** in galactic spectra. By 1925, following Edwin **Hubble**'s finding that the redshifts of galaxies increase with distance, he had measured the radial velocities of over 40 galaxies. Slipher also showed that light from certain nebulae was reflected from stars embedded in them, thus discovering reflection nebulae.

Sloan Digital Sky Survey (SDSS) A survey, including a **redshift survey**, begun in 1998, aimed at covering a quarter of the sky. Over the period of the project it is expected to obtain information about hundreds of millions of objects of all types, and plot the distribution of about a million galaxies and some 100,000 quasars in a volume of space over 30 times as large as in any previous survey. The SDSS has a dedicated 2.5-m (100-inch) reflector at **Apache Point Observatory**. It is named after Alfred Pritchard Sloan, Jr (1875–1966), whose foundation financed the project.

Small Magellanic Cloud SEE **Magellanic Clouds**.

SNC meteorites A group of achondritic meteorites (SEE **achondrites**) believed to have originated on Mars. Their ages (mostly 1.3 billion years or less) indicate that they originated on a sizeable planet that has been geologically active, and gases trapped in them are consistent with analyses of the Martian atmosphere made by spacecraft. It is unclear how they could have emerged as intact fragments from an impact of sufficient force to accelerate them to Mars's escape velocity. The name (pronounced 'snick') comes from the three subtypes: *shergottites*, *nakhlites* and *chassignites*.

SNR ABBREVIATION FOR **supernova remnant**.

SOFIA ABBREVIATION FOR **Stratospheric Observatory for Infrared Astronomy**.

SOHO ABBREVIATION FOR **Solar and Heliospheric Observatory**.

Solar and Heliospheric Observatory (SOHO) A European Space Agency probe, operated jointly with NASA, to study the corona of the Sun, oscillations of the Sun's surface and the solar wind. It was launched on 2 December 1995, and placed at the L_1 **Lagrangian point**, 1.5 million km (900,000 miles) from the Earth in the direction of the Sun, in March 1996. A main aim of the mission is to probe the Sun's interior structure. SOHO discovered many **sungrazer** comets.

solar apex, antapex SEE **apex**.

solar constant A measure of the amount of solar energy received by a body a certain distance from the Sun. For the Earth, the solar constant is defined as the solar power received per unit area, at the top of the atmosphere, at the average Earth–Sun distance of 1 AU; its value is about 1.35 kW/m^2. The 'constant' varies from day to day with sunspot activity, and also in the longer term with the **solar cycle**.

solar cycle The periodic fluctuation in the number of **sunspots** and the levels of other kinds of solar activity; the cycle lasts about 11 years. Over the course of a cycle, sunspots vary both in number and in latitude in a way which is neatly illustrated in the **butterfly diagram**. The variation in latitude is described by **Spörer's law**. The diagram here shows the variations in sunspot activity going back to the 17th century, as reckoned by the **relative sunspot number**. At *solar maximum*, when sunspot numbers are greatest, astronomers refer to the *active Sun*; *solar minimum* is also called the *quiet Sun*. There is some variation in the level of activity at solar maximum, and in the period up to 1715 – during the **Maunder minimum** – activity was low for about 70 years.

solar day SEE **day**.

solar eclipse An **eclipse** of the Sun by the Moon. Since the Moon's orbital plane is inclined to the plane of the ecliptic, a solar eclipse can occur only when the Moon is at conjunction (i.e. at new Moon) and at the same time is at or near one of its **nodes**; the Sun, Moon and Earth are then very nearly in a straight line. In addition, the Sun's angular distance from one of the Moon's nodes at conjunction will determine whether an eclipse can or cannot occur (SEE **ecliptic limits**).

The apparent diameters of the Sun and the Moon as seen in the sky are almost the same, but they do vary slightly, particularly the Moon's, whose maximum and minimum distances from the Earth differ by about 10%. A *total solar eclipse* is seen at places where the umbra of the Moon's shadow-cone falls on and moves over the Earth's surface; at the same time the eclipse will appear *partial* to observers on either side of the central track of totality. Shortly before and after totality the phenomena of **shadow bands** and **Baily's beads** are

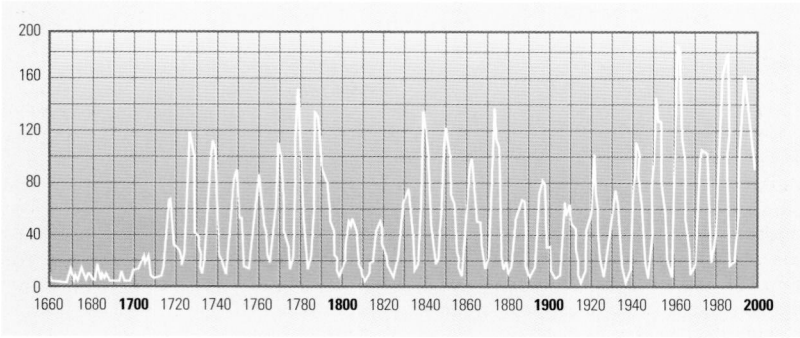

▲ **solar cycle** The number of sunspots observed peaks roughly every 11 years, although the peak intensity can vary widely, as can be seen in this diagram. During the second half of the 17th century there were very few spots; this period is known as the Maunder minimum.

observable. During the brief period of totality, the **corona** and any large **prominences** become visible. When the Moon is near apogee, the tip of its shadow-cone does not reach to the Earth and there is an *annular eclipse*, in which a rim of light is seen around the darkened disk of the Moon.

The overall duration of a solar eclipse from **first contact** to **fourth contact** can be as much as 4 hours; totality, from **second contact** to **third contact**, lasts at most $7\frac{1}{2}$ minutes. There are from two to five solar eclipses each year; if there are five, they will all be partial. The total number of solar and lunar eclipses in a year varies from two to seven; if there are only two, they will both be solar.

solar mass The mass of the Sun, used as a unit of mass for other stars and celestial objects.

solar parallax SEE **parallax**.

Solar System The Sun and all the celestial bodies which revolve around it: the nine **planets**, together with their **satellites** and ring systems, the thousands of **asteroids** and **comets**, meteoroids and other interplanetary material. The boundaries of the Solar System extend beyond the orbit of Pluto, which orbits the Sun at an average distance of 30 AU, to include the **Kuiper Belt**, which may extend to 100 AU, and the inferred **Oort Cloud** of comets, some of which may have aphelion distances of as much as 100,000 AU ($1\frac{1}{2}$ light years). Interplanetary space contains cosmic dust and extremely tenuous ionized gas (particles of the **solar wind**); SEE **interplanetary matter**.

All the planets have direct orbits: they revolve around the Sun in the same direction as the Sun itself rotates; the orbital motion of most of the satellites is direct too. The Sun is by far the most massive component of the Solar System, its mass being 330,000 times that of the Earth. The nine planets have a total mass equal to 448 times that of the Earth, of which Jupiter accounts for 70%.

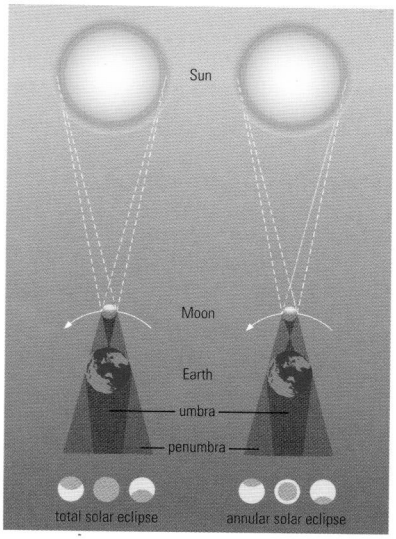

▲ **solar eclipse** In a total solar eclipse (left) the Sun's photosphere is completely obscured by the Moon's disk in the area covered by the cone of shadow. In an annular solar eclipse (right) the Moon is slightly farther away and so the cone of shadow does not reach the Earth.

The Solar System came into being nearly 5000 million years ago as the end-product of a contracting cloud of interstellar gas and dust, the *solar nebula* (SEE **cosmogony**). Similar systems centred on other stars, referred to as 'solar systems', without the capital S's (SEE **extrasolar planet**), and the existence of disks of matter surrounding stars (SEE **protoplanetary disk**), suggest that planets are by-products of star formation.

solar wind The steady flow of charged particles (mainly protons and electrons) from the solar corona into interplanetary space, controlled by the Sun's magnetic field. Solar wind particles are too energetic to be held back by the Sun's gravity. Some of them get trapped in planetary magnetic fields. At the Earth, some are trapped in the outer **Van Allen Belt**; others reach the Earth's upper atmosphere in oval regions surrounding the magnetic poles and cause **aurorae**. The solar wind carries away about 10^{-13} of the Sun's mass per year. Its intensity increases during periods of high solar activity. SEE ALSO **stellar wind**.

solstice The extreme northern or southern position of the Sun in its yearly path among the stars. At the solstices the Sun reaches its greatest declination, $23\frac{1}{2}$°N or $23\frac{1}{2}$°S. The *summer solstice* occurs around 21 June and the *winter solstice* is on or about 22 December.

solstitial colure The **great circle** that passes through the north and south celestial poles and the winter and summer **solstices**.

Sombrero Galaxy A spiral galaxy in the constellation Virgo, designated M104 or NGC 4594. The galaxy is seen nearly edge-on, and has the appearance of a sombrero hat. It is crossed by a thick dark lane of interstellar dust, and there is an unusually large central bulge containing numerous globular clusters. It lies about 30 million light years away.

South African Astronomical Observatory An observatory at Sutherland in the Karoo semi-desert north-east of Cape Town, founded in 1972. Its main instrument is a 1.88-m (74-inch) reflector originally at the Radcliffe Observatory, Pretoria. On the same site is the 11-m (36-ft) Southern African Large Telescope (SALT), the largest single-mirror instrument in the southern hemisphere, due to open in 2005. SALT, built in partnership with Poland, New Zealand, and universities in Germany and the USA, is a close copy of the **Hobby–Eberly Telescope**.

Southern Cross The popular name for the southern-hemisphere constellation **Crux**.

Southern Pleiades A bright open cluster in the southern constellation Carina, including the third-magnitude star Theta Carinae among its members. Designated IC 2602, the object takes its popular name from a description by Nicolas Louis de Lacaille. The Southern Pleiades lie at a distance of 480 light years.

space probe An uncrewed space vehicle sent to investigate the Moon, planets or interplanetary space.

spacetime A unified dimensional framework in which the three dimensions of space (length, breadth and height) and the dimension of time are considered together. An event in spacetime is specified by the three space coordinates and the time coordinate. The concept arose from **Einstein**'s Special and General Theories of Relativity. In the General Theory, gravity is a distortion of spacetime by matter.

space velocity The true velocity of a star in space with respect to the Sun. It is the hypotenuse of the right-angled triangle formed by its **tangential velocity** (obtained from observation of its proper motion) and its **radial velocity** (obtained by measurement of the Doppler shift in its spectrum).

Spacewatch A project to monitor the sky for **near-Earth asteroids** (NEAs), using the 0.9-m (36-inch) and 1.8-m (72-inch) telescopes at **Steward Observatory**. With most NEAs there is no risk of collision, but around a hundred come close enough to be classified as *potentially hazardous asteroids* (PHAs). The Spacewatch project has discovered several of these objects, as well as many NEAs and main-belt asteroids.

speckle interferometer SEE **optical interferometer**.

spectral classification The categorization of stars based on the characteristics of their spectra. In the 1860s Angelo **Secchi** made the first spectral classification of stars by dividing them into four groups according to colour and spectral lines. As spectroscopy improved, a more comprehensive system was developed at Harvard

SPECTRAL CLASSIFICATION: MAIN SPECTRAL TYPES

Type	Surface temperature (K)	Main features
O	50,000–30,000	Blue-white stars. Lines of ionized helium. Weak hydrogen lines.
B	30,000–10,000	Blue-white stars. Neutral helium lines, with hydrogen lines strengthening from B6 to B9.
A	9900–7500	White stars. Very strong hydrogen lines at A0, decreasing toward A9. Lines of ionized calcium increase in strength from A0 to A9.
F	7400–6100	Yellow-white stars. Ionized calcium continuing to increase in strength, and hydrogen weakening. Lines of other elements begin to strengthen.
G	6000–4600	Yellow stars. Strong lines of calcium, hydrogen weaker. Lines of iron become prominent.
K	4900–3250	Orange stars. Strong metallic lines. Molecular bands of CH and CN become prominent.
M	3500–2500	Orange-red stars. Strong absorption bands of titanium oxide and large numbers of metallic lines.

College Observatory. This was embodied in the ***Henry Draper Catalogue*** of stellar spectra, published in 1918–24. The sequence of spectral types was arranged according to the prominence or absence of certain lines in the spectra. There were

SPECTRAL CLASSIFICATION: LUMINOSITY CLASSES

Ia	Supergiants of high luminosity
Ib	Supergiants of lower luminosity
II	Bright giants
III	Normal giants
IV	Subgiants
V	Main-sequence stars (dwarfs)
VI	Subdwarfs

seven spectral types, labelled O, B, A, F, G, K and M (an order resulting from revisions of an earlier alphabetical sequence). This represents a temperature sequence, from the hottest stars, types O and B, which appear blue-white, to type M, the coolest stars, which appear orange-red. As a result of this ordering, O, B and A stars are called **early-type stars**; K and M stars are called **late-type stars**. Three new types were added when it was found that some cool stars had strong absorption bands not usually seen in other stars of the same colour. These were classes R and N, with strong bands of molecular carbon, and class S, with bands of zirconium oxide.

Stellar spectra can be classified into even finer divisions within these seven types, so decimal subdivisions were introduced. G5, for example, indicates a star midway in type between G0 and K0. Further refinements include the use of additional letters as suffixes to the spectral type, giving more information about the star, for example the existence of emission lines (e), metallic lines (m), broad lines due to rotation (n and nn) or a peculiar spectrum (p).

However, the Harvard system could not deal with stars of different luminosities at a given temperature (that is, dwarfs, giants and supergiants). In 1943 William Morgan, Philip Keenan and Edith Kellman of Yerkes Observatory redefined the spectral types and added a classification scheme for the luminosity (absolute magnitude) of stars. This is now known as the **Morgan–Keenan system**, shortened to *MK system*, and is now used universally. For example, a star classified in the MK system as O9.5 IV–V has a spectral type (and therefore a temperature) midway between

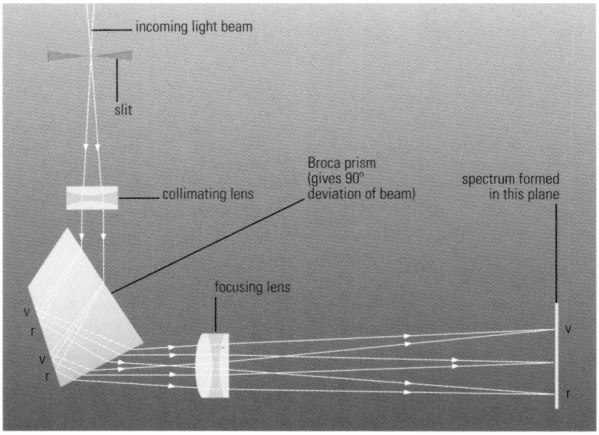

◀ spectroscope
Shown here is one type of traditional spectroscope, using a prism and lenses. The study of spectra, known as spectroscopy, is the key technique by which the physical properties of astronomical bodies are revealed.

that of an O9 and a B0 star, and a luminosity between that of a dwarf and a subgiant. The spectral types and luminosity classes of the MK system are listed in the tables on page 219.

The MK system is applicable to stars of normal chemical composition, which is about 95% of all stars. The various types of peculiar star are given their own special classification schemes. The Harvard R and N types are now combined into one **carbon star** class, the designations for which include a temperature type and a carbon band strength, as for example in C2,4. A similar classification is used for the **S stars**. A class of **L stars** has been added for cool stars emitting in the infrared, and, most recently, a class of **T stars** for the coolest brown dwarfs. White dwarfs are usually classified on a Harvard-type scheme, with D preceding the type.

spectral type A series of divisions, indicating surface temperature and hence colour, into which stars are classified according to the nature of their spectrum. SEE **spectral classification**, **A star**, **B star**, **carbon star**, **F star**, **G star**, **K star**, **L star**, **M star**, **O star**, **S star**, **T star**.

spectrogram A permanent record of the spectrum of an object, produced by a **spectrograph**.

spectrograph A **spectroscope**, fitted with a camera or an electronic detector such as a CCD, used to obtain a permanent record of the spectrum of a celestial object. Astronomical spectrographs are generally designed for use with a specific telescope. On telescopes with a **coudé** system or **Nasmyth focus** a large spectrograph can be mounted in a permanent position. Spectrographs generally operate in a band of wavelengths from the near-infrared to the near-ultraviolet. There are various designs for different purposes. *High-dispersion* instruments such as the **spectroheliograph** spread the spectral lines widely so that a narrow band of wavelengths can be studied in detail.

spectroheliograph A type of **spectrograph** used to image the Sun in the light of one particular wavelength only. A second slit placed in front of the photographic plate or detector images a narrow strip of the Sun's disk at a chosen wavelength. By moving the entrance slit and the second slit in tandem, the whole disk is scanned and an image of it, a *spectroheliogram*, is built up. The spectroheliograph was invented independently by George Ellery **Hale** and Henri Deslandres in the 1890s. The term is also used for a number of modern instruments of different design but working on the same principle.

spectrohelioscope A **spectroheliograph** adapted for visual use. The two slits are oscillated rapidly enough for the motion to be undetectable, so that by persistent vision the observer sees a steady, monochromatic image of the Sun's disk.

spectrometer A **spectroscope** equipped to measure accurately the positions and intensities of spectral lines, by means of a device such as a **Fabry–Pérot interferometer**. Spectrometers are also loosely referred to as **spectrographs**. Since spectra are now routinely recorded on CCDs, from which information in digital form can be analysed by computer, the distinction between modern spectrometers and spectrographs lies in their application rather than their design.

spectrophotometer An instrument that scans the lines in a spectrum recorded by a spectrograph and measures their intensities. It outputs a graph of intensity against wavelength called a *line profile*.

spectroscope An instrument for producing spectra for the analysis of electromagnetic radiation (light or other wavelengths). *Spectroscopy* is the use of such an instrument to probe the chemical composition and physical conditions (such as temperature and motion) of an object. As such it is the most important investigative technique available to astronomers.

In the general form of the instrument, a focused beam of light or other radiation is passed through a narrow slit, to prevent images of different wavelengths from overlapping and to ensure a sharply defined spectrum, and then collimated to produce a parallel beam. This beam is then split into its component wavelengths by a prism or – in most modern instruments – by a **diffraction grating**. After dispersion, the radiation is then focused onto the detector. In its original form, lenses were used for collimation and focusing, but modern equipment normally employs concave mirrors instead. In some instruments the dispersing element consists of a combined grating and prism (known as a *grism*).

All dispersing elements spread out the radiation into various *orders*; the overall dispersion varies between orders. Gratings can be designed to send most of the radiation into a specific chosen order. Particularly high spectral resolution over a wide range of wavelengths can be obtained by the use of a form known as an *echelle grating*, together with a *cross-disperser* (the two often combined in a grism), which separates the different orders and prevents them from overlapping.

Astronomical spectroscopes are known as **spectrographs** or **spectrometers**. Strictly, a spectrograph is a spectroscope equipped with a camera for recording a permanent record of a spectrum, whereas a spectrometer incorporates devices for accurately measuring the wavelengths and intensities of the spectral lines. In modern devices the detector is usually a **CCD**.

spectroscopic binary A **binary star** system whose two components are too close for them to be resolved visually as separate

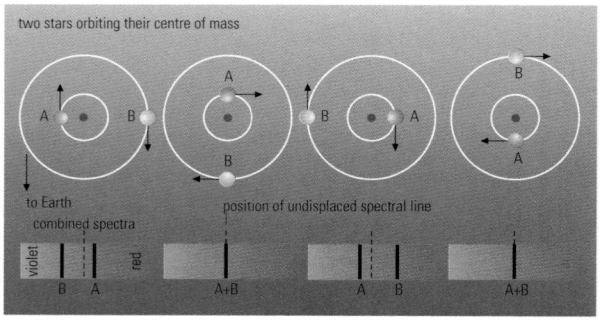

► **spectroscopic binary** Redshift is used to ascertain the relative motions of stars in close binary pairs.

▲ **Spitzer Space Telescope** This image of a star-forming 'bubble' of gas and dust (RCW 79) was captured by the Spitzer Telescope.

objects, but whose binary nature can be deduced from the periodic **Doppler shift** of the absorption lines in the combined spectrum. As the diagram on page 221 shows, when the two stars have reached positions in their orbits around each other where one is moving away from us and the other towards us, the spectral lines of the one are slightly redshifted, and those of the other slightly blueshifted, from their mean positions.

spectroscopic parallax A method of determining the distance of a star from its spectral type (SEE **spectral classification**) and apparent magnitude. Analysis of a star's spectrum reveals its spectral type. If the absolute magnitude of a star of that type is known, the star's distance can be calculated by comparing its absolute and apparent magnitudes. This is the commonest method of determining stellar distances. However, the scale must be calibrated against nearby stars whose distances can be measured directly by **trigonometric parallax**.

spectrum (plural **spectra**) The distribution of intensity of **electromagnetic radiation** with wavelength. When a beam of light from a source is deflected by a prism, or by another dispersing agent such as a diffraction grating, the radiation is fanned out into a spectrum because different wavelengths are deviated by different degrees.

There are various types of spectra . A *continuous spectrum* is an unbroken distribution of radiation over a broad range of wavelengths. White light, for example, is split into a continuous 'rainbow' band of colours from red to violet. A *line spectrum* contains lines corresponding to only certain wavelengths or frequencies which are characteristic of the chemical elements present in the source. An *emission line spectrum* contains lines emitted at particular wavelengths and is produced by substances in an incandescent state at low pressure. An *absorption line spectrum* contains dark lines at particular wavelengths superimposed on a continuous spectrum. It is produced when radiation from a hot source which is emitting a continuous spectrum passes through a layer of cooler gaseous material. The spectra of most stars are continuous spectra crossed by absorption lines, and sometimes by emission lines. Emission line spectra are typical of luminous nebulae. *Band spectra* contain absorption bands, wider than the spectral lines of atoms, and these are the spectral signatures of molecules.

The intensity of an emission line in a spectrum is a measure of the energy of the radiation at that wavelength. The width of the line indicates the wavelength band over which it is distributed. The intensity of the emission is greatest at the centre of the line and decreases towards the edges; the reverse is true for an absorption line. The pressure of the gas producing the spectral line can affect its width; this effect is known as *pressure broadening*. SEE ALSO **flash spectrum**, **hydrogen spectrum**, **Zeeman effect**.

spherical aberration An **aberration** (2) that results from the unequal deflection of light rays by different zones of a spherical lens or mirror. The rays are not brought to the same focus, causing a lack of definition about the edges of the image. Spherical aberration in lenses can be minimized by using a suitable combination of elements, and in reflecting telescopes it is overcome by the use of a **paraboloidal** mirror.

spheroidal Having the form of the surface formed by rotating an ellipse about one of its axes. If the minor axis is chosen, the result is an *oblate spheroid*. This is the shape of fluid bodies – stars and gas planets such as Jupiter – rotating at a sufficient speed, and also of solid but non-rigid bodies such as the Earth. Components of optical systems can also be spheroidal.

Spica The star Alpha Virginis, magnitude 0.98, distance 262 light years, luminosity 2200 times that of the Sun. It is a main-sequence star of type B1 and a spectroscopic binary.

spicule A narrow jet, like a tiny prominence, seen protruding from the Sun's chromosphere when it is viewed edge-on. Spicules are short-lived (about 5–10 minutes) and congregate at the edges of **supergranulation** cells. They are possibly caused by chromospheric gases shooting into the lower corona.

spiral galaxy A **galaxy** consisting of a nucleus of stars from which spiral arms emerge, winding around the nucleus and forming a flattened, disk-shaped region. The arms contain gas, dust and young stars, while the nucleus contains old stars.

Spitzer Space Telescope A spaceborne telescope developed by NASA, formerly known as the Space Infrared Telescope Facility (SIRTF), launched on 25 August 2003 to study the sky at infrared wavelengths between 3 and 180 μm with its 0.85-m (33-inch) telescope. It was placed into a solar orbit trailing behind the Earth.

sporadic meteor A meteor that is not part of a recognized meteor shower – one whose path cannot be traced back to a known **radiant**. The Earth is constantly sweeping up meteoroids, and 3–8 sporadic meteors should be visible each hour under ideal conditions. SEE ALSO **radar astronomy**.

Spörer's law The appearance of sunspots at lower latitudes over the course of the 11-year **solar cycle**. The average position drifts from mid-latitudes (about 30–40°) towards the equator as the cycle progresses from one minimum to the next. The phenomenon was first studied in detail by the German astronomer Gustav Spörer (1822–95), after whom it was later named. Despite it being called a 'law', it has no mathematical formulation. SEE ALSO **butterfly diagram**.

spring equinox SEE **vernal equinox**.

spring tide SEE **tides**.

Square Kilometre Array (SKA) A multi-national project to build a centimetre-wavelength radio telescope with an effective collecting area of 1 sq km (0.4 sq mi). It will synthesize an aperture with a diameter of about 1000 km (625 mi). A site is due to be selected in 2005/6.

SS Cygni star An alternative name for a **U Geminorum star**.

SS433 An unusual binary star, number 433 in a catalogue of stars with bright hydrogen emission lines compiled by American astronomers Bruce Stephenson and Nicholas Sanduleak. SS433 lies at the centre of a supernova remnant in Aquila called W50. It is notable for the variable **Doppler shift** in certain emission lines of its spectrum and for having lines that are

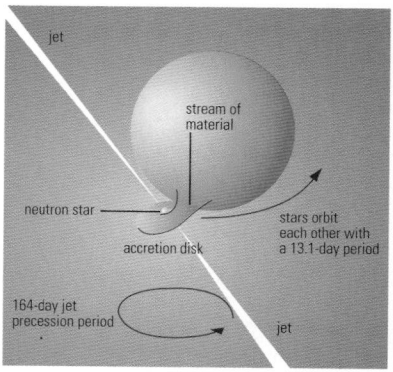

▲ **SS433** Shown here is a theoretical model of the unusual binary pair SS433, which is thought to consist of a hot, massive star orbited by a neutron star.

both red- and blueshifted. SS433 is thought to consist of a hot, massive star orbited every 13.1 days by a neutron star (presumably formed by the supernova that created W50). A stellar wind from the hot star, combined with orbital motion, produces the first set of emission lines, with variable Doppler shifts. Matter is transferred from the massive star to the neutron star via an **accretion disk**, but some of this material is ejected at high speed along two polar jets. The jets precess with a period of 164 days, producing the second set of variable lines with simultaneous red- and blueshifts.

S star A giant star whose surface temperature lies in the range for M stars, but whose spectrum contains absorption bands of zirconium oxide as well as the titanium oxide bands that are characteristic of M stars. The overabundance of zirconium, and also of carbon and many other heavy elements, is a result of convective mixing which brings to the surface products from nuclear reactions in the interior. As a result, many S stars show spectral lines of the radioactive element technetium. M stars with weak zirconium oxide bands are assigned the intermediate spectral type MS. S stars often show hydrogen emission lines in their spectra, and many are **long-period variables**.

standard epoch A fixed point in time, or **epoch**, to which star positions are referred in order to remove the effects of precession, proper motion and perturbation, together with a reference system in which the positions are measured. The standard epoch currently used for catalogues and star charts is designated 2000.0, meaning that all coordinates quoted are correct on noon 1 January 2000.

star A self-luminous ball of gas whose radiant energy is produced by nuclear reactions, mainly the conversion of hydrogen

STARS: THE BRIGHTEST (first magnitude or brighter)

Star	Name	Apparent magnitude	Spectral type	Distance (light years)
	Sun	−26.78	G2	—
α Canis Majoris	Sirius	−1.44	A0	8.60
α Carinae	Canopus	−0.62	A9	313
α Centauri	Rigil Kentaurus	−0.28**	G2+K1	4.39
α Boötis	Arcturus	−0.05 *v*	K2	36.7
α Lyrae	Vega	0.03 *v*	A0	25.3
α Aurigae	Capella	0.08 *v*	G6+G6	42.2
β Orionis	Rigel	0.18 *v*	B8	773
α Canis Minoris	Procyon	0.40	F5	11.4
α Eridani	Achernar	0.45 *v*	B3	144
α Orionis	Betelgeuse	0.45 *v*	M2	427
β Centauri	Hadar	0.61 *v*	B1	525
α Aquilae	Altair	0.76 *v*	A7	16.8
α Crucis	Acrux	0.77**	B0.5+B1	321
α Tauri	Aldebaran	0.87	K5	65.1
α Virginis	Spica	0.98 *v*	B1	262
α Scorpii	Antares	1.05 *v***	M1+B2.5	604
β Geminorum	Pollux	1.16	K0	33.7
α Piscis Austrini	Fomalhaut	1.16	A3	25.1
β Crucis	Mimosa	1.25 *v*	B0.5	353
α Cygni	Deneb	1.25 *v*	A2	3230
α Leonis	Regulus	1.36	B7	77.5
ε Canis Majoris	Adhara	1.50	B2	431

v variable ** *combined magnitude of double star*

STARS: THE NEAREST (closer than 12 light years)

Star	Apparent magnitude	Spectral type	Absolute magnitude	Distance (light years)
Sun	−26.78	G2	4.82	—
Proxima Centauri	11.01 *v*	M5	15.45	4.22
α Centauri A	−0.01	G2	4.34	4.39
α Centauri B	1.35	K1	5.70	4.39
Barnard's Star	9.54	M4	13.24	5.94
Wolf 359	13.44 *v*	M6	16.50	7.8
Lalande 21185	7.49	M2	10.46	8.31
Sirius A	−1.44	A0	1.45	8.60
Sirius B	8.44	DA2	11.34	8.60
UV Ceti A	12.54 *v*	M5.5	15.40	8.7
UV Ceti B	12.99 *v*	M6	15.85	8.7
Ross 154	10.37	M3.5	13.00	9.69
Ross 248	12.29	M5.5	14.79	10.3
ε Eridani	3.72	K2	6.18	10.50
HD 217987	7.35	M2	9.76	10.73
Ross 128	11.12 *v*	M4.5	13.50	10.89
L789–6	13.33	M5	15.64	11.2
61 Cygni A	5.20 *v*	K5	7.49	11.36
Procyon A	0.40	F5	2.68	11.41
Procyon B	10.7	DF	13.0	11.41
61 Cygni B	6.05 *v*	K7	8.33	11.43
HD 173740	9.70	M4	11.97	11.47
HD 173739	8.94	M3.5	11.18	11.64
GX Andromedae A	8.09 *v*	M1	10.33	11.64
GX Andromedae B	11.10	M4	13.35	11.64
G51–15	14.78	M6.5	16.98	11.8
ε Indi	4.69	K5	6.89	11.83
τ Ceti	3.49	G8	5.68	11.90
L372–58	13.03	M4.5	15.21	11.9

v variable

into helium. The temperatures and luminosities of stars depend on their masses (according to the **mass–luminosity relation**). The most massive known stars are about 130 solar masses (i.e. 130 times more massive than the Sun). Above about 150 solar masses a star is theoretically unstable and is likely to break up. Large stars can be very luminous and hot, and therefore appear blue. Medium-sized stars such as the Sun are yellow, while small stars are dull red. The smallest stars are about 0.08 solar masses; below this the temperature cannot become high enough for nuclear reactions to take place, and the object is instead a **brown dwarf**. The colours, and hence the temperatures, of stars give rise to the scheme of **spectral classification**. Most stars are born in pairs or groups (SEE **binary star, star cluster**). SEE ALSO **stellar evolution, stellar populations**.

starburst galaxy A **galaxy** undergoing a strong burst of star formation. Some starburst galaxies are detected by their enhanced optical, ultraviolet or radio emission, but most are characterized by strong emission at far-infrared wavelengths. Hot young stars emit strongly in the ultraviolet. This radiation is absorbed by dust clouds and re-emitted as infrared. Tidal interactions between galaxies can trigger starburst, as can radiation from massive stars and supernovae.

star cluster A group of physically associated stars. Star clusters are classified as globular or open. A **globular cluster** contains an extremely large number (10^5 to 10^7) of ancient Population II stars (SEE **stellar populations**) in a roughly spherical space. An *open cluster* contains fewer stars than a globular, in an irregular volume of space. Their members are young (**Population I**) stars and they are found in or near the plane of the Galaxy. SEE ALSO **association, stellar**; **moving cluster**.

star diagonal SEE **diagonal** (1).

Stardust A space probe to Comet Wild 2, launched on 7 February 1999 as part of NASA's **Discovery** programme. It encountered the comet in January 2004, making measurements and taking images. By using a collector covered with a special aerogel, it captured dust particles from the coma which will be ejected in a capsule when Stardust returns to Earth in January 2006.

Star of Bethlehem The 'star seen in east' mentioned in the biblical Gospel of St Matthew. It has been ascribed to various astronomical phenomena, including comets, novae and the conjunction of two or three planets. The three separate conjunctions of Jupiter and Saturn in Pisces in 7 BC are one possibility. However, the most remarkable astronomical event about that time was the apparent merging in Leo of Venus and Jupiter, the two brightest planets, in June 2 BC, which would have created a brilliant temporary 'star'. It is also possible that the Star of Bethlehem was a literary invention inserted into St Matthew's Gospel at a later date.

star streaming The overall effect of the proper motions of stars (corrected for the Sun's motion) not being random, giving rise to two preferred directions of motion in exactly opposite directions. One vertex is in the constellation Orion, the other in Scutum. Star streaming results from the motion of stars in their orbits around the Galaxy. It was discovered by the Dutch astronomer Jacobus Kapteyn in 1904.

star trailing The appearance of stellar images as curved lines, or trails, on images obtained by a stationary camera – one that is not driven to compensate for the rotation of the Earth. The lengths of the trails depend on the exposure time. A camera aimed at one of the celestial poles will record star trails as concentric arcs centred on the pole.

stationary point A point in the apparent path of a planet when it changes between easterly and westerly motion – SEE **retrograde** (2).

steady-state theory A cosmological theory put forward by Hermann **Bondi** and Thomas **Gold** in 1948, and further developed by Fred **Hoyle** and others. According to this theory the Universe has always existed; it had no beginning and will continue for ever. Although the Universe is expanding, it maintains its average density – its steady state – through the continuous creation of new matter. The steady-state theory was rejected because, unlike the rival **big bang** theory, it cannot naturally explain the **cosmic microwave background**, the abundances of helium and other light elements, or the observation that the appearance of the Universe has changed with time.

stellar association SEE **OB association**, **T association**.

stellar evolution The development of a star over its lifetime, which can range from thousands of years to thousands of millions of years, depending on its mass. The theory of stellar evolution is based on mathematical models of stellar interiors. Its account of the evolution of stars can be checked by observations of stars at each of the predicted stages.

Stars form when a cloud of gas and dust collapses under its own gravity. As the cloud collapses, its temperature rises because the energy of infalling atoms is released as heat when they collide. At this stage it is termed a *protostar*. A protostar will continue to contract under gravity until its centre is hot enough (about 15 million K) for nuclear reactions to commence, with the release of energy (SEE **carbon–nitrogen cycle**, **proton–proton reaction**). The outward flow of radiation from deep within halts the collapse (SEE **radiation pressure**). As this is happening, material is still being accreted. **T Tauri stars** are low-mass stars in this final stage of collapse.

Once nuclear fusion has commenced in the core of a star, the star adopts a stable structure with its gravity balanced by the radiation from its centre. This is known as the main-sequence phase of a star's life (SEE **Hertzsprung–Russell diagram**). It is the longest phase in the life of a star. The Sun has a main-sequence lifetime of some 10,000 million years, of which half has expired. Less massive stars are redder and fainter than the Sun, and stay on the main sequence for longer. More massive stars are bluer and brighter and have a shorter lifetime. As a star ages on the main sequence, nuclear fusion changes the composition of its core from mainly hydrogen to mainly helium.

Eventually, the hydrogen supply in the core of the star runs out. With the central energy source removed, the core collapses under gravity and heats itself further until hydrogen fusion is initiated in a spherical shell surrounding the core. As this change occurs the outer layers of the star expand considerably and the star becomes a **red giant** or, in the case of the most massive stars, a **supergiant**. While it is a red giant, the star's core temperature will reach 100 million K, hot enough for the fusion of helium to carbon to begin (the **triple-alpha process**). In low-mass stars the onset of helium fusion is sudden (the *helium flash*), but in high-mass stars it gets under way more gradually. After this stage high-mass stars can fluctuate in size, becoming **long-period variable** stars such as Mira.

When all of the helium fuel in the core has been converted into carbon, the core of the star again collapses and heats up. In a low-mass star the central temperature cannot rise far enough to initiate carbon fusion. The red giant sheds its outer layers as a **planetary nebula**, after which only a **white dwarf** remains. In a high-mass star the contraction of the carbon core leads to further episodes of nuclear fusion involving heavier elements, during which the star remains a supergiant. Once fusion has converted all of the core into elements with atomic weights close to that of iron, no further fusion is possible. At this point the core implodes and the star throws off its outer layers in a **supernova** explosion. The explosion takes place not at the centre but on the exterior of a dense core. The explosion expels several solar masses of gas at speeds of many thousands of kilometres per second, and also pushes inwards on the core with immense force. The effect of this implosion is to crush the already extraordinarily dense core still further, forming a **neutron star** or a **black hole**, both effectively end-points of stellar evolution.

stellar interferometer SEE **optical interferometer**.

stellar nomenclature Some stars have proper names based on names given to them by Greek, Roman and Arab astronomers, such as Sirius, Capella and Aldebaran. In addition to any proper name, naked-eye stars in each constellation are identified by a letter or number followed by the name of the constellation in the genitive (possessive) form (SEE the table of **constellations** on pages 48–9). The brightest stars are indicated by Greek letters, as in α (Alpha) Lyrae or ε (Epsilon) Eridani. These letters are known as *Bayer letters* because they were assigned by Johann **Bayer** in his star atlas *Uranometria* of 1603. Stars that do not have Bayer letters are prefixed by a number, as in 61 Cygni. These are known as *Flamsteed numbers* because they relate to the stars charted in John Flamsteed's *Historia coelestis britannica* of 1725. Fainter stars are referred to by their designation in any of a number of other catalogues. **Variable stars** have their own system of nomenclature.

▲ **stellar wind** The Bubble Nebula is caused by the strong stellar wind from the star at the bottom of the image. When young stars undergo a phase called the T Tauri phase their extremely strong stellar winds blow away much of the nebula in which they were formed.

stellar populations A classification of stars according to their age and location in the Galaxy. There are two principal populations. *Population I* consists of relatively young stars which are located in the plane of the Galaxy, in its spiral arms. The Sun is a Population I star. *Population II* consists of relatively old stars dispersed throughout the entire Galaxy, including **RR Lyrae stars** and subdwarfs, but prominently visible in its centre and halo, particularly in globular clusters. There is also a gradation of stars intermediate between these two extremes.

The distribution of populations in our Galaxy is explained in terms of the contraction of the Galaxy during its formation. The oldest population of stars was made during the infall, which is why Population II is distributed throughout the galactic halo and central regions of the Galaxy. Stars in these regions, including those in globular clusters, continue to have elliptical orbits around the Galaxy. After the Galaxy had developed its flat rotating disk and spiral arms, stars of Population I formed, moving in circular orbits. Between these two extremes lie intermediate populations showing progressively flatter distributions, representing successive stages in the contraction of the Galaxy towards a disk.

Chemical elements produced in stars are recycled back into the interstellar material by stellar winds and supernova explosions, so Population I stars have greater concentrations of heavier elements (**metals**) than Population II. The Sun's content of heavy elements puts it among the younger Population I stars. The chemical history of the Galaxy suggests that a Population III once existed, created when the Galaxy first formed, but which has now disappeared. This early population would have manufactured the smattering of heavy elements present in Population II before dying to form compact objects such as neutron stars.

stellar wind A stream of charged particles, mostly protons and electrons, from the surface of a star. The strength of the wind depends on the type of star, and its velocity can range from a few hundred to several thousand kilometres per second. Young stars evolving towards the main sequence have powerful stellar winds, up to a thousand times stronger than the Sun's **solar wind**. Old stars evolving into red giants also have strong stellar winds.

Stephano A small outer **satellite** of Uranus, discovered in 1999.

Stephan's Quintet A group of peculiar galaxies in the constellation Pegasus discovered in 1877 by the French astronomer Edouard Stephan (1837–1923). Four of the objects (NGC 7317, 7318A, 7318B and 7319) lie 270 million light years away, but the fifth (NGC 7320) is much closer, happening to lie in the same line of sight. The group, without NGC 7320, is sometimes called *Stephan's Quartet*.

Steward Observatory The observatory of the University of Arizona, Tucson. Its main telescope is a 2.3-m (90-inch) reflector, sited on Kitt Peak.

stony-iron meteorite A meteorite with roughly equal proportions of silicates (stony material) and metals, mostly nickel–iron. There are two main subtypes: pallasites and mesosiderites. *Pallasites*, which consist of olivine (a magnesium–iron silicate) mixed with nickel–iron, may have originated near the core/mantle interface of a planetary body. *Mesosiderites* are a much coarser combination of chunks of various silicates and nickel–iron, and could have been produced by impacts on a solid planetary surface.

stony meteorite A **meteorite** that consists mostly of silicates (stony material), with only a small amount of chemically uncombined metals (typically 5% nickel–iron). There are two subtypes: **chondrites** and **achondrites**.

Stratospheric Observatory for Infrared Astronomy (SOFIA) A joint venture between NASA and the German Space Agency to install a 2.7-m (105-inch) telescope aboard a Boeing 747-SP aircraft. It will fly at altitudes above 13 km (8 miles), where it will have unimpeded access to wavelengths of 0.3 to 1600 μm, from the visible to the microwave region. In this part of the spectrum, water vapour interferes with ground-based observations. SOFIA will be based at the Ames Research Center, where is expected to be ready for scientific flights in 2006.

strewnfield SEE **meteorite**, **tektites**.

Strömgren, Bengt George Daniel (1908–87) Swedish-born Danish astronomer. In 1940 he suggested the idea of what is now called a Strömgren sphere to explain how some nebulae shine. He also worked on photoelectric photometry, and the internal structure of stars.

Struve Russian-German family that produced four generations of astronomers. They included:

Struve, (Friedrich Georg) Wilhelm von (1793–1864) emigrated from Germany to Russia in 1833. In 1824 he had begun to observe double stars with a telescope made by Joseph von **Fraunhofer**, and in 1837 he published the first good catalogue of double stars, containing nearly 2500 doubles he had discovered himself. In 1840 he made the third good measurement of the parallax of a star (Vega).

Struve, Otto (Wilhelm) (1819–1905), son of Wilhelm. He studied double stars, and determined an accurate value for the rate of **precession**, taking into account the motion of the Sun with respect to nearby stars.

Struve, Otto (1897–1963), a grandson of the above, emigrated and became a naturalized American. A leading observer and astrophysicist of his time, he applied spectroscopy to the study of binary stars and stellar rotation, and discovered interstellar hydrogen and calcium regions.

Subaru Telescope An 8.3-m (327-inch) reflecting telescope for use at visible and infrared wavelengths, sited at **Mauna Kea**. It commenced operation in 1999. Subaru is the Japanese name for the Pleiades.

subdwarf A star that is less luminous by 1 to 2 magnitudes than main-sequence stars of the same **spectral type**. Subdwarfs are mainly of types F, G and K, and lie below the main sequence on the **Hertzsprung–Russell diagram**. Most are Population II stars (SEE **stellar populations**). They are placed in luminosity class VI, although an alternative designation is to prefix their spectral type with the letters 'sd'.

subgiant A star of smaller radius and lower luminosity than a normal giant star of the same **spectral type**. They are mainly of types G and K, and lie between the main sequence and the giants on the **Hertzsprung–Russell diagram**. Subgiants are placed in luminosity class IV.

submillimetre-wave astronomy Observations of **electromagnetic radiation** at wavelengths of less than a millimetre; in practice, between 0.3 and 3 mm. The techniques for detecting submillimetre radiation are similar to those used for radio wavelengths, but as it is absorbed by water vapour, submillimetre telescopes have to be sited at high altitudes. Important targets at these wavelengths are molecular clouds and regions of star formation.

subsolar point The point on the surface of the Earth or other body in the Solar System at which, at any given moment, the Sun is directly overhead. Similar points are defined for other pairs of bodies.

substellar object SEE **brown dwarf**, **extrasolar planet**.

summer solstice SEE **solstice**.

Summer Triangle The prominent triangle that is formed by the first-magnitude stars Altair (in Aquila), Vega (in Lyra) and Deneb (in Cygnus). The triangle is overhead on summer nights in northern temperate latitudes.

Sun The star at the centre of the Solar System, around which all other Solar System bodies revolve in their orbits. The apparent daily motion of the Sun across the sky and its annual motion along the **ecliptic** are caused by the Earth rotating on its axis and moving in its orbit. The Sun's light is occasionally blocked by the Moon (which, like the Sun, has an apparent diameter of about half a degree) in a **solar eclipse**. Data for the Sun are given in the table.

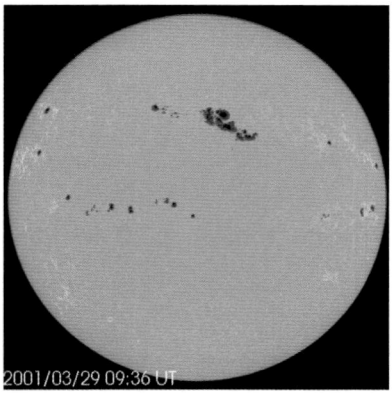

▲ **Sun** Although at a stable part of its life cycle, the Sun is not unchanging and undergoes a roughly 11-year cycle, during which its magnetic field 'winds up' and then declines. Sunspots are the most obvious result of these changes.

The Sun is a typical, average dwarf **G star**, on the **main sequence** of the Hertzsprung–Russell diagram. It consists of about 70% hydrogen (by weight) and 28% helium, with the remainder mostly oxygen and carbon. Its temperature, pressure and density increase towards the centre, where the values are about 15×10^6 K, 10^{11} bar and 150 g/cm^3, respectively. As with all stars, the Sun's energy is generated by nuclear fusion reactions that take place under the extreme conditions in the core, chiefly the **proton–proton reaction**, which converts hydrogen into helium. The mass converted into energy is 4.3 million tonnes per second, but even at this enormous rate the loss amounts to only 0.07% of the total mass in 10^{10} years. The amount of the Sun's energy normally reaching the Earth is quantified as the **solar constant**.

The Sun's core is about 400,000 km (250,000 miles) across. Energy released from the core passes up through the *radiative zone*, which is about 300,000 km (nearly 200,000 miles) thick, by a process of successive absorptions and re-emissions. It then passes through the 200,000-km (125,000-mile) thick *convective zone*, transported by rising and falling cells of gas, to the surface, the **photosphere**, from where it is radiated into space.

Most of the Sun's visible activity takes place in the 500-km (300-mile) thick photosphere. The pattern of convective cells that transport energy from below is visible as the **granulation** (SEE ALSO **supergranulation**). **Sunspots** are darker, cooler regions of the photosphere where the local magnetic field is enhanced. Their passage across the Sun's disk reveals **differential rotation** in the Sun's outer regions. Associated with sunspots are **flares** – sudden, violent releases of energy and material that lead to **aurorae** in the Earth's atmosphere. Other phenomena occurring in the photosphere are **faculae** and **plages**. There is a periodicity in level of solar activity known as the **solar cycle**.

Above the photosphere lies the turbulent **chromosphere**, which consists of hot gases and extends for thousands of kilometres. This is the realm of **spicules** and **prominences**, huge magnetically constrained clouds of material above the Sun's limb. Extending outward from the chromosphere for millions of kilometres is the extremely tenuous solar **corona**. Material escapes continuously from the corona, producing the **solar wind**. Transient large-scale eruptions of material from the corona are known as **coronal mass ejections**. The solar wind and the Sun's magnetic field dominate a region of space called the **heliosphere**, which extends to the boundaries of the Solar System.

The Sun must never be viewed directly through any optical instrument. There are two ways for amateurs to observe the Sun's disk. A *full-aperture filter* has a special metallic coating which cuts out harmful radiation before the Sun's light enters the telescope. Alternatively the Sun's image can be projected onto a white screen. Temporary or permanent blindness can result from direct viewing or from the use of unsuitable filters. Professional instruments for solar work include the **coronagraph** and the **spectroheliograph**.

Many probes and satellites, including **Ulysses**, the **Solar and Heliospheric Observatory** and **Yohkoh**, have been launched to study the Sun.

SUN: DATA

Distance from the Earth	
mean (the astronomical unit)	149.6×10^6 km
maximum (at aphelion)	152.1×10^6 km
minimum (at perihelion)	147.1×10^6 km
Diameter	1.391×10^6 km
Density (mean)	1.409 g/cm^3
Mass	1.989×10^{30} kg
Volume	1.412×10^{18} km^3
Period of axial rotation	
at equator	25.4d
at poles	about 35d
Inclination of axis of rotation to pole of ecliptic	7° 15′
Surface gravity (Earth = 1)	28
Spectral type	G2V
Luminosity	3.86×10^{26} W
Magnitude (mean visual)	−26.86 (apparent) +4.71 (absolute)
Rotational velocity (mean)	1.9 km/s
Escape velocity	617 km/s
Temperature of surface	5780 K
Temperature of core	15 million K

sundial A simple timekeeping device in which a shadow cast by the Sun falls on a dial graduated in hours. The shadow is cast by a short pillar called a *style* or *gnomon* standing out from the dial, which may be mounted vertically or horizontally. A sundial shows **apparent solar time**.

sundog SEE **parhelion**.

sungrazer A **comet** which at perihelion passes through the Sun's corona. Many of the bright long-period comets of the 19th century were sungrazers. Often, such comets do not survive their close passage of the Sun. Hundreds of small sungrazers have been detected using the coronagraphs on the **SOHO** spacecraft.

sunrise The moment when the Sun's upper limb appears above the horizon in the morning. It is defined as the moment when the Sun's **zenith distance** is 90° 50′, and decreasing. This value is arrived at by making allowance for the Sun's semi-diameter (16′) and for atmospheric refraction (34′).

sunset The moment when the Sun's upper limb disappears below the horizon in the evening. Similarly to **sunrise**, it is defined as the moment when the Sun's zenith distance is 90° 50′, and increasing.

sunspot A region in the Sun's **photosphere** that is cooler than its surroundings and therefore appears darker. A sunspot consists of a dark central region, the *umbra*, and a lighter outer region, the *penumbra*. They vary in size from around 1000 to 50,000 km (600 to 30,000 miles), and occasionally up to about 200,000 km

(125,000 miles), at which size they are visible to the naked eye. Their duration varies from a few hours to a few weeks, or months for the very biggest. Sunspots are seen to move across the face of the Sun as it rotates. The number of spots visible depends on the stage of the 11-year **solar cycle**.

Sunspots occur where there is a local strengthening of the Sun's magnetic field, which cools the area to around 4000 K. The stronger field is believed to suppress the convection of hotter gases from lower levels. Spots are usually found in pairs of opposite magnetic polarity, between latitudes 30° and 40° north or south of the equator (SEE ALSO **following, preceding**).

The spectrum of a sunspot differs from that of the photosphere as a result of the spot's lower temperature. The radial outflow of gases in the penumbra was detected by a Doppler shift of spectral lines (known as the *Evershed effect*). The **Zeeman effect** in the spectra of some sunspots demonstrates the presence of strong magnetic fields, the polarity of which reverses for each solar cycle.

supercluster SEE **cluster of galaxies**.

supergiant An extremely luminous star of large diameter and low density. Supergiants can be of any spectral type, from O to M. Rigel (type B) and Betelgeuse (type M) are examples. The luminosities of supergiants are several magnitudes greater than those of giant stars of the same spectral type, and so they lie at the top of the **Hertzsprung–Russell diagram**. Red (M-type) supergiants such as Betelgeuse have the largest diameters, around 1000 times that of the Sun. Supergiants are assigned to luminosity class I. The brightest are often given the separate class Ia, and the others placed in class Ib.

supergranulation A pattern of convective cells distributed fairly uniformly over the Sun's **photosphere**, and much larger – perhaps 30,000 km (19,000 miles) to a side – than ordinary photospheric granules (SEE **granulation**). Material has been detected flowing from the centre to the edge of the cells, where most of the magnetic flux coming from the photosphere is concentrated, and it is believed that it is the magnetic field at the edges of the cells that leads to the formation of **spicules**.

superior conjunction SEE **conjunction**.

superior planet Any of the planets Mars, Jupiter, Saturn, Uranus, Neptune and Pluto, all of whose orbits lie outside the Earth's.

supernova (plural **supernovae**) A stellar explosion in which virtually an entire star is disrupted. For a week or so, a supernova may outshine all the other stars in its galaxy. The luminosity (absolute magnitude up to about −19) is some 23 magnitudes (1000 million times) brighter than the Sun, and the energy released in the explosion is the same as is released over the star's entire previous life. A supernova is about 1000 times brighter than a **nova**. The diagram shows the light curve of Supernova 1989B, in the galaxy M66. Supernovae are designated by the year in

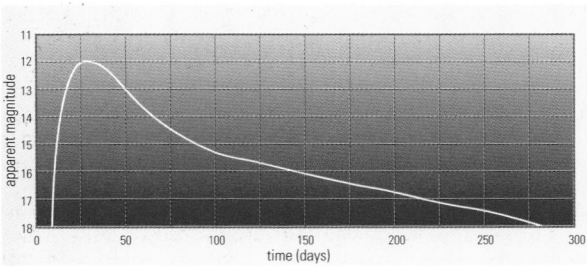

◀ **supernova** The light curve of a supernova shows a sudden brightening as the outer atmosphere is thrown off after the stellar core has collapsed. This peak is followed by a steady decline for about two months, after which this fading slows further.

which they are observed, plus one or more letters to indicate the order of discovery in that year.

Analysis of ancient records has identified several supernovae in our own Galaxy, observed before the invention of the telescope. One was seen in 1054 by Chinese astronomers. The **Crab Nebula** and its associated pulsar are the remains of this event. In 1572 Tycho Brahe and others saw a new star as bright as Venus in the constellation Cassiopeia (SEE **Tycho's Star**). The last naked-eye supernova in our Galaxy erupted in 1604 in Ophiuchus, and was observed for over a year (SEE **Kepler's Star**). There is probably about one supernova every 30 years in a galaxy like our own, but most of them are concealed by dust.

Supernovae are of two main types, classified by their light curves and spectra. *Type II* supernovae have hydrogen lines in their spectra, whereas *Type I* supernovae do not. Type I supernovae have been found in all kinds of galaxies, but Type II have never been found in ellipticals. Type I supernovae are further classified into subtypes. Type Ia supernovae are believed to be formed by the explosion of the white dwarf component of a binary star. Hydrogen from the companion leaks onto the white dwarf, where it accumulates and drives the white dwarf over the critical mass of 1.4 solar masses (the **Chandrasekhar limit**), causing it to explode. Types Ib and Ic and Type II supernovae are thought to be caused by the core collapse and explosion of massive stars at the end of their life, when the nuclear fuel inside them is used up. In Types Ib and Ic, the progenitor star, probably a **Wolf–Rayet star**, is less than about 8 solar masses. In Type II supernovae, the collapse of the core of a star greater than about 8 solar masses forms a **neutron star** or a **black hole**; the collapse releases energy which is picked up by the outer layers, and these layers are ejected into space at about 5000 km/s (3000 mile/s).

After a couple of years the supernova has expanded so much that it becomes thin and transparent. For hundreds or thousands of years the ejected material remains visible as a **supernova remnant**.

Supernova 1987A A bright supernova that flared up in the Large Magellanic Cloud in February 1987, reaching magnitude 2.8 – the first naked-eye supernova since Kepler's Star of 1604. A blue supergiant designated Sanduleak −69° 202 was identified as the star that had exploded.

supernova remnant (SNR) A gaseous emission nebula, the expanding shell of matter thrown off into space by the outburst of a **supernova**. These remnants are often strong radio and X-ray sources. The ejected layers of the supernova collide with the surrounding interstellar gas and heat up to perhaps a million degrees K, emitting X-rays. Electrons spiral in the magnetic fields in the compressed gas and emit **synchrotron radiation** in the form of radio waves. The most intense radio source in the sky, **Cassiopeia A**, is a supernova remnant. Supernova remnants usually appear as hollow shells. However, some (including the **Crab Nebula**) are filled balls of radio and X-ray emission. Such a remnant is termed a *plerion*, meaning 'filled'. This implies that they contain electrons produced by an active pulsar, formed in the supernova explosion.

Surveyor A series of seven US spacecraft, five of which made successful soft landings on the Moon between May 1966 and January 1968 in preparation for the Apollo landings. They took photographs and made chemical analyses of the surface.

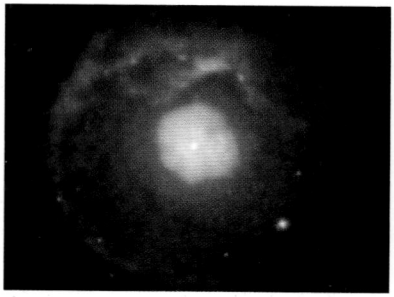

▲ **supernova remnant** The central cloud of electrons and surrounding shell of hot gas are well seen in this image of G21.5-0.9 obtained by the Chandra Observatory.

SU Ursae Majoris star SEE **U Geminorum star**.

Swan Nebula SEE **Omega Nebula**.

Swift Gamma Ray Burst Explorer A multi-wavelength observatory launched on 20 November 2004 and equipped to study gamma-ray bursts and their afterglows at gamma-ray, X-Ray, ultraviolet and optical wavelengths. Swift is a NASA mission with participation from the UK and Italy. It will determine the origins and classifications of gamma-ray bursts and determine how the blast wave evolves and interacts with its surroundings.

Swift–Tuttle, Comet A **comet** discovered in 1862 by American astronomers Lewis Swift (1820–1913) and Horace Tuttle (1837–1923), subsequently identified with Comet Kogler of 1737. It is the second-brightest short-period comet, Halley's Comet being the brightest. The large number of jets seen emerging from the nucleus at its return in 1992 suggest that the comet's changing period (at present 130 years) may be caused by **non-gravitational forces**. It is the parent comet of the **Perseid** meteor shower.

Sycorax An outer **satellite** of Uranus in an eccentric, **retrograde** (1) orbit, discovered in 1997.

symbiotic variable A **binary star** consisting of two stars of widely differing surface temperatures, such as a cool red giant and a hot dwarf. Gas from the cool star falls onto the smaller star and heats it. Symbiotic variables are also known as *Z Andromedae stars*, after their prototype, which has a range from magnitude 8.3 to 12.4.

synchronous orbit An orbit in which a satellite's period of revolution is the same as the primary's period of axial rotation. From the planet's surface the satellite appears to hover over one point, never rising nor setting. An example is **Charon**'s orbit around Pluto.

synchronous rotation The axial rotation of a celestial body in the same period as its period of revolution. Consequently, it always presents the same face towards the body about which it revolves, as in the case of the Moon orbiting the Earth. Tidal friction (SEE **tides**) has locked the Moon into this condition, which is also known as *captured rotation*. Other examples of bodies with synchronous rotation are the regular satellites of Jupiter and Saturn.

synchrotron radiation The electromagnetic radiation emitted by charged particles (usually electrons) that are accelerated by a strong magnetic field to speeds which are a significant fraction of the speed of light. The higher the energy of the particles, the shorter the wavelength of the radiation they emit. Synchrotron radiation is so called because it was first observed in particle accelerators called snychrotrons. Cosmic sources include **radio galaxies** and supernova remnants such as the **Crab Nebula**.

synodic month The period between two identical phases of the Moon; it is the same as the duration of one **lunation**. Its length is 29.53059 mean solar days.

synodic period The period of apparent revolution of one body about another as observed from the Earth, for example from one opposition or conjunction to the next. COMPARE **sidereal period**.

Syrtis Major A dark, triangular feature near the Martian equator, and the planet's most conspicuous feature in telescopic views from the Earth. It was first recorded by Christiaan **Huygens** in 1659. It is a sloping, cratered area, officially named Syrtis Major Planum.

syzygy The approximate alignment of three celestial bodies; in particular, the alignment of the Earth, the Sun, and the Moon or another planet. Thus syzygy occurs at full Moon and new Moon, and at planetary oppositions and conjunctions.

T

TAI ABBREVIATION FOR **International Atomic Time** (from its name in French).

tail A long train of dust or gas particles expelled from the nucleus of a **comet**.

tangential velocity The component of a star's velocity at right angles to the line of sight. It is also known as *transverse velocity*. The tangential velocity can be found from measurements of the star's annual **proper motion** and its distance in parsecs. If the **radial velocity** is also known, then the **space velocity** of the star can be calculated.

Tarantula Nebula An emission nebula faintly visible to the naked eye on the south-eastern edge of the Large Magellanic Cloud; also known as NGC 2070 or 30 Doradus. It is 1000 light years across and 500,000 solar masses, bigger and brighter than any nebula in our Galaxy or in any other nearby galaxy. It has a complex filamentary structure, and has a dense cluster of stars, designated R136, at its centre.

T association A region of recent and active star formation consisting of low-mass stars known as **T Tauri stars**. The members of T associations offer a series of snapshots of stars like our Sun at various stages of its infancy. They are still surrounded by the dusty, obscuring material of the cloud from which they are forming, and consequently they are often brightest in the infrared. COMPARE **OB association**.

Taurids A meteor shower showing steady, low activity between mid-October and late November, with a broad maximum during the first week of November. At maximum, the **ZHR** is little more than 10. In early November, the shower's two radiants lie close to the Pleiades and west of the Hyades star clusters in Taurus. The Taurids are produced by material from Comet **Encke**. Meteors from the shower are slow, and can sometimes be bright.

Taurus A constellation of the zodiac, representing the bull into which Zeus changed himself to carry off Princess Europa. It contains two of the most famous open clusters in the sky, the **Pleiades** and the **Hyades**, and also the **Crab Nebula**. Its brightest star is **Aldebaran**. Beta (Elnath), magnitude 1.7, and Zeta, magnitude 3.0, mark the tips of the bull's horns. Lambda is an eclipsing binary (range 3.4 to 3.9, period 3.95 days). There is also the prototype of the **T Tauri stars**.

tectonics Name given to the various processes by which the surface of a planetary body is deformed as a result of heating from within. Tectonic processes operate on bodies that have undergone **differentiation** and so possess a surface crust with a molten layer below. Large-scale movements of the molten layer affect the crust, producing faults and folds, and lifting mountains. Several satellites (e.g. **Ariel** and **Europa**) have surface features which appear to have been produced by tectonic activity. *Plate tectonics* operates, as far as is known uniquely, on the **Earth**.

▲ **Tarantula Nebula** Located in the Large Magellanic Cloud, the Tarantula Nebula is much the brightest of the external galaxies visible from Earth.

tektites Small, glassy objects, typically centimetre-sized, found scattered across certain specific areas (*strewnfields*) of the Earth's surface. They consist mainly of silica, with small quantities of metallic oxides, and are from 600,000 to 65 million years old. All tektites have clearly solidified rapidly from a temperature of around 1500–2000 K. Their origin is uncertain, though their flight markings, which are characteristic of solidification while flying through the air at high speed, suggest that they originated as terrestrial rock was vaporized by the impact of large meteorites. Tektites are named after the location of the strewnfield: for example, *Australites* are from Australia.

telescope An instrument for collecting and magnifying light or other electromagnetic radiation from a distant object. There are two basic types of optical telescope: the **refracting telescope**, in which the light-collecting element is a lens, and the **reflecting telescope**, in which it is a mirror; **catadioptric** systems are hybrids of the two. The image formed by the primary (main) lens or mirror is magnified by an **eyepiece**. The eyepiece may be replaced by a camera, photographic plate or **CCD** for imaging (SEE ALSO **Schmidt camera**); a **photometer** for measuring brightness; or a **spectrometer** for examining spectra. A telescope is supported (see **mounting**) in such a way that its weight is supported and it is easily aimed at celestial objects.

The **focal ratio** of a telescope is equal to the focal length of the primary lens or mirror divided by its diameter. Refractors have longer focal ratios than reflectors, which, aperture for aperture, are thus shorter. The most compact designs are short-focus instruments like the **Schmidt–Cassegrain telescope**. The image produced by an astronomical telescope is inverted, as opposed to that formed in a terrestrial telescope, which is erect (upright).

The refractor installed in the Yerkes Observatory in 1897, which has an objective 1 m (40 inches) in diameter, is still the largest of its kind. Large telescopes built since then have all been reflectors, because large mirrors are easier to make and support than large lenses. The world's largest reflectors are currently the **Keck Telescopes**, which each have a composite mirror 10 m (396 inches) across. With the development of spacecraft, it has become possible to send instruments such as the **Hubble Space Telescope** into space above the Earth's atmosphere, and so make observations unaffected by atmospheric absorption. SEE ALSO **drive**, **radio telescope**.

Telescopium A faint southern constellation representing a telescope, introduced by **Lacaille**. Alpha, its brightest star, is magnitude 3.5, but it contains little of interest.

Telesto A small **satellite** of Saturn, discovered by Bradford Smith and others in 1980 between the two Voyager encounters. Telesto is irregular in shape, and is **co-orbital** with Tethys and **Calypso**.

telluric line An **absorption line** in the spectrum of a celestial object produced by molecules such as oxygen and water in the Earth's atmosphere.

Tempel 1, Comet A faint periodic comet discovered in 1867 by Wilhem Tempel (1821–89). The comet has an orbital period of 5.5 years and was the target of the Deep Impact spacecraft in 2005.

terminator The boundary between the illuminated and unilluminated hemispheres of a planet or satellite. Viewed from Earth, Mercury, Venus, Mars and the Moon show phases and therefore their terminators are visible. The mountainous lunar surface gives the Moon a visibly irregular terminator.

Terrestrial Planet Finder Two NASA spacecraft, one planned for launched in 2014 and the other by 2020, to study the formation of disks of dust and gas around newly formed stars and the presence and features of mature planets orbiting the nearest stars. The number, various sizes and locations of planets will be studied with a view to ascertaining their capacity as abodes for life.

terrestrial planets The planets Mercury, Venus, Earth and Mars, so called because they have rather similar characteristics in respect of size and density, and few or no satellites.

Tethys A medium-sized **satellite** of Saturn, diameter 1058 km (657 miles), discovered by Giovanni Cassini in 1684. The whole of Tethys's surface is densely cratered, but evidence of past geological activity is provided by some less heavily cratered areas which must have undergone resurfacing. The two dominant features are Odysseus, a crater 440 km (270 miles) in diameter, and Ithaca Chasma, a 100-km (60-mile) wide trough running three-quarters of the way round the satellite; their formation may be linked. Tethys shares its orbit with two **co-orbital** satellites, **Calypso** and **Telesto**.

Thalassa One of the small inner **satellites** of Neptune discovered in 1989 during the Voyager 2 mission.

Thales of Miletus (*c*.625–*c*.550 BC) Greek philosopher. He provided the first theory of the Universe's origin, in which everything developed from a primordial mass of water. He knew of the **saros** and is credited with having used it to predict the eclipse of the Sun that occurred probably on 28 May 585 BC, during a battle between the Lydians and the Medes.

Thebe One of the small inner **satellites** of Jupiter discovered in 1979 during the Voyager missions. It is irregular, measuring about 110 × 90 km (70 ×55 miles).

theory of everything (TOE) SEE **grand unified theory**.

third contact In a total **solar eclipse**, the moment when the trailing edge of the Moon's disk begins to uncover the Sun's disk, and totality ends. In a **lunar eclipse** it is the moment when the Moon begins to leave the umbra of the Earth's shadow. The term is also used for the corresponding stage in eclipses involving other bodies.

▲ **Tethys** Ithaca Chasma can be seen to the right in this Voyager 2 image of Saturn's satellite Tethys. Younger craters are superimposed on the giant trough, indicating that it is an ancient feature.

Thirty Meter Telescope (TMT) A US–Canadian proposal for a 30-m (100-ft) optical/infrared reflecting telescope with a segmented primary mirror. It incorporates elements of previous plans for instruments in this class, including the *Giant Segmented Mirror Telescope* and *California Extremely Large Telescope* proposals.

three-body problem A fundamental problem in celestial mechanics: to determine the motions of three bodies under the influence only of their mutual gravitational attractions. There is no exact general solution, only solutions for special cases, but highly accurate approximations can be achieved by using modern computers. Work on the three-body problem was stimulated originally by the need to understand the orbit of the Moon, under the gravitational influence of the Earth and the Sun, and more recently by the need to calculate the orbits of artificial satellites. SEE ALSO **many-body problem**.

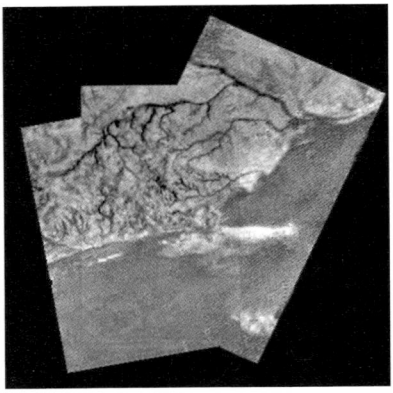

▲ **Titan** River channels on Titan's surface are clearly visible in this composite of images obtained by Huygens as it descended to the surface on 14 January 2005.

tides Distortions induced in a celestial body by the gravitational attraction of one or more others. The gravitational attraction a body experiences is greatest on the side nearest the attracting body, and least on the side farthest away, causing the body to elongate slightly in the direction of the attracting force and thus acquiring a *tidal bulge* on each side. If the deformed body is orbiting the attracting body, different parts of its surface periodically experience tidal distortion as it rotates. These tides have a significant effect on a gaseous atmosphere or a fluid ocean, but the solid crust of a planet such as the Earth, which is supported by a fluid mantle, is able to flex and so also experiences tidal distortion, but to a much lesser degree. The enormous gravitational force exerted by Jupiter on its satellite **Io** produces a large amount of flexure and interior heating, and this *tidal heating* is what makes Io so volcanically active. Smaller and completely rigid bodies can under certain circumstances be broken up by tidal forces.

The familiar ocean tides on the Earth are raised by the gravitational attraction of the Moon and the Sun, the Moon's influence being about three times the Sun's. There are two high tides and two low tides each day. When the Sun and the Moon are exerting a pull in the same direction (at new Moon and full Moon) their effects are additive and high tides are higher (*spring tides*). When the pull of the Sun is at right angles to that of the Moon (at first quarter and last quarter), high tides are lower (*neap tides*). *Tidal friction* caused by the rotation of the Earth beneath the oceanic bulges acts as a brake on the Earth's rotation, causing the length of the day to increase by about one millisecond per century. The angular momentum which is thus being lost is transferred to the Moon (SEE **secular acceleration**).

time The continual passage of existence, perceived as an ordering of events. Time may also be defined as the dimension of **spacetime** in which different events that have identical spatial dimensions are distinguished.

timescale A scale for measuring the passage of time by reference to some regularly recurring phenomenon. From remote antiquity, timescales have been based on the passage of the Earth once around its orbit, giving the unit of the **year**, and the rotation of the Earth on its axis, giving the unit of the **day**. However, the rotation of the Earth is not constant, and so it cannot be used as an accurate and invariable standard of time. Modern timescales are now based on a continuous count of time units defined by atomic resonators (**International Atomic Time**). The basis of civil timekeeping is **Greenwich Mean Time** (known technically as **Universal Time**). In addition, astronomers reckon time by the stars: **sidereal time**. SEE ALSO **apparent solar time**, **equation of time**, **local time**, **mean solar time**, **time zone**.

time zone One of 24 divisions of the Earth's surface, each 15° of longitude wide, within which the time of day is reckoned to be the same. The meridian of Greenwich serves as the zero of longitude. Standard time in each successive zone west of Greenwich is one hour behind that in the preceding zone (with some local variations). Large territories such as the USA and Russia span several time zones. There is a discrepancy at longitude 180° (the *International Date Line*), which is resolved by omitting one day from the calendar when crossing from west to east, or repeating one day if the crossing is from east to west.

Titan The largest **satellite** of Saturn, and the second largest in the Solar System, discovered by Christiaan Huygens in 1655. Titan is unique among planetary satellites in having a substantial atmosphere. At 1.88 g/cm^3 Titan is the densest of Saturn's large satellites, and is composed of rock and water ice in roughly equal proportions. Titan's atmosphere consists mostly of nitrogen, with some methane and traces of other hydrocarbon compounds, and exerts a surface pressure 1.5 times that on the Earth. Photochemical reactions driven by sunlight produce a largely opaque reddish cloud layer 200 km (125 miles) above the surface. The surface temperature is 95 K, at which methane can exist as solid, liquid or gas; methane can play the role that water does on Earth, forming, clouds, rain, and pools or more substantial bodies of liquid.

During its descent to Titan's surface on 14 January 2005, the Huygens probe (SEE **Cassini-Huygens**) imaged features which are clearly river channels, caused by flows of liquid methane. The surface is believed to have been shaped by the same erosional processes that have operated on Earth. Huygens landed on a surface with the consistency of wet sand, covered by a thin harder crust. Substantial quantities of methane are apparently bound up in the topsoil as 'groundwater'; heat generated by the landing released some of this methane. On the surface, Huygens' cameras imaged a landscape strewn with 'pebbles' of dirty water ice over which a light fog of methane or ethane was present.

Titania The largest of the five main **satellites** of Uranus, discovered by William **Herschel** in 1787. It is 1580 km (980 miles) in diameter, and its density of 1.68 g/cm^3 indicates that it consists of a mixture of rock and water ice. Titania is Oberon's twin in size, but its surface resembles that of **Ariel**, with similar features and an apparently similar geological history.

Tombaugh, Clyde William (1906–97) American astronomer. In 1930, nearly a year into a search based on predictions by Percival **Lowell**, he discovered the planet **Pluto**. He continued to search for other planets for more than ten years, discovering in the process star clusters, clusters of

► **Titania** The surface of Titania is heavily cratered and traversed by faults. This Voyager 2 picture shows details down to 7 km (4 miles) across.

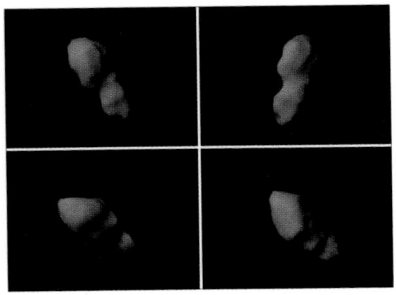

▲ **Toutatis** These four images of the asteroid Toutatis were constructed using radar information from NASA's Goldstone and Arecibo radar telescopes. Toutatis appears to consist of two irregularly shaped, cratered objects.

galaxies, a comet and hundreds of asteroids. After World War II he developed telescopic cameras for tracking rockets after launch.

topocentric Describing observations made from a point on the surface of the Earth, e.g. topocentric coordinates as opposed to geocentric coordinates (which are measured from the centre of the Earth).

total eclipse SEE **lunar eclipse, solar eclipse**.

Toutatis Asteroid no. 4197, discovered by Christian Pollas in 1989. Radar observations suggest that it consists of two bodies, 4 km (2.5 miles) and 2.5 km (1.5 miles) in diameter, in close contact. Toutatis is an **Apollo asteroid**, and one of the largest **near-Earth asteroids**.

transient lunar phenomenon (TLP) SEE **lunar transient phenomenon**.

transit (1) The passage of a celestial body across the observer's **meridian**. Such transits, observed with a **transit instrument**, were once important in the measurement of time and for navigational purposes.

transit (2) The passage of a body directly between the Earth and the Sun. The planets Mercury and Venus do so on occasion, when they appear as a black spot crossing the solar disk. Accurate measurements of such transits helped to establish the scale of the Solar System.

transit (3) The passage of a planetary satellite across the planet's disk. Jupiter's four main satellites frequently transit its disk, and Saturn's inner satellites are occasionally seen in transit. Observations of transits of Jupiter's satellites by Ole **Römer** established the enormous but finite speed of light.

transit (4) The passage of a surface or atmospheric feature of a body across its **central meridian** as it rotates, the timing of which gives a means of measuring the body's rotation period.

transit instrument A telescope that is mounted on a horizontal east–west axis and movable only in the vertical plane of the meridian. Transit instruments were once used to time the **transits** (1) of bodies across the local meridian and measure their altitudes. Such measurements were used for precision timekeeping, but have been superseded by devices such as the **photographic zenith tube**.

Transition Region and Coronal Explorer (TRACE) A NASA satellite launched on 2 April 1998 to study the Sun's **corona** and the transition region – the border between the chromosphere and the corona. The primary mirror of its 300-mm (12-inch) Cassegrain telescope has different coatings on its four quadrants to enable TRACE to make images at different ultraviolet wavelengths.

trans-Neptunian object SEE **Kuiper Belt**.

transverse velocity SEE **tangential velocity**.

Trapezium Popular name for the multiple star Theta Orionis in the Orion Nebula. Its four main components, magnitudes

5.1, 6.7, 6.7 and 8.0, are arranged in the shape of a trapezium. The stars are very hot, and light up the nebula.

Triangulum A small but easily found constellation between Andromeda and Aries. Its brightest star, Beta (magnitude 3.0), forms a well-marked triangle with Alpha (3.4) and Gamma (4.0). The main feature of interest is the spiral galaxy known as the Pinwheel or Triangulum Galaxy (M33), 2.8 million light years away in our **Local Group**, visible with binoculars under clear, dark skies.

Triangulum Australe A southern constellation representing a triangle, marked by the stars Alpha (magnitude 1.9), Beta and Gamma (both 2.9). The open cluster NGC 6025 is visible with binoculars.

Trifid Nebula An emission nebula in the constellation Sagittarius, apparently divided into three main sectors by dark lanes of dust. It is also known as M20 or NGC 6514. At its centre is an 8th-magnitude double star. The Trifid lies 6700 light years away.

trigonometric parallax A means of determining the distance of a star by triangulation. As the Earth revolves around the Sun, a nearby star shifts its position slightly with respect to more distant stars. This change in position can be measured from images obtained six months apart, i.e. when the Earth is on opposite sides of the Sun. As the diagram shows, the angular displacement is a direct measure of the star's parallax (π). Since the baseline (the distance from the Sun to the Earth) is known, the distance d of the star can be determined by trigonometry.

The first star to have its distance measured by this method was 61 Cygni, by Friedrich **Bessel** in 1838. Parallax is measured in arc seconds. The inverse of the parallax is the distance in **parsecs**. Alpha Centauri is the star with the largest known parallax, 0.752 arc seconds, which corresponds to a distance of 1.3 parsecs. The error in measuring parallax from individual images is typically about 10 milliarcseconds, but by combining images this can be reduced to about 4 milliarcseconds. The parallax of a star at 25 parsecs is 40 milliarcseconds, so the uncertainty is around 10%. The most accurate parallax determinations have been made by the **Hipparcos** satellite.

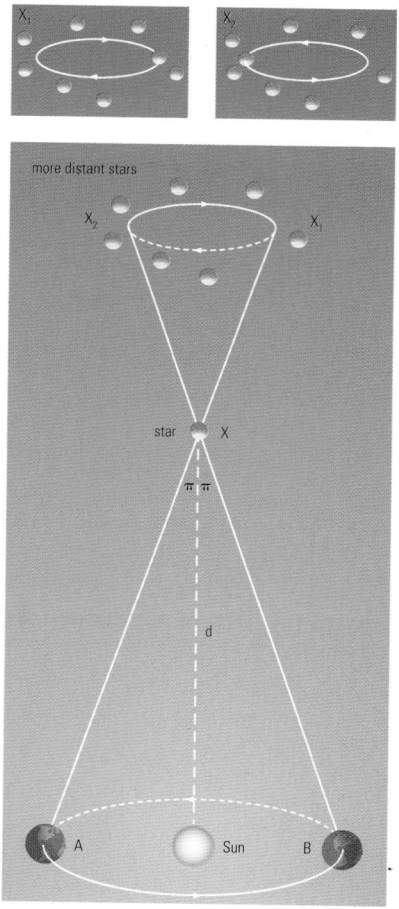

▲ **trigonometric parallax** If a nearby object (X) is observed from opposite sides of the Earth's orbit (A and B), it will appear in different positions against the stellar background (X_1 and X_2). As the distance between A and B is known, the distance, d, to X can be calculated.

triple-alpha process A nuclear reaction in which three alpha particles (helium nuclei) are transformed into carbon with the release of energy. It takes place after all of the hydrogen in a star's core has been exhausted. The core contracts and its temperature rises until it exceeds 100 million K, when the triple-alpha reaction begins. In each reaction two helium nuclei combine to form a beryllium nucleus which, in turn, captures a further helium nucleus to form a carbon nucleus. Additional helium captures can produce oxygen, neon and a number of heavier elements. The reaction is believed to be the dominant energy-producing process in red giants. It is also known as the *Salpeter process* after the US physicist Edwin Salpeter (1924–), who described it.

triplet A compound lens consisting of an assembly of three component lenses, which may be air-spaced or cemented.

Triton The largest **satellite** of Neptune, discovered in 1846 by William Lassell. It is in an inclined, **retrograde** (1) orbit and has almost certainly been captured by Neptune, probably from the **Kuiper Belt**. Its diameter is 2705 km (1680 miles), and its density of 2.07 g/cm^3 is consistent with a composition largely of three parts rock to one of water ice. Triton is the coldest and most distant world so far visited: Voyager 2's instruments measured a surface temperature of 35 K (−238°C). Its surface is complex and varied. There is much evidence of very recent volcanic activity in a variety of features formed by the upwelling of fluid material. In the south polar region, which is covered with a thin polar cap of nitrogen ice, dark eruptive plumes send fine dark material rising to altitudes of 8 km (5 miles) before being carried downwind for 100 km (60 miles) or more in the tenuous nitrogen atmosphere.

Trojan asteroid A member of one of two groups of **asteroids** sharing Jupiter's orbit. One group lies ahead of Jupiter, and the other behind, oscillating about the L_4 and L_5 **Lagrangian points** of Jupiter's orbit around the Sun. The Trojans ahead of Jupiter, about two-thirds of the total, are known as the *Achilles group*, after the first one to be found, **Achilles** in 1906; those behind form the *Patroclus group*, named after the second Trojan to be discovered, also in 1906. About 1700 were known by late 2004. The term is also used for asteroids sharing the orbits of other planets; six possible Martian Trojans and one Neptunian have been found.

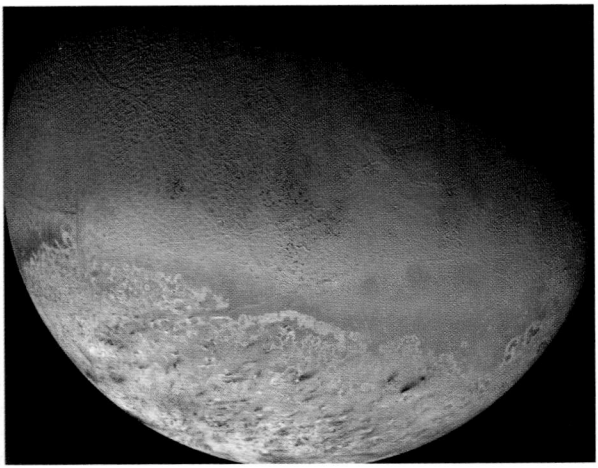

◀ **Triton** Voyager 2 imaged Triton in 1989, and this mosaic shows some of the varied terrains. The broad area at the bottom of the picture is the southern polar ice cap, which probably consists mainly of nitrogen ice.

tropical month The time taken for one revolution of the Moon around the Earth relative to the vernal **equinox**. It equals 27.32158 days.

tropical year The time taken for the Earth to orbit the Sun relative to the vernal equinox. Its length is 365.24219 mean solar days. Because of the precession of the vernal equinox, which moves along the ecliptic in a direction opposite to the Sun's motion, the tropical year is about 20 minutes shorter than the sidereal year. It is sometimes also known as the *solar year*.

Tropic of Cancer The parallel of latitude on Earth, $23\frac{1}{2}^{\circ}$ north, that marks the most northerly declination reached by the Sun at the summer **solstice**, on or about 21 June.

Tropic of Capricorn The parallel of latitude on Earth, $23\frac{1}{2}^{\circ}$ south, that marks the most southerly declination reached by the Sun at the winter **solstice**, on or about 22 December.

true anomaly SEE **anomaly**.

Trumpler classification The classification of open star clusters devised by the Swiss-American astronomer Robert Trumpler (1886–1956) on the basis of three characteristics: the number of stars in the cluster, the concentration of stars towards the centre of the cluster and the range of brightness within each cluster.

T star An extremely low-temperature **brown dwarf** star characterized by methane lines in its spectrum. Methane molecules can form only below about 1500 K. Like the **L stars**, no thermonuclear processes are occurring in T stars and they are detectable only because they derive some energy from their continued gravitational contraction.

T Tauri star A very young star still settling onto the main sequence and belonging to a class of irregular variables, named after the prototype, T Tauri. T Tauri stars are found in nebulae or young clusters and are characterized by high-velocity infall or outflow of gas which occurs as they adjust to the onset of nuclear reactions. Emission lines in their spectra indicate an extended atmosphere. Most known T Tauri stars are less massive than the Sun. Heavier stars either pass through the T Tauri stage while they are still obscured in the nebulae from which they were born, or have a different appearance at the same stage of life. SEE ALSO **Hertzsprung–Russell diagram**, **T association**.

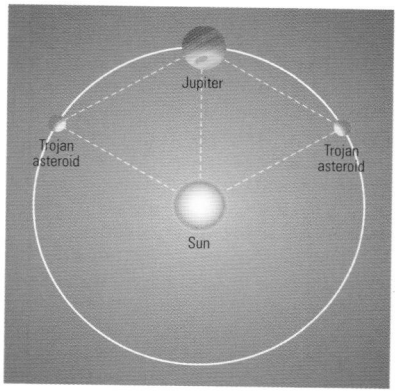

▲ **Trojan asteroid** The asteroids known as the Trojans orbit at the Lagrangian points in front of or behind Jupiter. They oscillate around these points over a period of 150–200 years.

Tucana A far southern constellation representing a toucan. Its overall faintness is redeemed by the presence of the Small Magellanic Cloud and the superb globular cluster 47 Tucanae (NGC 104). At magnitude 4 and half a degree across, 47 Tucanae is inferior only to Omega Centauri. It lies 15,000 light years away. NGC 362 is another globular, visible in binoculars.

Tunguska Event A huge aerial explosion on 30 June 1908, just north of the Stony Tunguska River in Siberia, which devastated the surrounding forest. The explosion, which was heard more than

800 km (500 miles) away and was recorded seismographically all around the world, was preceded by a fireball as bright as the Sun. Trees up to 40 km (25 miles) away were felled, and in the central 1000 sq km (400 sq miles) trees were incinerated.

The absence of an impact crater or meteoritic fragments puzzled earlier investigators. The best current theory is that a **near-Earth asteroid** roughly 50 m (150 ft) in diameter entered the atmosphere obliquely, and shattered and vaporized at an altitude of about 8 km (5 miles), blanketing the area with dust. Dust particles preserved in resin taken from trees in the area have the composition of known stony meteorites, and the damage pattern is consistent with the impactor having had an asteroidal density.

tuning fork diagram A diagram of **galaxy** types, originated by Edwin Hubble. The 'handle' of the fork consists of elliptical galaxies (E), numbered according to their degree of elongation, while the two prongs of the fork are made up of ordinary (S) and barred (SB) spirals, designated a, b or c according to how tightly wound the spiral arms are. The classification was originally interpreted as a sequence of evolution, with ellipticals developing into spirals, but this is now known to be incorrect. SEE the diagram at **Hubble classification**.

twenty-one centimetre line The emission line of neutral hydrogen in interstellar clouds. It lies in the radio spectrum at a wavelength of about 21 cm; the frequency is 1420 MHz. Its existence was predicted by Hendrik van de **Hulst** in 1944 and discovered by him and others in 1951. SEE ALSO **H I region**.

twilight The period during which the illumination of the sky gradually increases before sunrise, and decreases after sunset. The phenomenon is caused by the scattering of sunlight by molecules and particles of dust in the Earth's atmosphere. The duration of twilight depends on the steepness of the Sun's apparent path with respect to the horizon, so that twilight lasts longer at higher latitudes. Three forms of twilight are distinguished. *Civil twilight* ends or begins when the centre of the Sun's disk is 6° below the horizon, and is regarded as a period during which normal daytime activities cease to be possible. *Nautical twilight* ends or begins when the centre of the Sun's disk is 12° below the horizon, and the marine horizon is no longer visible. *Astronomical twilight* ends or begins when the centre of the Sun's disk is 18° below the horizon, and is the time when the faintest stars can be seen with the naked eye.

Tycho A large lunar crater, 85 km (53 miles) in diameter, in the Moon's southern uplands. It is the centre of the most conspicuous system of lunar **rays**, which stretch across much of the Moon's nearside, evidence that Tycho is a relatively young crater.

Tycho Brahe SEE **Brahe, Tycho**.

Tycho's Star A supernova in Cassiopeia in 1572 that was observed and described by Tycho **Brahe**. It was brighter than Venus at maximum and visible during daytime. Its remnant has been detected as a radio and X-ray source. Its light curve identifies it as a Type I supernova.

U

UBV system A system of three-colour photometry devised by Harold Johnson and William Morgan of Yerkes Observatory. Stellar magnitudes are determined at three different wavelength bands through three colour filters: ultraviolet (U) peaking at 360 nm; blue (B) peaking at 440 nm; and yellow (V for visual) peaking at 550 nm. SEE ALSO **colour index**.

U Geminorum star A member of a class of dwarf novae which show sudden outbursts of between 2 and 6 magnitudes, followed by a slower return to minimum, where they remain until the next outburst. Periods range from 10 days to several years. They are sometimes known also as *SS Cygni stars*. The best-known examples are SS Cygni, whose light curve is shown in the diagram, and U Geminorum. All are close binaries, comprising a subgiant or dwarf star of type K or M, comparable in mass to the Sun, which has filled its **Roche lobe**, and a white dwarf surrounded by an **accretion disk** of infalling matter. Almost all show rapid irregular flickering of about half a magnitude at minimum. Because they are binaries, some show eclipses at minimum, such as U Geminorum itself. As well as normal maxima, the *SU Ursae Majoris* subclass have 'supermaxima' which last about five times as long and are twice as bright. It has been suggested that they evolve from **W Ursae Majoris stars**. They are closely related to **Z Camelopardalis stars**.

Uhuru The first X-ray astronomy satellite, launched by the USA on 12 December 1970. Many X-ray sources bear numbers prefixed with the letter U, being their designation in the Uhuru catalogue.

ultraviolet astronomy The study of **electromagnetic radiation** from space with wavelengths between those of the visible spectrum and X-rays, i.e. from about 350 nm down to about 90 nm. Apart from the longest wavelengths (called the *near ultraviolet*), ultraviolet (UV) radiation does not penetrate the Earth's atmosphere, so observations must be made from rockets and satellites. The **International Ultraviolet Explorer** (IUE) was launched in 1978. Coverage was extended to the shortest ultraviolet wavelengths by **Rosat** in

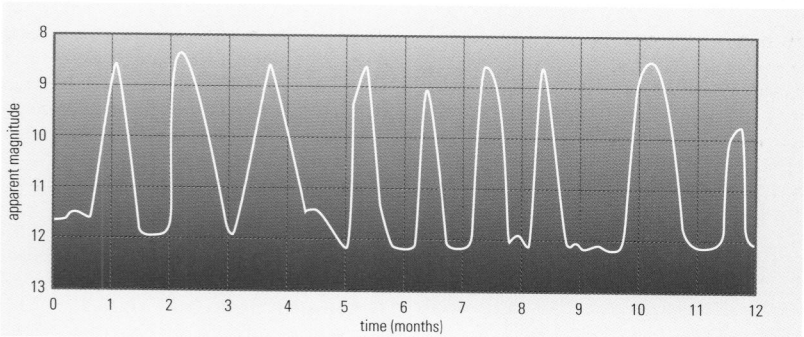

▲ **U Geminorum star** The light curve of a typical U Geminorum star shows repeated outbursts at intervals ranging from approximately ten days to several years. During outburst, the star's brightness climbs rapidly, with a slower decline. Outbursts occur as a result of mass transfer to an accretion disk around the smaller, more massive star in these close binary systems.

1990 and the **Extreme Ultraviolet Explorer** (EUVE) in 1992. Recent satellites include the **Far Ultraviolet Spectroscopic Explorer**, launched in 1999, and the **Galaxy Evolution Explorer**, in 2003.

Unlike explorations at other wavelengths, such as radio, infrared and X-rays, UV astronomy has discovered few new sources. Instead, the main application is in spectroscopy. Many of the abundant atoms and ions in celestial sources have their strongest spectral lines in the UV. Observations at UV wavelengths have been significant for studies of the composition and motions of interstellar gas. Hot stars of spectral types O and B emit strongly in the UV, and this emission excites surrounding nebulae. In the Solar System, UV spectroscopy is used to study planetary atmospheres and the gas in comets. Farther off in the Universe, UV astronomy is important in the study of luminous galaxies and quasars, which contain hot stars and large quantities of gas.

Ulugh Beg (1394–1449) Mongol ruler and astronomer. He established an observatory at Samarkand, and compiled the first star catalogue to surpass those of Hipparchus and **Ptolemy** in precision.

▲ **Umbriel** This pole-on view of Umbriel was obtained by Voyager 2 on 24 January 1986. The smallest details that can be resolved are 10 km (6 miles) across, and the heavily cratered surface is clearly visible.

Ulysses A European Space Agency probe launched on 6 October 1990 to study the polar regions of the Sun. After a flyby of Jupiter in February 1992, it entered a solar orbit which took it over the Sun's south pole 1994, and over the north pole a year later. In the late 1990s, Ulysses studied the evolution of **coronal mass ejections** from a position back near the orbit of Jupiter. It passed over the south and north poles again in 2000 and 2001.

umbra (1) The inner part of the shadow cast by a celestial body illuminated by an extended source such as the Sun. An observer in the umbral region of a shadow sees all of the source obscured, as, for example, in a total **solar eclipse**. SEE ALSO **lunar eclipse**. COMPARE **penumbra** (1).

umbra (2) The dark central area of a sunspot. COMPARE **penumbra** (2).

Umbriel One of the five main **satellites** of Uranus, discovered by William **Lassell** in 1851. Its diameter is 1170 km (727 miles), and its density of 1.51 g/cm^3 is consistent with a composition of mostly water ice and rock. Umbriel is a geologically inactive, heavily cratered satellite with a predominantly dark surface. The only bright spot is the floor of the 140-km (85-mile) diameter crater named Wunda.

United States Naval Observatory The US government observatory with headquarters at Washington, DC. Its main telescope there is a 0.66-m (26-inch) refractor. USNO has an observing station at Flagstaff, Arizona, with a 1.55-m (61-inch) reflector.

Universal Time (UT) Standard timescale used for scientific purposes throughout the world, popularly known as **Greenwich Mean Time** (GMT). There are several versions of Universal Time. *UT0* is based on the observed rotation of the Earth relative to the stars. *UT1* is UT0 corrected for a slight wandering of the Earth's geographical poles, which affects the longitude of the place of observation.

Coordinated Universal Time (UTC; the abbreviation is of the term in French) is based on the caesium atomic clock, and is used for radio time signals. As the Earth's rotation is gradually slowing, the time shown by an atomic clock gradually diverges from UT1. To keep atomic time in step with UT1 to the nearest second, time signals are retarded when necessary by one **leap second** at midnight on 30 June or 31 December. UTC is what is generally known as GMT.

Universe All of space and everything contained in it. SEE **cosmology**.

upper culmination SEE **culmination**.

Uranus The seventh major planet from the Sun, and the second-smallest of the four giant planets. With a mean magnitude of 5.5, Uranus is visible to the naked eye under good conditions. Its maximum apparent diameter is 4.1 arc seconds, and through a telescope it appears as a small, featureless, greenish blue disk. In size, mass, atmosphere and colour, Uranus resembles Neptune. The main data for Uranus are given in the table.

▲ **Uranus** Cloud features on Uranus are fairly elusive. This heavily processed Voyager 2 image from 1986 shows some structure in the planet's atmosphere at lower right.

URANUS: DATA	
Globe	
Diameter (equatorial)	51,118 km
Diameter (polar)	49,947 km
Density	1.29 g/cm^3
Mass (Earth = 1)	14.53
Volume (Earth = 1)	62.18
Sidereal period of axial rotation (equatorial)	17h 14m (retrograde)
Escape velocity	21.3 km/s
Albedo	0.51
Inclination of equator to orbit	97° 52′
Temperature at cloud-tops	55 K
Surface gravity (Earth = 1)	0.79
Orbit	
Semimajor axis	19.22 AU = 2871×10^6 km
Eccentricity	0.046
Inclination to ecliptic	0° 46′
Sidereal period of revolution	84.01y
Mean orbital velocity	6.81 km/s
Satellites	27

Uranus was the first planet to be discovered since ancient times. Although it had been observed on several occasions (and once had even been catalogued as a star), its non-stellar nature was first recognized by William **Herschel**, who observed its disk on 13 March 1781, during a telescopic survey of faint naked-eye stars. Uranus's axis of rotation is steeply inclined, and lies close to the ecliptic plane. Its poles thus spend around half of its 84-year orbit in sunlight, and half in darkness, and highly exaggerated seasonal variations are experienced by both the planet and its satellites. The flyby of the Voyager 2 probe in 1986 provided most of our current knowledge of the planet.

The upper atmosphere is about 83% molecular hydrogen and 15% helium, the other 2% being mostly methane. Strong enhancement of Voyager images revealed faint banding and a haze over the Sun-facing pole. Near-infrared images obtained from the ground and from the Hubble Space Telescope have since revealed dark spots and bright clouds similar to those visible when Voyager flew past **Neptune**. It may be that Uranus's 'weather' was unusually calm at the time of the Voyager visit. The planet has insufficient internal heat to drive a dynamic meteorology like Neptune's. **Differential rotation** operates on Uranus, atmospheric features taking 16 to 17 hours to rotate, compared with the internal rotation period of over 17 hours. Wind speeds vary from about 150 to 600 km/h (90 to 370 mile/h).

Uranus's interior structure remains conjectural. There may be an iron–silicate core, surrounded by a deep, possibly semi-liquid 'mantle' consisting of water, ammonia and methane ices in which the planet's magnetic field (comparable in strength to Saturn's) originates. The magnetic axis is tilted by 59° to the axis of rotation, and is displaced from the planet's centre. The magnetic field gives rise to an appreciable **magnetosphere**.

The five largest **satellites** were known before the Voyager encounter, which led to the discovery of 11 more. A total of 27 Uranian satellites were known at the end of 2004. The main components of Uranus's ring system were discovered in 1977 when the planet occulted a star (SEE **occultation**). Others were imaged by the Voyager spacecraft. The brightest and outermost is the Epsilon Ring, while the innermost is a very diffuse sheet of material. Between the two are nine narrow, darkish rings. SEE ALSO **ring, planetary**.

Ursa Major A famous northern constellation, the Great Bear, whose main pattern is known as the Plough or Big Dipper . This feature consists of seven stars: Alpha (Dubhe), magnitude 1.8; Beta (Merak), 2.3; Gamma (Phekda), 2.4; Delta (Megrez), 3.3; Epsilon (Alioth), 1.8; Zeta (**Mizar**, a famous double), 2.2; and Eta (Alkaid), 1.9. Dubhe and Merak are the 'pointers' to the pole star. Five of the Plough stars make up a **moving cluster**; the exceptions are Dubhe and Alkaid. There are many galaxies in Ursa Major, including several Messier objects (M81, M82, M101, M108 and M109), as well as the **Owl Nebula**.

Ursa Minor The constellation that contains the north celestial pole. Its brightest star is Alpha (**Polaris**), the north pole star, magnitude 2.0. The constellation's seven main stars make a pattern rather like a faint and distorted Plough (Big Dipper), and is known as the Little Dipper.

UT ABBREVIATION FOR **Universal Time**.

UV Ceti star Another name for a **flare star**.

V

Valhalla A 4000-km (2500-mile) diameter impact basin – the largest such structure in the Solar System – on Jupiter's satellite **Callisto**, surrounded by a series of concentric rings representing 'ripples' from the impact.

Valles Marineris SEE **Mariner Valley**.

Van Allen Belts Two zones of plasma (high-energy charged particles) in the Earth's **magnetosphere**. The concentric, torus-shaped belts trap charged particles which then spiral around magnetic field lines, back and forth between the two magnetic poles. The diagram, which is roughly to scale, shows the form of the belts. The outer belt contains mainly electrons from the **solar wind**, and the inner belt mainly protons from the same source. Within the inner belt is a radiation belt consisting of particles produced by interactions between the solar wind and heavier **cosmic ray** particles. The intensity of the Van Allen Belts varies with the level of solar activity (SEE **solar cycle**). Because the Earth's magnetic axis is offset from its rotational axis, the belts are not a uniform distance from the Earth's surface. The inner belt comes very close to the surface over the South Atlantic Ocean; this is called the *South Atlantic Anomaly*. The two main belts were discovered by American physicist James Van Allen (1914–) and his collaborators during investigations using Geiger counters on board the first US Explorer satellites in 1958. The cosmic ray belt was found by Russian Cosmos satellites in 1991.

variable star A star whose brightness varies with time. *Intrinsic variables* are stars that vary because of some inherent process, such as pulsations in size or events in their atmosphere. In *extrinsic variables*, external factors such as eclipses or obscuring dust affect the amount of light reaching us from the star. There are five main classes of variable star. **Eclipsing binaries** are extrinsic variables in which members of a binary system periodically pass in front of each other as seen from Earth. **Pulsating stars** are intrinsic variables that expand and contract in size, either regularly or irregularly. **Cepheid variables** and **RR Lyrae stars** are examples of pulsating stars with regular cycles of variation. Less regular variation is found among the **long-period variables** (or *Mira stars*), **semiregular variables** and **irregular variables**. **Cataclysmic variables** include **supernovae**, **novae**, **dwarf novae** and **symbiotic variables**. **Eruptive variables** are a very diverse class, including **flare stars**, **R Coronae Borealis stars** and **shell stars**. Finally, **rotating variables** include various types where a star's shape or non-uniform surface brightness

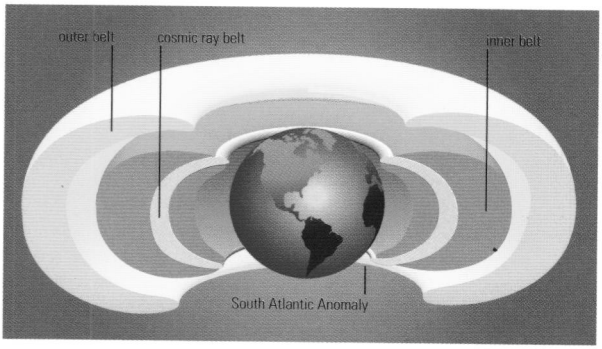

◀ **Van Allen Belts**
Earth is surrounded by radiation belts, which contain energetic particles trapped by the magnetic field. The inner belt dips low over the South Atlantic, where its particle population can present a hazard to satellites.

produces the variations in magnitude. Each main group also has its own subclassification.

Variable stars have their own peculiar naming system. The letter R is assigned to the first variable to be discovered in a particular constellation (unless it already has an existing designation with a greek or roman letter). Subsequent discoveries are given the letters S to Z. Then come RR to RZ, followed by SS to SZ, and similarly to ZZ. After that come AA to AZ, BB to BZ, and so on to QZ (but the letter J is never used). This scheme caters for 334 variable stars in a constellation; subsequently discovered variables are numbered V335, V336, and so on.

variation A perturbation in the Moon's motion caused by the Sun's changing gravitational pull on the Moon as the Moon orbits the Earth.

Varuna A large Kuiper Belt object discovered by R.S. McMillan on 28 November 2000. It has an estimated diameter of 900 km (560 miles) and takes 285 years to orbit the Sun at a mean distance of 43 AU.

Vega The star Alpha Lyrae, fifth-brightest in the sky, magnitude 0.03, distance 25 light years, luminosity 50 times that of the Sun. Vega is surrounded by a disk of gas and dust extending out to 85 AU from it, possibly a **protoplanetary disk**.

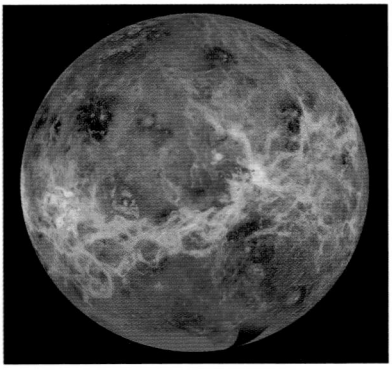

▲ **Venus** Centred on longitude 180°, this image is a composite view of Venus prepared from Magellan Imaging Radar results in 1991.

Veil Nebula The brightest part of the **Cygnus Loop**, a large supernova remnant in Cygnus.

Vela A southern constellation, once part of the larger grouping of Argo Navis, the ship Argo, and representing the ship's sails. Its leading star is Gamma, magnitude 1.8 but slightly variable; this is the brightest **Wolf–Rayet star**, with a luminosity 40,000 times that of the Sun. Vela is a rich region; the stars Delta (magnitude 1.9) and Kappa (2.5) make up the **False Cross** with Iota and Epsilon Carinae. Other features of the constellation are the **Gum Nebula** and the **Vela Pulsar**.

Vela Pulsar A pulsar with a period of 89 milliseconds, discovered in 1968 in the constellation Vela; known as PSR 0833-45. The optical counterpart, discovered in 1977, flashes with the same period.

Venera A series of space probes launched by the Soviet Union to the planet Venus between 1961 and 1983. Venera 7 in 1970 was the first craft to land successfully on Venus, although it transmitted for only 23 minutes before succumbing to the extreme temperatures and pressures at the surface. Venera 8 in 1972 proved more resilient and transmitted for 50 minutes. Veneras 9 and 10 in 1975 obtained the first images of the surface. Veneras 11 and 12 landed in 1978 and sent back information about the atmosphere, but no pictures. Veneras 13 and 14 landed in 1982, obtaining colour images of the surface and sampling the rocks. Veneras 15 and 16, both radar mappers, went into orbit around the planet in October 1983 and operated until the following July.

Venus The second major planet from the Sun, and the second-largest of the terrestrial planets. Visible around dawn or dusk as the so-called *Morning Star* or *Evening Star*, it is the most conspicuous celestial object after the Sun and the Moon. At its brightest (magnitude −4.7) it is even visible to the naked eye by day in a clear sky. At **greatest elongation** Venus is 45° to 47° from the Sun and has an apparent diameter of 25 arc seconds. Like the Moon and Mercury, it exhibits **phases**. **Transits** of Venus are rare, and occur in pairs eight years apart, with over a century between each pair. The most recent was on 7 June 2004; and the next is on 5–6 June 2012. The main data for Venus are given in the table.

A telescope shows the planet's dazzling yellowish white cloud cover. In ultraviolet images, visually elusive markings show clearly as a Y-shaped feature, enabling the clouds' rotation period of just under 4 days to be measured. The **ashen light**, **cusp caps** and **Schröter effect** may be watched for by amateur observers; their causes have been long debated, but they are most probably real atmospheric phenomena, and not optical effects.

Spacecraft, including the two US **Pioneer Venus** probes and **Magellan**, have revealed information about Venus's surface. Later Soviet **Venera** landers were built to withstand the extreme surface conditions – a temperature of 750 K and a pressure of over 90 bars – for long enough to transmit measurements and images. Magellan produced highly detailed radar maps of the surface. **Venus Express** is due to reach the planet in 2006.

A gently undulating plain covers two-thirds of Venus. Highlands account for a further quarter, and depressions and chasms the remainder. The two principal highland regions are Ishtar Terra, in the northern hemisphere, and Aphrodite Terra, largely in the southern. The plains are cratered, but there are no craters less than a few kilometres across because smaller **meteoroids** would have been destroyed during passage through the dense atmosphere. Most of the surface features are volcanic in origin. There are large shield volcanoes such as Rhea Mons, and calderas. The volcanic peak Maat Mons is surrounded by lava flows that may be no more than two decades old, suggesting that Venus is still volcanically active. There are many other types of near-circular volcanic feature, which have been given names such as *arachnoids*, *coronae* and *ovoids*.

VENUS: DATA

Globe	
Diameter	12,104 km
Density	5.20 g/cm^3
Mass (Earth = 1)	0.8149
Volume (Earth =1)	0.8568
Sidereal period of axial rotation	243d 0h 30m (retrograde)
Escape velocity	10.4 km/s
Albedo	0.76
Inclination of equator to orbit	177° 20′
Surface temperature	750 K
Surface gravity (Earth =1)	0.90
Orbit	
Semimajor axis	0.723 AU = 108.2×10^6 km
Eccentricity	0.007
Inclination to ecliptic	3° 24′
Sidereal period of revolution	224.701d
Mean orbital velocity	35.02 km/s
Satellites	0

The atmosphere is 96% carbon dioxide and 3½% nitrogen, with traces of helium, argon, neon and krypton. Its great density has produced the very high surface temperature by an extreme **greenhouse effect**. At various altitudes there are haze layers of sulphuric acid, sulphur dioxide, water vapour and sulphur. A chemically active atmosphere is probably the only agent for 'weathering' of surface features. Wind speeds in the atmosphere, which are 350 km/h (220 mile/h) at the cloud-tops, fall to 10 km/h (6 mile/h) at the surface.

The internal structure is largely unknown. Venera measurements indicate a low-density crust. By analogy with the similar-sized Earth, there may be a silicate mantle and iron–nickel core. Certainly there is no appreciable magnetic field, although the solar wind does interact strongly with Venus's ionosphere to produce a well-defined **bow shock** 2000 km (1300 miles) ahead of the planet.

Venus Express A European Space Agency Venus orbiter due for launch in October or November 2005. It will reach the planet in spring 2006 and conduct the first global investigation of the Venusian atmosphere and its plasma environment.

vernal equinox (spring equinox) The point at which the Sun crosses the celestial equator from south to north, or the time at which this occurs. This point, also known as the *First Point of Aries*, is the zero point of the celestial coordinate called right ascension. The vernal equinox moves westward by about one-seventh of an arc second daily because of the effect of **precession**.

Very Large Array (VLA) The most complex radio telescope on a single site in the world. It consists of 27 dishes, each 25 m (82 ft) in diameter, which can be moved along a giant Y-shaped track on the plains near Socorro, New Mexico. Each arm of the VLA can be made up to 21 km (13 miles) long, using the technique of **aperture synthesis** to produce a virtual dish some 34 km (21 miles) wide.

Very Large Telescope (VLT) A group of four 8.2-m (232-inch) telescopes at Paranal Observatory, Chile, built by the **European Southern Observatory**, the last of which was completed in 2001. With all four telescopes linked together, the VLT has a light-gathering power equivalent to that of a single 16-m (630-inch) mirror. The four main instruments can be linked with several smaller ones to form an **optical interferometer**.

Very Long Baseline Array (VLBA) An array of ten 25-m (82-ft) radio telescopes spread across the continental USA and Hawaii, giving a baseline for **very long baseline interferometry** 8000 km (5000 miles) long. Resolutions of 1 milliarcsecond are possible. The instrument began operation in 1993. SEE ALSO **radio interferometer**.

very long baseline interferometry (VLBI) A technique in radio astronomy that extends the concept of **aperture synthesis** by combining signals from radio telescopes which may be thousands of kilometres apart to achieve a high accuracy of positional measurements, of 1 milliarcsecond for the **Very Long Baseline Array**, for example. By combining signals from radio astronomy satellites and ground-based radio telescopes, baselines of around 40,000 km (25,000 miles) have been achieved.

Vesta Asteroid no. 4, diameter 501 km (311 miles), discovered in 1807 by Heinrich Olbers. At magnitude 6.4 it is the brightest asteroid and the only one ever visible (under ideal conditions) to the naked

◀ **Very Large Array**
One of the world's premier radio observatories, the Very Large Array is located at an elevation of 2120 km (6970 ft) on the plains of San Augustin, near Socorro.

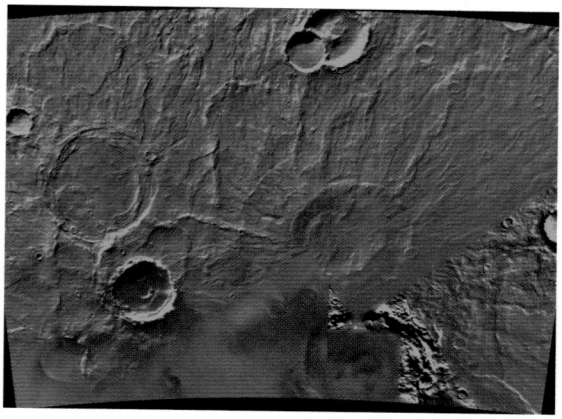

► **Viking** Compiled from images obtained by the Viking orbiters, this composite shows the Amphitrites Patera region of Mars. Radial ridges extend northwards for about 400 km (250 miles) from this old volcano.

eye. Vesta's spectrum suggests that its surface, unique among the larger asteroids, was once molten (SEE **achondrite**). There is a huge, 460-km (285-mile) impact basin at the south pole.

Viking Two NASA space probes to Mars, launched on 20 August and 9 September 1975. Each was a combined orbiter and lander. Viking 1 went into orbit around Mars in June 1976 and the lander landed on 20 July 1976; Viking 2 went into orbit in August 1976 and its lander reached the surface on 3 September. Both landers photographed their surroundings, studied the weather and soil, and conducted experiments which found no indications of life at the landing sites. The orbiters photographed the surface of the planet and its two satellites.

Virgo The largest constellation of the zodiac, lying on the celestial equator, and representing the goddess of justice. Its brightest star is **Spica**. Gamma is a splendid binary with equal components of magnitude 3.6, period 169 years, at their closest in the year 2005. Virgo is one of the richest areas for faint galaxies since it includes the **Virgo Cluster**. Closer to us than this cluster is the **Sombrero Galaxy**.

Virgo Cluster (Coma–Virgo Cluster) A rich concentration of galaxies in the direction of the constellation Virgo, although some lie across the border in Coma Berenices (hence the alternative name). The centre of the cluster is about 55 million light years away, and is marked by the giant elliptical galaxy M87, also known as the radio source Virgo A, which is ejecting a jet of gas. In all there are 16 Messier objects in the cluster, which contains over 2000 galaxies in all.

visible spectrum The range of wavelengths in the electromagnetic spectrum perceptible by the human eye. It varies slightly from one person to another, but extends typically from 385 nm (violet) to 700 nm (red). SEE ALSO **electromagnetic radiation**.

visual binary A **binary** system that can be observed as a double star with a telescope, as distinct from a spectroscopic binary.

visual magnitude (symbol m_v) The **apparent magnitude** of a celestial object in the colour region to which the human eye is most sensitive, i.e. about 560 nm.

VLA ABBREVIATION FOR **Very Large Array**.

VLBA ABBREVIATION FOR **Very Long Baseline Array**.

VLBI ABBREVIATION FOR **very long baseline interferometry**.

VLT ABBREVIATION FOR **Very Large Telescope**.

Volans A small, faint southern constellation adjacent to Carina, representing a flying fish. Its brightest stars are of 4th magnitude.

volatile A substance that solidifies at a low temperature. Substances such as water and carbon dioxide, familiar as gases on Earth, are examples. Bodies in the Solar System that formed nearer the Sun have a lower volatile content than those that formed farther away. Volatiles in comets, which formed in the outer reaches of the Solar System, evaporate to form the gas tail as the comet approaches the Sun.

volcanism The eruption of molten material at the surface of a planetary body. Volcanism in which molten silicate rock erupts, cools, and becomes new surface material has taken place on Venus and Mars as well as the Earth, and similar material produced the lunar maria (SEE **mare**) by volcanic flooding. In the outer Solar System a mixture of water ice and ammonia ice can behave as lava in a process called *cryovolcanism*, and has produced, for example, the volcanic floodplains on **Ariel**. Sulphur volcanism operates on **Io**, the most volcanically active world known.

Voyager Two NASA space probes to the outer planets. Launched in 1977, on 20 August (Voyager 2) and 5 September (Voyager 1), they passed Jupiter in March and July 1979, respectively, photographed the planet and satellites, and discovered Jupiter's ring system. Voyager 1 reached Saturn in November 1980, followed by Voyager 2 in August 1981; they photographed the rings, as well as the satellites. Voyager 1's path was bent out of the ecliptic by its close approach to Titan, but Voyager 2 proceeded to Uranus, reaching it in January 1986, and Neptune, in August 1989. By early 2005 both Voyager probes were on their way out of the Solar System, their last task to locate the heliopause, the boundary of the **heliosphere**.

Vulpecula A dim constellation next to Cygnus, representing a fox. Its brightest stars are of 4th magnitude. It contains the **Dumbbell Nebula**, reputedly the easiest planetary nebula to see in the sky.

W

weakly interacting massive particle SEE **WIMP**.

Weizsäcker, Carl Friedrich von (1912–) German theoretical physicist and astrophysicist. In 1938 he and Hans **Bethe** independently proposed a detailed theory for the production of energy in the Sun and other stars (SEE **carbon–nitrogen cycle**) in which hydrogen is converted into helium by nuclear fusion. In 1944 he revived the *nebular hypothesis* of the origin of the Solar System originally proposed by Immanuel Kant and Pierre Simon de **Laplace**.

▲ **Whirlpool Galaxy** The magnificent Whirlpool Galaxy takes its name from its pronounced spiral structure. The small companion galaxy at top, NGC 5195, appears to be interacting with the Whirlpool.

West, Comet A bright comet discovered by Danish astronomer Richard West (1941–) in November 1975. The following March it became a prominent naked-eye object with a fan-shaped tail. The comet was extremely active, and the nucleus broke into at least four fragments as it passed within 30 million km (20 million miles) of the Sun.

Westerbork Radio Observatory A radio astronomy observatory located near Groningen in the Netherlands, operated by the Netherlands Foundation for Research in Astronomy. Its major instrument is an **aperture synthesis** telescope consisting of 14 dishes 25 m (82 ft) in diameter, along an east–west baseline 2.7 km (1.7 miles) long.

Whipple, Fred Lawrence (1906–2004) American astronomer. He was best known for his 'dirty snowball' theory of comets, proposed in 1949 and shown to be correct in 1986 when space probes were sent to Halley's Comet. Whipple discovered several comets and worked on cometary orbits; he also studied planetary nebulae, stellar evolution and the Earth's upper atmosphere.

Whirlpool Galaxy A well-defined spiral galaxy of type Sc which appears face-on and has very prominent arms; also known as M51 and NGC 5194. It lies in Canes Venatici, not far from the end of the tail of Ursa Major, and was the first galaxy in which spiral structure was noted, by Lord **Rosse** in 1845. A small companion galaxy, NGC 5195, is connected to it by an extension of one of the spiral arms. The Whirlpool Galaxy is 25 million light years away.

white dwarf A type of star about the size of the Earth, but with a mass about that of the Sun. As a result, its density is enormously greater than that of any terrestrial material (0.1 to 100 tonnes/cm^3). This is

▲ **white dwarf** This close-up view from the Hubble Space Telescope shows part of the globular cluster M4 in Scorpius. While most of the objects seen here are old, red stars, several white dwarfs (circled in blue) have also been found.

because the normal atomic structure is broken down completely, with electrons and nuclei packed tightly together (the state known as **degenerate matter**). A white dwarf cannot have a mass of more than about 1.4 solar masses (the **Chandrasekhar limit**). For larger masses, gravity overwhelms the pressure of the electrons and the star collapses under its own weight, forming a **neutron star** or **black hole**. White dwarfs are of low luminosity and gradually cool down to become cold, dark objects. They represent the final stage in the evolution of low-mass stars after they have lost their outer layers. The first white dwarf to be discovered was the companion of **Sirius**.

Widmanstätten pattern The distinctive pattern on a sectioned and polished surface of an **iron meteorite** that has been etched with acid. It is named after Count Aloys Joseph von Widmanstätten (1754–1849), who noted it in 1808. The pattern is only seen in meteoritic material, and consists of intersecting plates of the iron–nickel minerals kamacite and taenite.

Wild 2, Comet A periodic comet in a 6.2 year orbit, discovered by the Swiss astronomer Paul Wild in 1978. The **Stardust** spacecraft made a close (236 km; 148 miles) flyby in January 2004, imaging the comet's nucleus and collecting material for return to Earth. The nucleus was found to be a 5-km (3-mile) diameter, more or less spherical body with heavy cratering; its surface topography was suggestive of greater mechanical strength than had previously been assumed for comets. Numerous gas jets were seen to be emerging from the nucleus.

Wild Duck Cluster An open cluster in the constellation Scutum, also known as M11 and NGC 6705. Its wedge shape resembles wild ducks in flight. The cluster lies 6200 light years away.

Wilkinson Microwave Anisotropy Probe (WMAP) A NASA spacecraft launched on 30 June 2001 to study the small-scale variations (anisotropy) in the **cosmic microwave background** (CMB). It is positioned at the Earth's L_2 **Lagrangian point**, from where it can map the CMB at high resolution. It is named after David Wilkinson, a member of the mission's science team.

William Herschel Telescope (WHT) A reflecting telescope with a 4.2-m (165-inch) mirror at the **Roque de los Muchachos Observatory** in the Canary Islands. It has facilities for high-dispersion and multi-object spectroscopy. The WHT began operation in 1987.

Wilson, Robert Woodrow (1936–) American physicist. In 1964, with Arno **Penzias**, he detected the **cosmic microwave background** radiation that provides the strongest evidence for the **big bang**. They shared the 1978 Nobel Prize for Physics for their discovery.

Wilson effect The foreshortening of a **sunspot** when it is near the Sun's limb, accompanied by a widening of the penumbra on the side nearest the limb, and a narrowing on the side farthest from the limb. The phenomenon was discovered by the Scottish astronomer Alexander Wilson (1714–86), who took it as an effect of perspective, indicating that sunspots are saucer-shaped depressions. However, not all spots show the Wilson effect. The effect is now thought to arise from sunspots being more transparent than the surrounding photosphere, and the umbra more transparent than the penumbra.

WIMP Abbreviation for 'weakly interacting massive particle', a hypothetical atomic particle whose existence has been invoked to account for the **missing mass** of the Universe. WIMPs would interact with each other and with ordinary matter only via the weak nuclear force and gravity, and so would take no part in nuclear fusion reactions.

winter solstice SEE **solstice**.

WMAP ABBREVIATION FOR **Wilkinson Microwave Anisotropy Probe**.

Wolf, Maximilian Franz Joseph Cornelius ('Max') (1863–1932) German astronomer. He was a pioneer of photographic methods in astronomy. In 1891 he made the first photographic discovery of an asteroid (Brucia, no. 323); in all he made 232 asteroid discoveries. He detected several new nebulae, including the **North America Nebula**, the nebulosity around the Pleiades, numerous dark nebulae and the first cluster of galaxies (in Coma Berenices) to be identified as such, as well as the **Virgo Cluster**.

Wolf number SEE **relative sunspot number**.

Wolf–Rayet star (WR star) A highly luminous star whose spectrum contains extremely strong emission lines. WR stars are divided into two kinds. In the WN type, emission lines from nitrogen dominate the spectrum. In the WC type, emission lines of carbon and oxygen predominate. Both types have strong lines of helium and a few have moderate or weak lines of hydrogen as well. All the emission lines are broad. WR stars are very hot, with surface temperatures between 25,000 and 50,000 K, luminosities between 100,000 and 1 million times the Sun's, and masses from 10 to 50 solar masses. They have strong **stellar winds** that typically carry away 3 solar

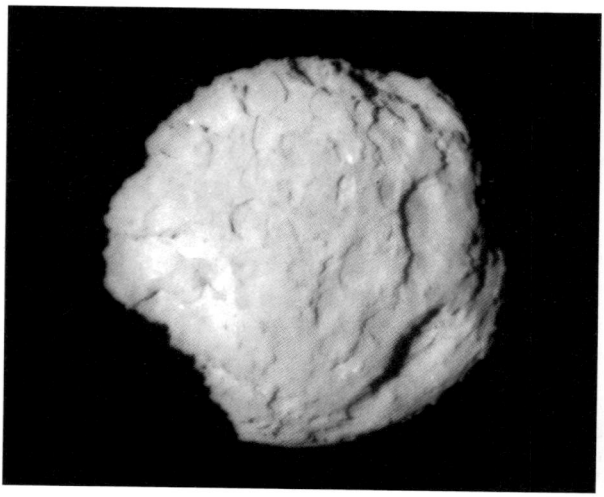

► **Wild 2, Comet** The heavily cratered nucleus of Comet Wild 2, as imaged by the Stardust spacecraft on January 2, 2004.

masses per million years. Many central stars of planetary nebulae are WR stars. The type of WR star that emerges after the original hydrogen-rich envelopes have been lost depends on how far the star's evolution had progressed. Wolf–Rayet stars are named after their French discoverers, Charles Wolf (1827–1918) and Georges Rayet (1839–1906).

wormhole A hypothetical tunnel in **spacetime**. In 1916 Albert Einstein discovered that one of the solutions to equations describing a black hole showed that there could be a connection between different regions of spacetime. In principle, this would make time travel possible. Recent cosmological theories suggest that on very small scales spacetime has a foam-like structure, pervaded by wormholes. These wormholes could lead to the formation of 'baby universes', which would have profound implications for our understanding of the **big bang**.

wrinkle ridge A winding ridge on the surface of a lunar mare, having sloping sides rising to a typical height of 200 m (650 ft), and several hundred kilometres long. They are often associated with **rilles**, and may have originated when the lava that formed the maria cooled and contracted. Wrinkle ridges are also found on Venus.

W Ursae Majoris star A member of a class of eclipsing binary stars, almost in contact as they orbit each other. They differ from **Beta Lyrae stars** in that the two components are smaller, less luminous stars of nearly identical brightness.

W Virginis star A member of a group of pulsating variable stars superficially similar to **Cepheid variables**. They are giant stars, typically in the spectral range G0 to M0 and with absolute magnitudes from −1 to −4, on the instability strip in the **Hertzsprung–Russell diagram**. Their masses can be as low as 0.5 solar masses, suggesting that they have evolved from low-mass main-sequence stars. Light curves of W Virginis stars can be distinguished from those of classical Cepheids by their less regular shape and a double-peaked maximum. They are Population II stars (SEE **stellar populations**) with periods that range from about 1 to about 100 days, and their **period–luminosity law** is distinctly different from that of normal Cepheids; however, they can still be used as distance estimators for galactic and extragalactic objects. Occasionally W Virginis stars show small period changes but in one case, RU Camelopardalis, the variations suddenly stopped in 1964 for about three years. It is not understood how pulsations can stop, or start, in such a short time.

XYZ

X-ray astronomy The study of celestial objects emitting **electromagnetic radiation** of short wavelength (about 0.01 to 10 nm), shorter than ultraviolet radiation but longer than gamma-rays. The Earth's atmosphere is opaque to X-rays, so this radiation must be studied by means of instruments aboard rockets or satellites. For astronomical purposes X-rays are usually described according to the energy associated with the X-ray photons; this energy is expressed in **electronvolts**.

X-ray astronomy began in 1949 with the discovery that the Sun emits X-rays. Other celestial X-ray sources were detected by rocket flights in the 1960s, followed by the first all-sky X-ray surveys by the US satellite Uhuru, launched in 1970, and later the UK satellite Ariel V. These surveys revealed the existence of **X-ray binaries**, in which one component is a neutron star or a black hole, as well as X-ray emission from supernova remnants, active galaxies, quasars and hot gas pervading clusters of galaxies. An **X-ray telescope** aboard the Einstein satellite, launched in 1978, produced the first true images of X-ray sources, rather than simply counting X-rays as previous instruments had done. Objects in nearby galaxies such as M31 and the Magellanic Clouds were resolved and studied for the first time. It was also discovered that almost all types of ordinary star emit X-rays. Subsequent X-ray satellites have included **Rosat**, **Yohkoh**, the **Chandra X-ray Observatory** and **Newton**.

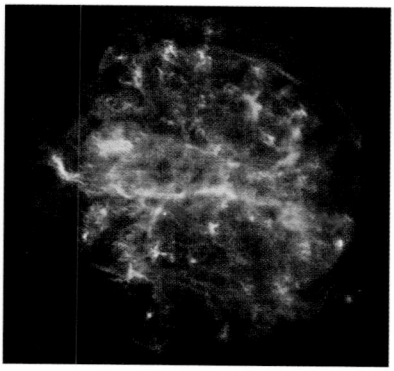

▲ **X-ray astronomy** This Chandra X-ray Observatory image shows a young supernova remnant in Centaurus. The rapidly expanding shell of gas contains large amounts of oxygen, neon, magnesium, sulphur and silicon. It has a pulsar at its centre.

▲ **X-ray binary** The strong X-ray source Cygnus X-3 is a close binary system in which matter from a normal star is being drawn into a neutron star or black hole. The X-ray emission from Cygnus X-3 varies regularly with a 4.8 hour period, as the compact star circles its companion star.

X-ray binary A binary system consisting of a normal star and a collapsed star – a **neutron star**, **black hole** or, in less intense sources, a **white dwarf**. The two components are very close together. Where the normal star is a giant or supergiant, a stellar wind blows directly onto the compact companion, causing the X-ray emission. In other cases, expansion of the normal star in the course of its evolution results in a flow of gas towards the collapsed star. This matter forms an **accretion disk** of hot material which emits X-rays. SEE ALSO **X-ray burster**.

X-ray burster A source of intense flashes of X-rays. These X-ray bursts have rise times of approximately 1 second, fall times of approximately 60 seconds and total luminosities equivalent to one week's energy output of the Sun. The intervals between bursts are irregular, ranging from hours to days, while many sources undergo burst-inactive phases that can last for weeks or even months. Burst sources are associated with old Population II objects (SEE **stellar populations**) and emit no pulsations. They are thought to be binaries containing an old neutron star with no magnetic field, which explains the absence of pulsations. The neutron star is paired with an old low-mass star, and individual bursts come from thermonuclear explosions after material accreted from the companion star has exceeded a critical mass on the neutron star's surface. *X-ray transients* are nova-like outbursts of X-rays believed to come from similar binaries in which the mass transfer is very uneven. SEE ALSO **pulsar**.

X-ray Multi-Mirror Mission (XMM) SEE **Newton**.

X-ray telescope An instrument for imaging X-ray sources. All X-ray telescopes are satellite-borne because X-rays do not penetrate the Earth's atmosphere. Unlike less energetic electromagnetic radiation, X-rays cannot be focused by reflection from a conventional concave mirror (as in a **reflecting telescope**) because, for all but the shallowest angles of incidence, they penetrate the surface. In *grazing incidence* X-ray telescopes the focusing element is a pair of coaxial surfaces, one **paraboloidal** and the other **hyperboloidal**, from which incoming X-rays are reflected at a very low 'grazing' angle towards a focus. The detecting element is usually a **CCD** adapted for X-ray wavelengths. An alternative instrument is the *microchannel plate detector*, in which the incident radiation falls on a plate which is made up of many fine tubes, rather like a short, wide fibre-optic bundle. The plate is charged, so that the radiation generates electrons which are accelerated down the tubes and together form an image that can be read off.

X-ray transient SEE **X-ray burster**.

Yagi antenna A basic form of antenna used in simple **radio telescopes**. It consists of several parallel elements mounted on a straight member, and often forms the basis of cheap arrays used in **aperture synthesis**; it is also a familiar form of TV aerial. It was developed by the Japanese engineer Hidetsugu Yagi (1886–1976).

year The time taken by the Earth to complete one revolution around the Sun. Various years, defined according to the choice of reference point, are given in the table. The *civil year* (**calendar** year) averages 365.2425 **mean solar days**. SEE ALSO the entries for each type of year listed in the table.

TYPES OF YEAR	
anomalistic year	365.25964 mean solar days
eclipse year	346.62003 mean solar days
sidereal year	365.25636 mean solar days
tropical year	365.24219 mean solar days

Yerkes Observatory The observatory of the University of Chicago, at Williams Bay, Wisconsin. It was founded by George Ellery **Hale**. Its main instrument is a 1-m (40-inch) refractor, opened in 1897 and still the largest in the world.

Ymir A small outer **satellite** of Saturn, discovered in 2000.

Yohkoh A satellite launched on 30 August 1991 to study the Sun, in particular solar flares, at X-ray and gamma-ray wavelengths. The mission was a Japanese–British–American collaboration; Yohkoh is Japanese for 'sunbeam'. The satellite returned many X-ray images of the Sun, the first obtained from orbit since Skylab missions in 1973. Its instruments revealed much about the rapid changes

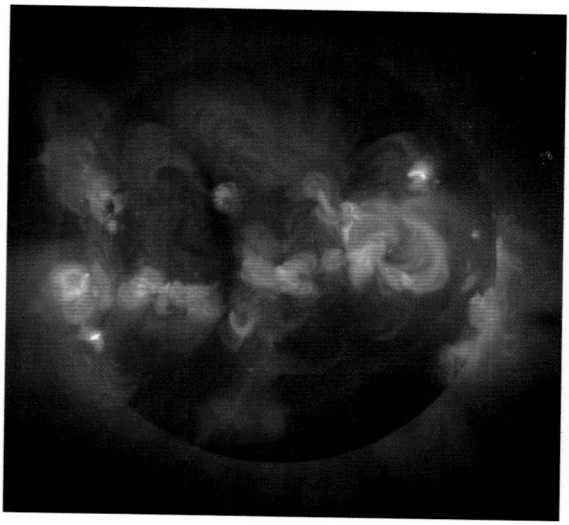

► **Yohkoh** This Yohkoh image of the active Sun at X-ray wavelengths shows hot coronal plasma in magnetic loops above active regions. Cooler, dark regions above the poles are described as coronal holes; these extend to lower solar latitudes at sunspot minimum.

that occur in the corona, and how flares originate and evolve.

Z Andromedae star SEE **symbiotic variable**.

Z Camelopardalis star A member of a small subgroup of **dwarf novae** similar to **U Geminorum stars**, except that they experience occasional 'standstills' (called 'stillstands' in North America), remaining more or less constant at some intermediate brightness. The onset of the standstills, and their duration, ranging from a few days to many months, are unpredictable.

Zeeman effect The splitting of a spectral line into several components by a strong magnetic field. Where these components cannot be resolved, the effect is apparent as a widening of the original line. The Zeeman effect occurs in the spectra of sunspots and stars. It demonstrates the existence of magnetic fields in celestial bodies and, since the field strength depends on the degree of splitting, it allows the fields to be measured. The phenomenon was predicted by Hendrik Lorentz, and discovered in 1896 by Pieter Zeeman (1865–1943).

Zelenchukskaya Observatory An observatory, also known as the Special Astrophysical Observatory, on Mount Pastukhov in the Caucasus Mountains of southern Russia, the site of the 6-m (236-inch) Large Azimuthal Telescope, which was opened in 1975. Also at Zelenchukskaya is the RATAN-600 radio telescope, which consists of a ring of reflecting panels 600 m (2000 ft) in diameter.

zenith The point on the **celestial sphere** that is vertically above the observer, 90° from the horizon. The point diametrically opposite, beneath the observer's feet, is called the *nadir*.

zenithal hourly rate (ZHR) The number of meteors per hour that would be seen at the maximum of a particular **meteor shower** under ideal conditions if the radiant were immediately overhead. The ZHR is obtained by applying a number of correction factors to the observed hourly rate of meteors in the shower. The observed hourly rate is always less than the ZHR. There are short-term and long-terms variations in showers' ZHRs as the associated **meteor streams** evolve.

zenith distance The angular distance of a celestial body from the zenith. It is equal to 90° minus the altitude of the body above the horizon.

zero-age main sequence (ZAMS) The **main sequence** on the Hertzsprung–Russell diagram as defined by stars of zero age, i.e. before they have undergone any substantial evolution.

ZHR ABBREVIATION FOR **zenithal hourly rate**.

zodiac A belt on the celestial sphere, about 8° either side of the ecliptic, which forms the background for the motions of the Sun, Moon and planets (except Pluto). The zodiac is divided into 12 *signs*, each 30° long, which are named after the constellations they contained at the time of the ancient Greeks: Aries, Taurus, Gemini, Cancer, Leo, Virgo, Libra, Scorpio, Sagittarius, Capricorn, Aquarius and Pisces. Because of **precession** these signs no longer coincide with the constellations of the same names. The ecliptic also passes through the constellation Ophiuchus, which is not a zodiacal sign.

zodiacal light A cone of faint light, usually fainter than the Milky Way, visible at all seasons from dark locations in the tropics in the absence of moonlight. It stretches along the ecliptic from the western horizon after evening twilight, or from the eastern horizon before morning twilight. At the point in the sky opposite the Sun, the zodiacal light brightens a little; this is the **gegenschein**. The spectrum of the zodiacal light resembles the Sun's, indicating that the phenomenon results from the scattering of sunlight by particles in the plane of the ecliptic. These particles constitute the *zodiacal dust cloud*, and probably originate partly from matter ejected by the Sun, and partly from the decay of comets and asteroids.

zone of avoidance A region of the sky along the plane of the Milky Way where almost no galaxies can be seen because of absorption by interstellar gas and dust in our Galaxy. It is between 10° and 40° wide.

Zwicky, Fritz (1898–1974) Swiss-American astronomer, born in Bulgaria. In 1934 he and Walter **Baade** suggested that what is left after a supernova explosion is a neutron star; this was confirmed in 1968 with the discovery of the pulsar in the **Crab Nebula**. In the mid-1960s, using the newly developed **Schmidt camera**, Zwicky discovered and examined many supernovae in other galaxies, and began a long study of **clusters of galaxies**.

Appendix

THE GREEK ALPHABET

α	A	alpha	ν	N	nu
β	B	beta	ξ	Ξ	xi
γ	Γ	gamma	ο	O	omicron
δ	Δ	delta	π	Π	pi
ε	E	epsilon	ρ	P	rho
ζ	Z	zeta	σ	Σ	sigma
η	H	eta	τ	T	tau
θ	Θ	theta	υ	Y	upsilon
ι	I	iota	φ	Φ	phi
κ	K	kappa	χ	X	chi
λ	Λ	lambda	ψ	Ψ	psi
μ	M	mu	ω	Ω	omega

CONVERSION FACTORS

Distances

1 nm = 10 Å
1 inch = 25.4 mm
1 mm = 0.03937 inch
1ft = 0.3048 m
1 m = 39.37 inches = 3.2808 ft
1 mile = 1.6093 km
1 km = 0.6214 mile
1 km/s = 2237 mile/h
1 pc = 3.0857×10^{13} km = 3.2616 l.y. = 206,265 AU
1 l.y. = 9.4607×10^{12} km = 0.3066 pc = 63,240 AU

Temperatures (to the nearest degree)

°C to °F:	×1.8, +32
°C to K:	+273
°F to °C:	−32, ÷1.8
°F to K:	÷1.8, +255
K to °C:	−273
K to °F:	×1.8, −460

Note: To convert temperature *differences*, rather than points on the temperature scale, ignore the additive or subtractive figure and just multiply or divide.

SYMBOLS FOR UNITS, CONSTANTS AND QUANTITIES

a	semimajor axis
Å	angstrom unit
AU	astronomical unit
c	speed of light
d	distance
e	eccentricity
E	energy
eV	electron-volt
f	following
F	focal length, force
g	acceleration due to gravity
G	gauss
G	gravitational constant
h	hour
h	Planck constant
H_0	Hubble constant
Hz	hertz
i	inclination
IC	*Index Catalogue*
Jy	jansky
k	Boltzmann constant
K	degrees kelvin
L	luminosity
L_n	Lagrangian points (n = 1 to 5)
l.y.	light year
m	metre, minute
m	apparent magnitude, mass
m_{bol}	bolometric magnitude
m_{pg}	photographic magnitude
m_{pv}	photovisual magnitude
m_v	visual magnitude
M	absolute magnitude, mass (stellar)
N	newton
p	preceding
P	orbital period
pc	parsec
q	perihelion distance
q_0	deceleration parameter
Q	aphelion distance
r	radius, distance
R	Roche limit
s	second
t	time
T	temperature (absolute), epoch (time of perihelion passage)
T_{eff}	effective temperature
v	velocity
W	watt
y	year
z	redshift
α	constant of aberration, right ascension
δ	declination
λ	wavelength
μ	proper motion
ν	frequency
π	parallax
ω	longitude of perihelion
Ω	observed/critical density ratio, longitude of ascending node
°	degree
′	arc minute
″	arc second

MULTIPLES AND SUBMULTIPLES USED WITH SI UNITS

Multiple	Prefix	Symbol	Submultiple	Prefix	Symbol
10^3	kilo-	k	10^{-3}	milli-	m
10^6	mega-	M	10^{-6}	micro-	μ
10^9	giga-	G	10^{-9}	nano-	n
10^{12}	tera-	T	10^{-12}	pico-	p
10^{15}	peta-	P	10^{-15}	femto-	f
10^{18}	exa-	E	10^{-18}	atto-	a

Acknowledgements

Abbreviations

AURA – Association of Universities for Research in Astronomy, Inc.
CXC – Chandra X-ray Observatory Center, Harvard-Smithsonian Center for Astrophysics
DLR – Deutschen Zentrum für Luft- und Raumfahrt
ESA – European Space Agency
ESO – European Southern Observatory
GSFC – Goddard Space Flight Center
JHU – Johns Hopkins University
JPL – Jet Propulsion Laboratory
JSC – Johnson Space Center
MPE – Max-Planck-Institut für extraterrestrische Physik, Garching
NASA – National Aeronautics and Space Administration
NOAO – National Optical Astronomy Observatory
NRAO – National Radio Astronomy Observatory
NSF – National Science Foundation
SOHO – Solar and Heliospheric Observatory. SOHO is a project of international cooperation between ESA and NASA
STScI – Space Telescope Science Institute
ST-ECF – Space Telescope European Coordinating Facility
USGS – US Geological Survey

5 Nigel Sharp/NOAO/AURA/NSF; **6** NOAO/AURA/NSF; **7** ESO; **8** NASA/JPL; **13** NASA/David R. Scott; **14** National Astronomy and Ionosphere Center/Cornell Univ./NSF; **15** NASA/JPL; **21** ESO; **23** NASA/ESA/Hubble Heritage Team (STScI/AURA); **25** A. Dupree Harvard-Smithsonian Center for Astrophysics/NASA/STScI; **27** Big Bear Solar Observatory; **30** ESO; **37** NASA/JPL/Space Science Institute; **38** NASA/CXC/GSFC/U.Hwang et al; **39** NASA/ESA/HEIC/Hubble Heritage Team (STScI/AURA); **41** NOAO/AURA/NSF; **43** NASA/JPL/USGS; **51** Courtesy of the TRACE consortium, Stanford-Lockheed Institute for Space Research/NASA; **53** DMR/COBE/NASA/Two-Year Sky Map; **56** ESO; **59** NASA; **62** NASA/JPL; **64** Bill Schoening/NOAO/AURA/NSF; **66** NASA/ESA/Hubble Heritage Team (STScI/AURA); **72** NOAO/AURA/NSF; **73t** NASA/JPL/Space Science Institute, **73b** NASA/JPL/Space Science Institute; **75** NASA/JHU Applied Physics Laboratory; **77** NASA/JPL/DLR; **79** Courtesy SOHO/MDI Consortium; **80** SEC/NOAA – National Oceanic and Atmospheric Administration/US Department of Commerce; **81** Courtesy SOHO EIT Consortium; **83** Nigel Sharp/NOAO/AURA/NSF; **86** NASA/JPL/DLR; **87** NASA/JPL/Univ. Arizona; **88** ESO; **89t** NASA/JPL/USGS; **89b** Gemini Observatory/NSF/Univ. Hawaii Institute for Astronomy; **90** Nigel Sharp/NOAO/AURA/NSF; **91t** ESA/MPE, **91b** Michael Rich, Kenneth Mighell and James D. Neill (Columbia Univ.)/Wendy Freedman (Carnegie Observatories)/NASA; **93** A. Fruchter and the ERO Team (STScI, ST-ECF)/NASA; **94** ESO; **95** NASA/JPL/Cornell Univ.; **100** (X-ray) NASA/UMass/D. Wang et al (optical) NASA/HST/D. Wang; **101** NOAO/AURA/NSF; **102** ESO; **105** ESA; **106** T.A.Rector (NOAO/AURA/NSF) and Hubble Heritage Team (STScI/AURA/NASA); **109** NASA/JSC; **111t** NASA/JPL, **111b** NASA/JPL/Space Science Institute; **112** NASA/GSFC/Langley Research Center/JPL/Multi-angle Imaging SpectroRadiometer Team; **113** IRAS/IPAC/JPL-Caltech/NASA; **114** ESA/ISOCAM/ISO-GAL Team; **115** NASA; **116** ESA; **118** NASA/JPL/Lunar and Planetary Laboratory; **120** Ian Morison/Jodrell Bank Observatory; **122** NASA/JPL/Univ. Arizona; **125** NOAO/AURA/NSF; **128** NASA/JPL; **129** NOAO/AURA/NSF; **140** ESO; **141** NASA/JPL/USGS; **142t** NASA/JPL/USGS; **142b** NASA/ASU/Cornell; **143** NASA/JPL/Malin Space Science Systems; **144** NASA/JHU Applied Physics Team; **145** NASA/JPL; **150** NASA/JPL/Space Science Institute; **151** NASA/JPL/USGS; **153** NASA/JPL/USGS; **158** NASA/JPL; **160** NASA/North Carolina School of Science and Mathematics/C. Olbert et al; **164** NASA/JPL; **166** Todd Boroson/NOAO/AURA/NSF; **169** Mark McCaughrean/ESO; **170** Karen Kwitter (Williams College), Ron Downes (STScI), You-Hua Chu (University of Illinois) and NOAO/AURA/NSF; **172** Mir 27 Crew © Centre National d'Études Spatiales, France; **174** NASA/JPL/Malin Space Science Systems; **175** NASA/JPL/Space Science Institute; **176** ESO; **177** Bill Schoening/NOAO/AURA/NSF; **179** NASA/ESA/Hubble Heritage Team (STScI/AURA); **180** NASA, ESA and AURA/Caltech; **181** Dr R. Albrecht ESA/ESO/ST-ECF/NASA; **185** NASA/JPL; **186** NASA/JPL; **187** NASA/CXC/Smithsonian Astrophysical Observatory; **188** John Bahcall, Institute for Advanced Study, Princeton/Mike Disney (Univ. Wales)/NASA; **190** NASA/JPL; **192** NRAO/Associated Universities, Inc.; **194** NASA/JPL/USGS; **196** NASA/Hubble Heritage Team (STScI/AURA); **199** NASA/JPL/Space Science Institute; **201** NASA, ESA and the Hubble Heritage Team (AURA/STScI); **205** NASA/JPL/Space Science Institute; **213** NOAO/AURA/NSF; **222** NASA/JPL-Caltech/E. Churchwell (University of Wisconsin-Madison); **228** NASA/Donald Walter (S. Carolina State Univ.)/ Paul Scowen and Brian Moore (Arizona State Univ.); **230** Courtesy SOHO/MDI Consortium; **233** NASA/CXC/U.Manitoba/H.Matheson & S.Safi-Harb; **235** ESO; **237** NASA/JPL; **238** ESA/NASA/JPL/University of Arizona; **239** NASA/JPL; **240** NASA/NSSDC; **242** NASA/JPL/USGS; **246** NASA/JPL; **247** NASA/JPL; **250** NASA/JPL; **252** NRAO/AU/NSF; **253** NASA/JPL/USGS; **255** T.A. Rector and Monica Ramirez/NOAO/AURA/NSF; **256** Harvey Richer (Univ. British Columbia)/STScI/NASA; **257** NASA/JPL-Caltech; **259b** NASA/CXC/Rutgers/J. Hughes et al; **259t** NASA/SRON/MPE; **261** Institute of Space and Astronautical Science, Japan.

Artwork © Philip's